Glencoe McGraw-Hill

Study Guide and Intervention Workbook

Geometry

Glencoe

To the Student

This *Study Guide and Intervention Workbook* gives you additional examples and problems for the concept exercises in each lesson. The exercises are designed to aid your study of mathematics by reinforcing important mathematical skills needed to succeed in the everyday world. The materials are organized by chapter and lesson, with two Study Guide and Intervention worksheets for every lesson in *Glencoe Geometry*.

Always keep your workbook handy. Along with your textbook, daily homework, and class notes, the completed *Study Guide and Intervention Workbook* can help you in reviewing for quizzes and tests.

To the Teacher

These worksheets are the same ones found in the Chapter Resource Masters for *Glencoe Geometry*. The answers to these worksheets are available at the end of each Chapter Resource Masters booklet as well as in your Teacher Wraparound Edition interleaf pages.

The McGraw·Hill Companies

Glencoe

Send all inquiries to:
Glencoe/McGraw-Hill
8787 Orion Place
Columbus, OH 43240

ISBN 13: 978-0-07-890848-4
ISBN 10: 0-07-890848-5

Geometry Study Guide and Intervention Workbook

Printed in the United States of America

5 6 7 8 9 10 REL 14 13 12 11 10

Contents

NAME ______________________ DATE ____________ PERIOD ____________

1-1 Study Guide and Intervention

Points, Lines, and Planes

Name Points, Lines, and Planes In geometry, a **point** is a location, a **line** contains points, and a **plane** is a flat surface that contains points and lines. If points are on the same line, they are **collinear**. If points on are the same plane, they are **coplanar**.

Example **Use the figure to name each of the following.**

a. a line containing point *A*

The line can be named as ℓ. Also, any two of the three points on the line can be used to name it.
$\overleftrightarrow{AB}$, $\overleftrightarrow{AC}$, or $\overleftrightarrow{BC}$

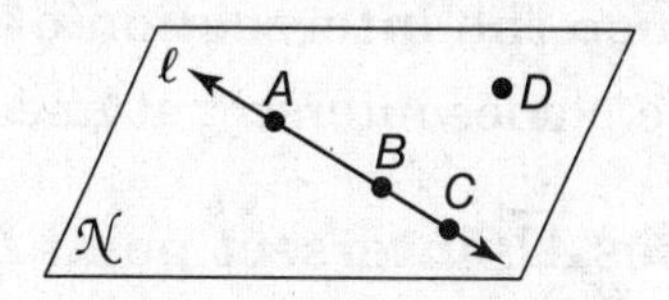

b. a plane containing point *D*

The plane can be named as plane $\mathcal{N}$ or can be named using three noncollinear points in the plane, such as plane *ABD*, plane *ACD*, and so on.

Exercises

Refer to the figure.

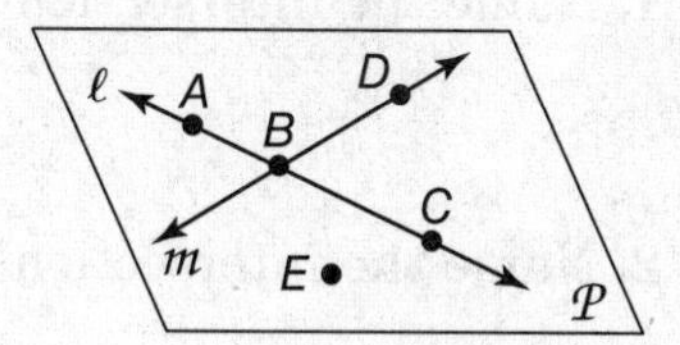

1. Name a line that contains point *A*.

2. What is another name for line *m*?

3. Name a point not on $\overleftrightarrow{AC}$.

4. What is another name for line ℓ?

5. Name a point not on line ℓ or line *m*.

Draw and label a figure for each relationship.

6. $\overleftrightarrow{AB}$ is in plane *Q*.

7. $\overleftrightarrow{ST}$ intersects $\overleftrightarrow{AB}$ at *P*.

8. Point *X* is collinear with points *A* and *P*.

9. Point *Y* is not collinear with points *T* and *P*.

10. Line ℓ contains points *X* and *Y*.

NAME ______________________ DATE ____________ PERIOD ____________

1-1 Study Guide and Intervention *(continued)*

Points, Lines, and Planes

Points, Lines, and Planes in Space **Space** is a boundless, three-dimensional set of all points. It contains lines and planes. The **intersection** of two or more geometric figures is the set of points they have in common.

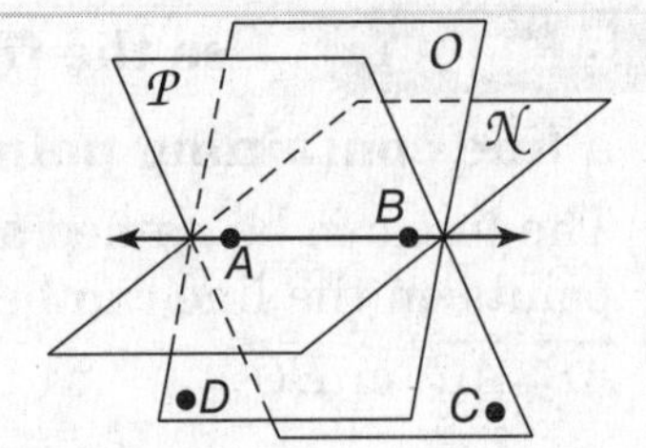

Example

a. Name the intersection of the planes O and $\mathcal{N}$.

The planes intersect at line $\overleftrightarrow{AB}$.

b. Does $\overleftrightarrow{AB}$ intersect point D? Explain.

No. $\overleftrightarrow{AB}$ is coplanar with D, but D is not on the line $\overleftrightarrow{AB}$.

Exercises

Refer to the figure.

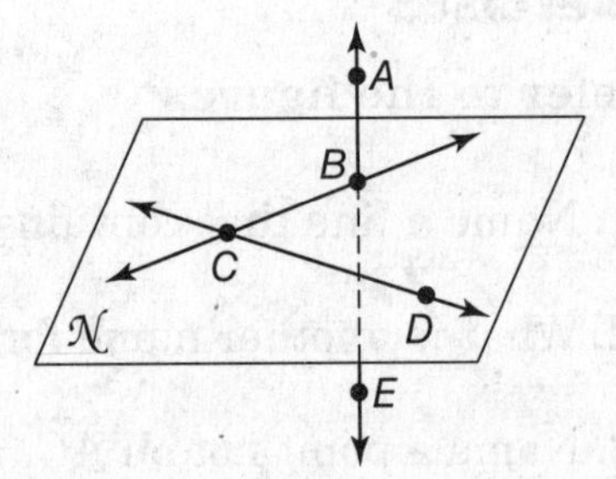

1. Name the intersection of plane $\mathcal{N}$ and line $\overleftrightarrow{AE}$.

2. Name the intersection of $\overleftrightarrow{BC}$ and $\overleftrightarrow{DC}$.

3. Does $\overleftrightarrow{DC}$ intersect $\overleftrightarrow{AE}$? Explain.

Refer to the figure.

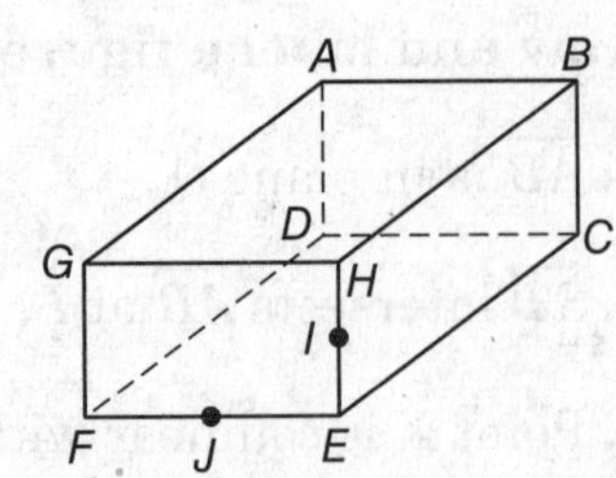

4. Name the three line segments that intersect at point A.

5. Name the line of intersection of planes GAB and FEH.

6. Do planes GFE and HBC intersect? Explain.

NAME ______________________________ DATE ____________ PERIOD ____________

1-2 Study Guide and Intervention

Linear Measure

Measure Line Segments A part of a line between two endpoints is called a **line segment.** The lengths of $\overline{MN}$ and $\overline{RS}$ are written as MN and RS. All measurements are approximations dependant upon the smallest unit of measure avaliable on the measuring instrument.

Example 1 **Find the length of $\overline{MN}$.**

M N

cm 1 2 3 4

The long marks are centimeters, and the shorter marks are millimeters. There are 10 millimeters for each centimeter. The length of $\overline{MN}$ is about 34 millimeters.

Example 2 **Find the length of $\overline{RS}$.**

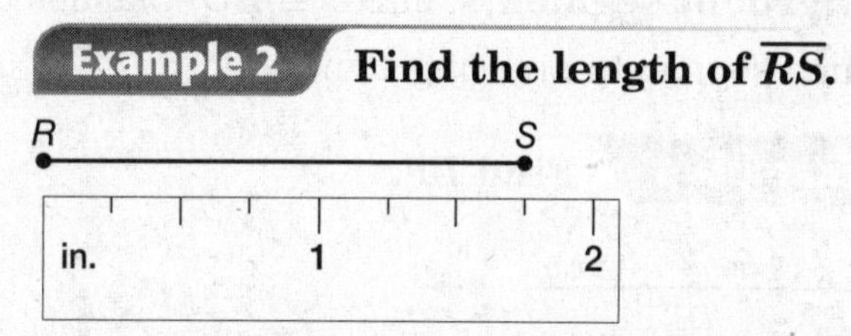

The long marks are inches and the short marks are quarter inches. Points S is closer to the $1\frac{3}{4}$ inch mark. The length of $\overline{RS}$ is about $1\frac{3}{4}$ inches.

Exercises

Find the length of each line segment or object.

1.

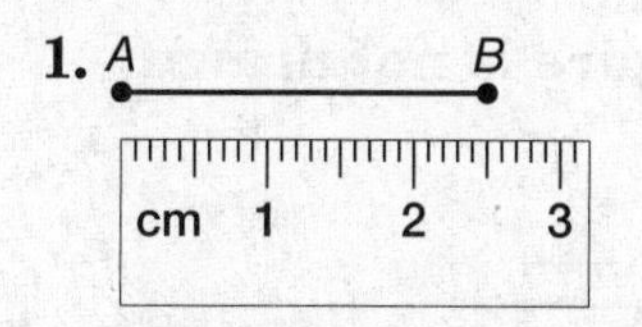

2.

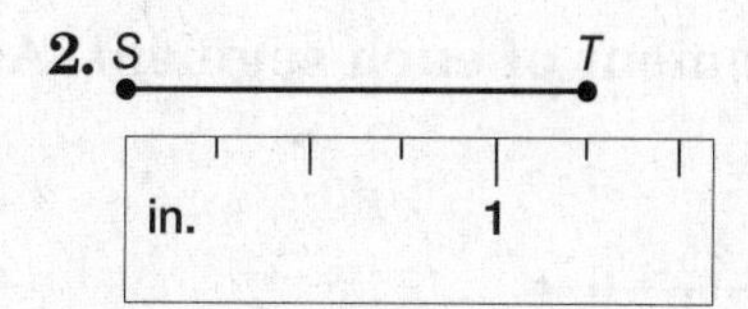

3.

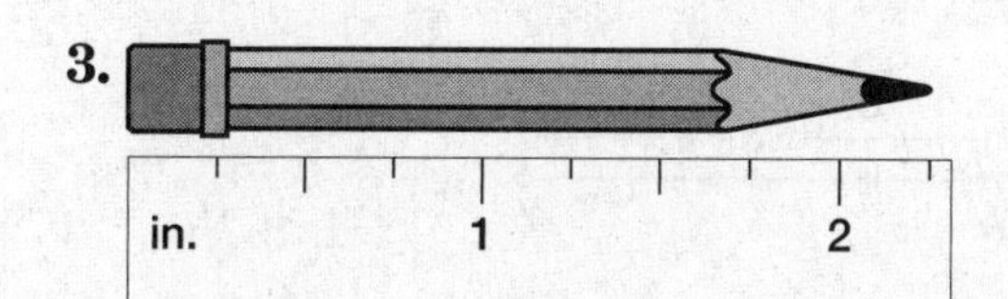

4.

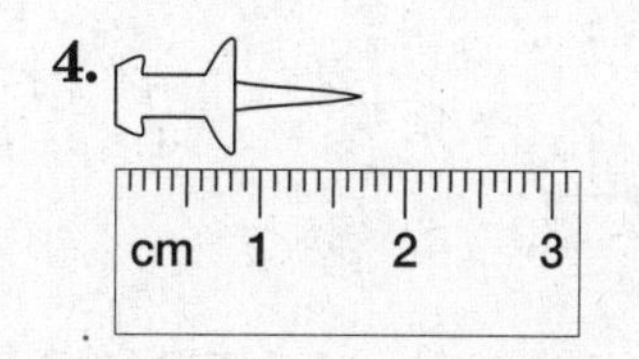

5.

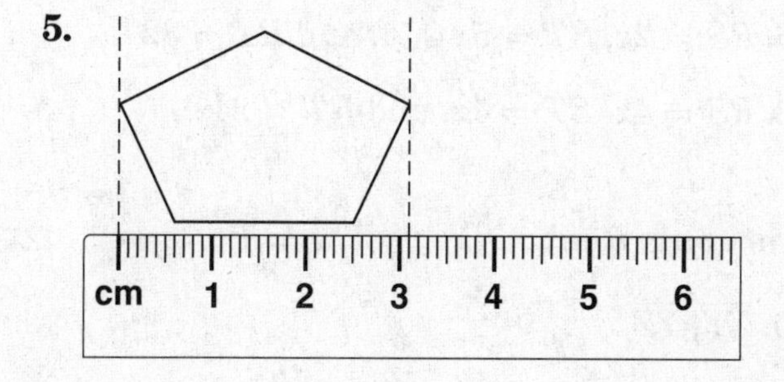

6.

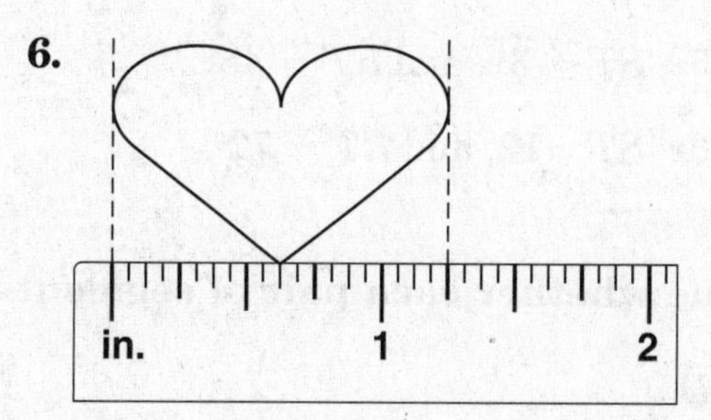

1-2 Study Guide and Intervention (continued)

Linear Measure

Calculate Measures On $\overleftrightarrow{PQ}$, to say that point M is between points P and Q means P, Q, and M are collinear and $PM + MQ = PQ$.

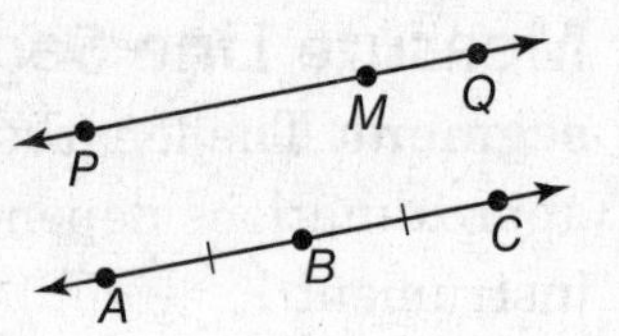

On $\overleftrightarrow{AC}$, AB = BC = 3 cm. We can say that the segments are **congruent segments**, or $\overline{AB} \cong \overline{BC}$. Slashes on the figure indicate which segments are congruent.

Example 1 **Find EF.**

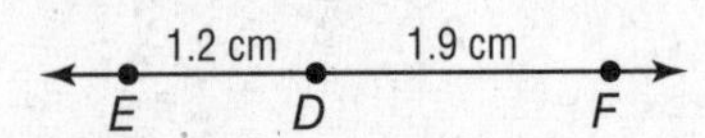

Point D is between E and F. Calculate EF by adding ED and DF.

$ED + DF = EF$	Betweenness of points
$1.2 + 1.9 = EF$	Substitution
$3.1 = EF$	Simplify.

Therefore, EF is 3.1 centimeters long.

Example 2 **Find x and AC.**

$2x + 5$

A x B $2x$ C

B is between A and C.

$AB + BC = AC$	Betweenness of points
$x + 2x = 2x + 5$	Substitution
$3x = 2x + 5$	Add $x + 2x$.
$x = 5$	Simplify.

$AC = 2x + 5 = 2(5) + 5 = 15$

Exercises

Find the measurement of each segment. Assume that each figure is not drawn to scale.

1. RT 2.0 cm 2.5 cm R S T

2. BC 6 in. A $2\frac{3}{4}$ in. B C

3. XZ $3\frac{1}{2}$ in. $\frac{3}{4}$ in. X Y Z

4. WX 6 cm W X Y

ALGEBRA Find the value of x and RS if S is between R and T.

5. $RS = 5x$, $ST = 3x$, and $RT = 48$

6. $RS = 2x$, $ST = 5x + 4$, and $RT = 32$

7. $RS = 6x$, $ST = 12$, and $RT = 72$

8. $RS = 4x$, $ST = 4x$, and $RT = 24$

Determine whether each pair of segments is congruent.

9. $\overline{AB}$, $\overline{CD}$

A 11 cm D
5 cm 5 cm
B 11 cm C

10. $\overline{XY}$, $\overline{YZ}$

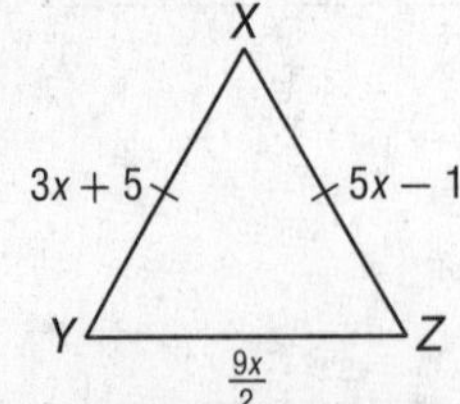

1-3 Study Guide and Intervention

Distance and Midpoints

Distance Between Two Points

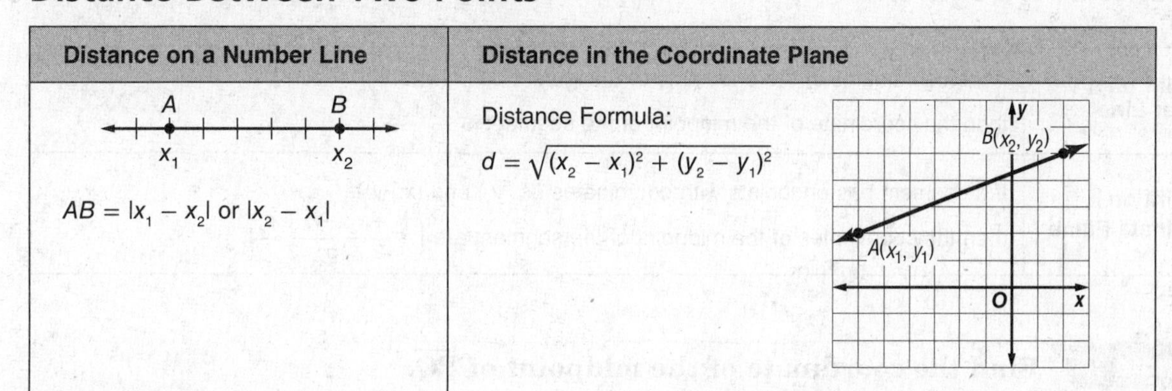

Distance on a Number Line	Distance in the Coordinate Plane
$AB = \lvert x_1 - x_2 \rvert$ or $\lvert x_2 - x_1 \rvert$	Distance Formula: $d = \sqrt{(x_2 - x_1)^2 + (y_2 - y_1)^2}$

Example 1 **Use the number line to find *AB*.**

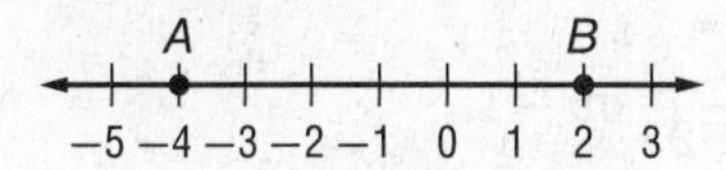

$$AB = |(-4) - 2| = |-6| = 6$$

Example 2 **Find the distance between *A*(−2, −1) and *B*(1, 3).**

Distance Formula

$$d = \sqrt{(x_2 - x_1)^2 + (y_2 - y_1)^2}$$

$$AB = \sqrt{(1 - (-2))^2 + (3 - (-1))^2}$$

$$AB = \sqrt{(3)^2 + (4)^2}$$

$$= \sqrt{25}$$

$$= 5$$

Exercises

Use the number line to find each measure.

A B C D E F G
−10 −8 −6 −4 −2 0 2 4 6 8

1. *BD* **2.** *DG*

3. *AF* **4.** *EF*

5. *BG* **6.** *AG*

7. *BE* **8.** *DE*

Find the distance between each pair of points.

9. *A*(0, 0), *B*(6, 8) **10.** *R*(−2, 3), *S*(3, 15)

11. *M*(1, −2), *N*(9, 13) **12.** *E*(−12, 2), *F*(−9, 6)

13. *X*(0, 0), *Y*(15, 20) **14.** *O*(−12, 0), *P*(−8, 3)

15. *C*(11, −12), *D*(6, 2) **16.** *K*(−2, 10), *L*(−4, 3)

NAME ______________________ DATE __________ PERIOD __________

1-3 Study Guide and Intervention *(continued)*

Distance and Midpoints

Midpoint of a Segment

Midpoint on a Number Line	If the coordinates of the endpoints of a segment are x_1 and x_2, then the coordinate of the midpoint of the segment is $\frac{x_1 + x_2}{2}$.
Midpoint on a Coordinate Plane	If a segment has endpoints with coordinates (x_1, y_1) and (x_2, y_2), then the coordinates of the midpoint of the segment are $\left(\frac{x_1 + x_2}{2}, \frac{y_1 + y_2}{2}\right)$.

Example 1 **Find the coordinate of the midpoint of $\overline{PQ}$.**

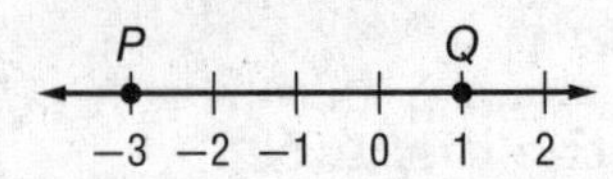

The coordinates of P and Q are −3 and 1.

If M is the midpoint of $\overline{PQ}$, then the coordinate of M is $\frac{-3 + 1}{2} = \frac{-2}{2}$ or −1.

Example 2 **Find the coordinates of M, the midpoint of $\overline{PQ}$, for $P(-2, 4)$ and $Q(4, 1)$.**

$M = \left(\frac{x_1 + x_2}{2}, \frac{y_1 + y_2}{2}\right) = \left(\frac{-2 + 4}{2}, \frac{4 + 1}{2}\right)$ or (1, 2.5)

Exercises

Use the number line to find the coordinate of the midpoint of each segment.

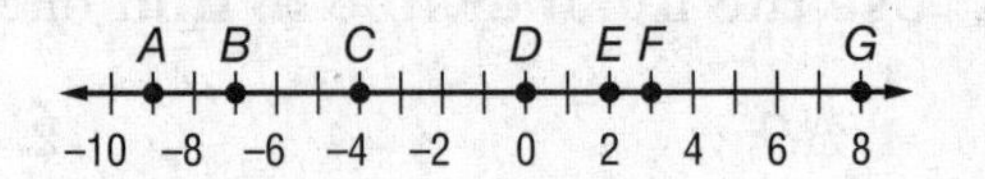

1. $\overline{CE}$

2. $\overline{DG}$

3. $\overline{AF}$

4. $\overline{EG}$

5. $\overline{AB}$

6. $\overline{BG}$

7. $\overline{BD}$

8. $\overline{DE}$

Find the coordinates of the midpoint of a segment with the given endpoints.

9. $A(0, 0)$, $B(12, 8)$

10. $R(-12, 8)$, $S(6, 12)$

11. $M(11, -2)$, $N(-9, 13)$

12. $E(-2, 6)$, $F(-9, 3)$

13. $S(10, -22)$, $T(9, 10)$

14. $K(-11, 2)$, $L(-19, 6)$

NAME _______________ DATE _______ PERIOD _______

1-4 Study Guide and Intervention

Angle Measure

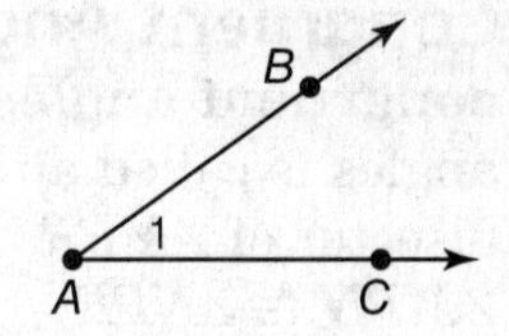

Measure Angles If two noncollinear **rays** have a common endpoint, they form an **angle.** The rays are the **sides** of the angle. The common endpoint is the **vertex.** The angle at the right can be named as $\angle A$, $\angle BAC$, $\angle CAB$, or $\angle 1$.

A **right angle** is an angle whose measure is 90. An **acute angle** has measure less than 90. An **obtuse angle** has measure greater than 90 but less than 180.

Example 1

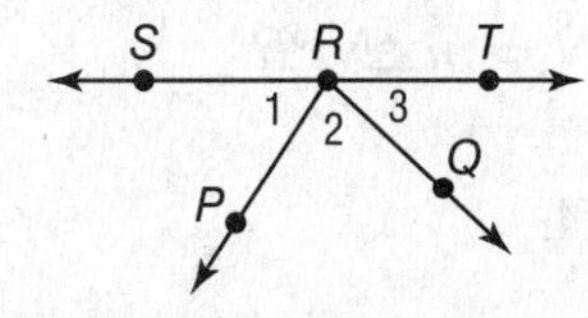

a. Name all angles that have *R* as a vertex.

Three angles are $\angle 1$, $\angle 2$, and $\angle 3$. For other angles, use three letters to name them: $\angle SRQ$, $\angle PRT$, and $\angle SRT$.

b. Name the sides of $\angle 1$.

$\overrightarrow{RS}$, $\overrightarrow{RP}$

Example 2 **Classify each angle as *right, acute,* or *obtuse.* Then use a protractor to measure the angle to the nearest degree.**

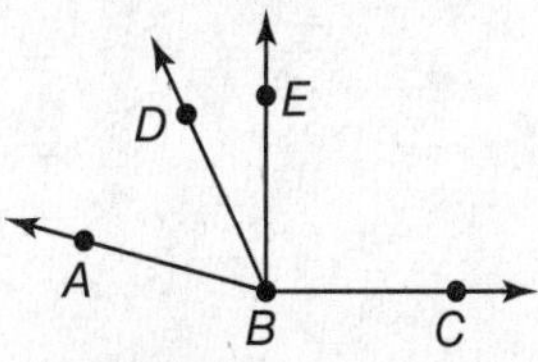

a. $\angle ABD$

Using a protractor, $m\angle ABD = 50$.
$50 < 90$, so $\angle ABD$ is an acute angle.

b. $\angle DBC$

Using a protractor, $m\angle DBC = 115$.
$180 > 115 > 90$, so $\angle DBC$ is an obtuse angle.

c. $\angle EBC$

Using a protractor, $m\angle EBC = 90$.
$\angle EBC$ is a right angle.

Exercises

Refer to the figure at the right.

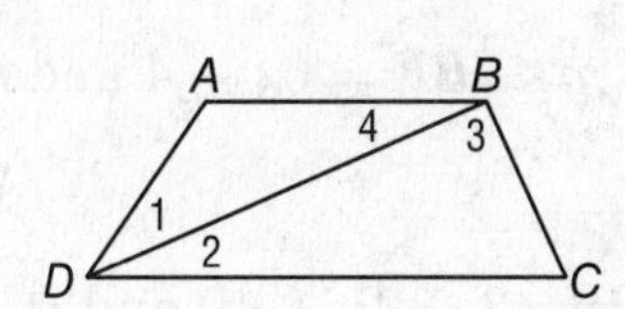

1. Name the vertex of $\angle 4$.

2. Name the sides of $\angle BDC$.

3. Write another name for $\angle DBC$.

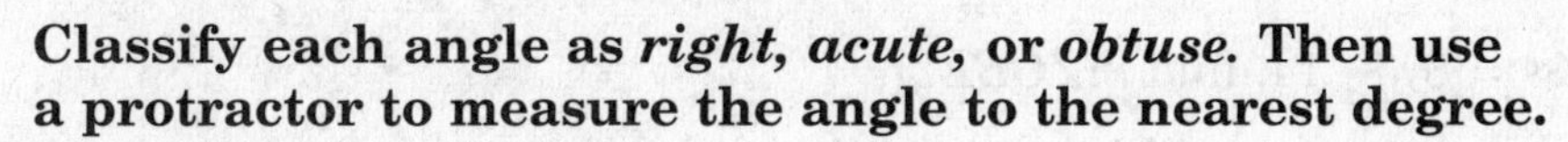

Classify each angle as *right, acute,* or *obtuse.* Then use a protractor to measure the angle to the nearest degree.

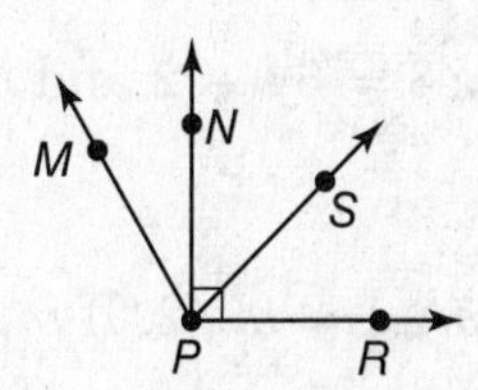

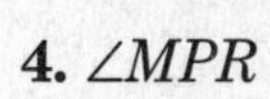

4. $\angle MPR$

5. $\angle RPN$

6. $\angle NPS$

1-4 Study Guide and Intervention *(continued)*

Angle Measure

Congruent Angles Angles that have the same measure are **congruent angles.** A ray that divides an angle into two congruent angles is called an **angle bisector.** In the figure, $\overrightarrow{PN}$ is the angle bisector of $\angle MPR$. Point N lies in the interior of $\angle MPR$ and $\angle MPN \cong \angle NPR$.

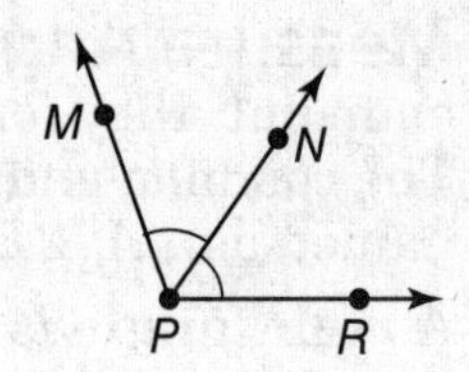

Example **Refer to the figure above. If $m\angle MPN = 2x + 14$ and $m\angle NPR = x + 34$, find x and find $m\angle NPR$.**

Since $\overrightarrow{PN}$ bisects $\angle MPR$, $\angle MPN \cong \angle NPR$, or $m\angle MPN = m\angle NPR$.

$$\begin{aligned} 2x + 14 &= x + 34 \\ 2x + 14 - x &= x + 34 - x \\ x + 14 &= 34 \\ x + 14 - 14 &= 34 - 14 \\ x &= 20 \end{aligned}$$

$$\begin{aligned} m\angle NPR &= 2x + 14 \\ &= 2(20) + 14 \\ &= 40 + 14 \\ &= 54 \end{aligned}$$

Exercises

ALGEBRA In the figure $\overrightarrow{QP}$ and $\overrightarrow{QR}$ are opposite rays. $\overrightarrow{QS}$ bisects $\angle PQT$.

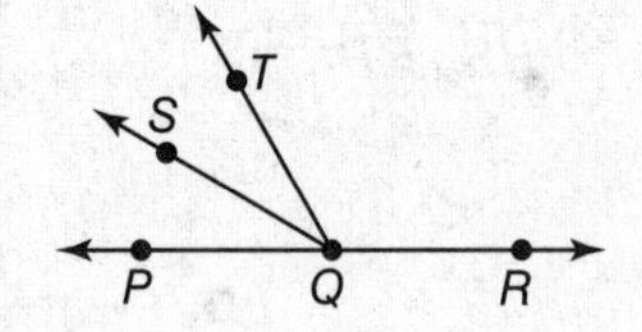

1. If $m\angle PQT = 60$ and $m\angle PQS = 4x + 14$, find the value of x.

2. If $m\angle PQS = 3x + 13$ and $m\angle SQT = 6x - 2$, find $m\angle PQT$.

ALGEBRA In the figure $\overrightarrow{BA}$ and $\overrightarrow{BC}$ are opposite rays. $\overrightarrow{BF}$ bisects $\angle CBE$.

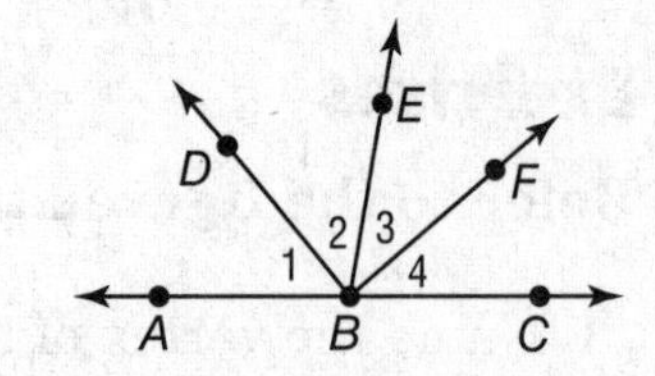

3. If $m\angle EBF = 6x + 4$ and $m\angle CBF = 7x - 2$, find $m\angle EBF$.

4. If $m\angle 3 = 4x + 10$ and $m\angle 4 = 5x$, find $m\angle 4$.

5. If $m\angle 3 = 6y + 2$ and $m\angle 4 = 8y - 14$, find $m\angle CBE$.

6. Let $m\angle 1 = m\angle 2$. If $m\angle ABE = 100$ and $m\angle ABD = 2(r + 5)$, find r and $m\angle DBE$.

NAME ______________________ DATE __________ PERIOD __________

1-5 Study Guide and Intervention

Angle Relationships

Pairs of Angles **Adjacent angles** are two angles that lie in the same plane and have a common vertex and a common side, but no common interior points. A pair of adjacent angles with noncommon sides that are opposite rays is called a **linear pair**. **Vertical angles** are two nonadjacent angles formed by two intersecting lines.

Example **Name an angle or angle pair that satisfies each condition.**

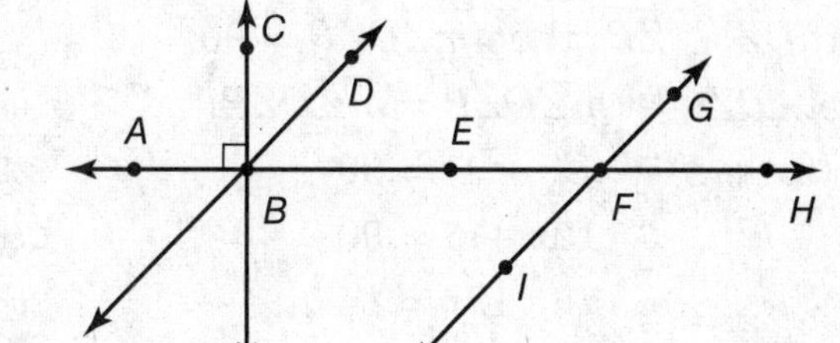

a. two vertical angles

$\angle EFI$ and $\angle GFH$ are nonadjacent angles formed by two intersecting lines. They are vertical angles.

b. two adjacent angles

$\angle ABD$ and $\angle DBE$ have a common vertex and a common side but no common interior points. They are adjacent angles.

c. two supplementary angles

$\angle EFG$ and $\angle GFH$ form a linear pair. The angles are supplementary.

d. two complementary angles

$m\angle CBD + m\angle DBE = 90$. These angles are complementary.

Exercises

Name an angle or angle pair that satisfies each condition.

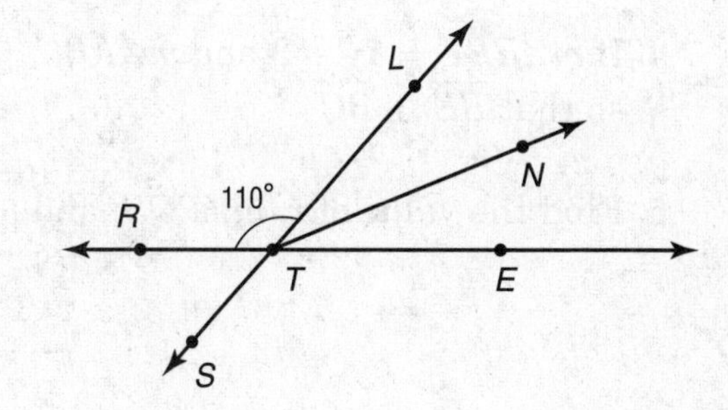

1. two adjacent angles

2. two acute vertical angles

3. two supplementary adjacent angles

4. an angle supplementary to $\angle RTS$

For Exercises 5–7, use the figure at the right.

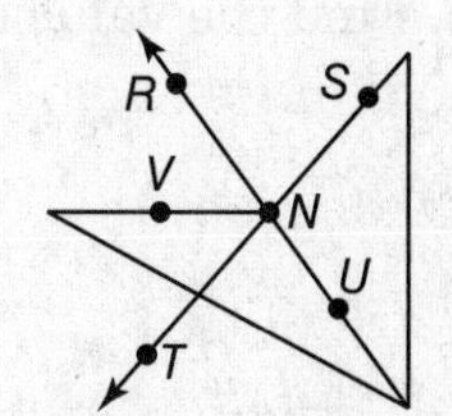

5. Identify two obtuse vertical angles.

6. Identify two acute adjacent angles.

7. Identify an angle supplementary to $\angle TNU$.

8. Find the measures of two complementary angles if the difference in their measures is 18.

NAME ______________________________ DATE ____________ PERIOD ________

1-5 Study Guide and Intervention *(continued)*

Angle Relationships

Perpendicular Lines Lines, rays, and segments that form four right angles are **perpendicular.** The right angle symbol indicates that the lines are perpendicular. In the figure at the right, $\overleftrightarrow{AC}$ is perpendicular to $\overleftrightarrow{BD}$, or $\overleftrightarrow{AC} \perp \overleftrightarrow{BD}$.

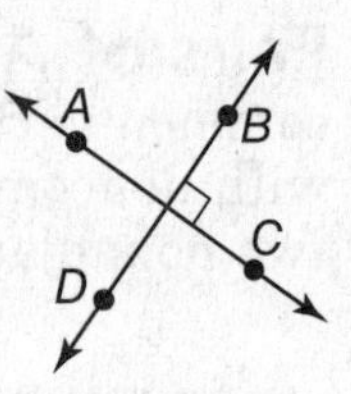

Example **Find x so that $\overrightarrow{DZ}$ and $\overrightarrow{ZP}$ are perpendicular.**

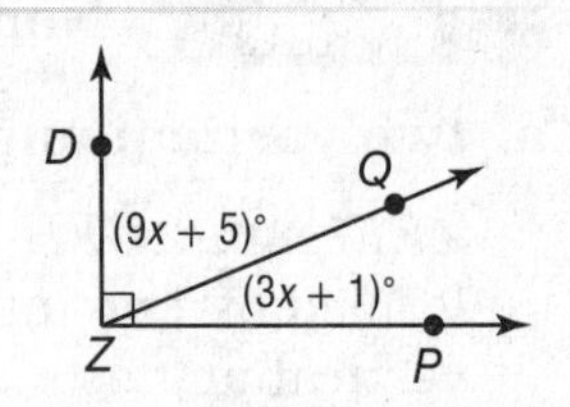

If $\overrightarrow{DZ} \perp \overrightarrow{ZP}$, then $m\angle DZP = 90$.

$m\angle DZQ + m\angle QZP = m\angle DZP$	Sum of parts = whole
$(9x + 5) + (3x + 1) = 90$	Substitution
$12x + 6 = 90$	Combine like terms.
$12x = 84$	Subtract 6 from each side.
$x = 7$	Divide each side by 12.

Exercises

1. Find the value of x and y so that $\overleftrightarrow{NR} \perp \overleftrightarrow{MQ}$.

2. Find $m\angle MSN$.

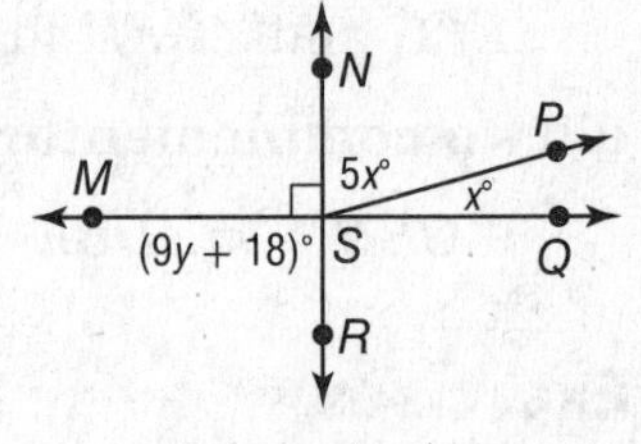

3. $m\angle EBF = 3x + 10$, $m\angle DBE = x$, and $\overrightarrow{BD} \perp \overrightarrow{BF}$. Find the value of x.

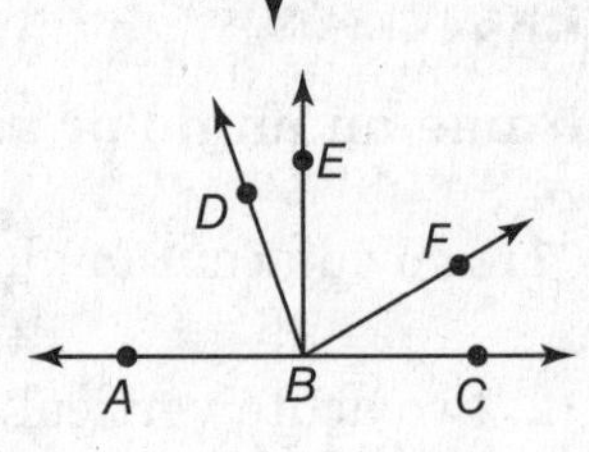

4. If $m\angle EBF = 7y - 3$ and $m\angle FBC = 3y + 3$, find the value of y so that $\overrightarrow{BE} \perp \overrightarrow{BC}$.

5. Find the value of x, $m\angle PQS$, and $m\angle SQR$.

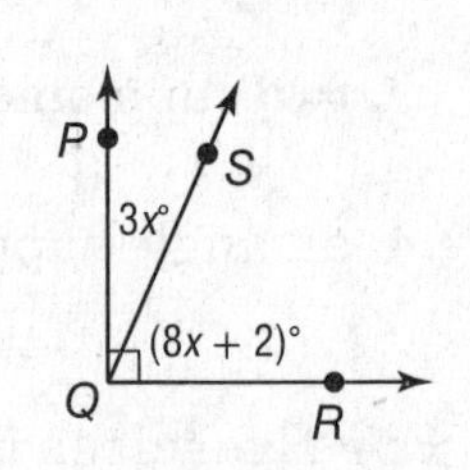

6. Find the value of y, $m\angle RPT$, and $m\angle TPW$.

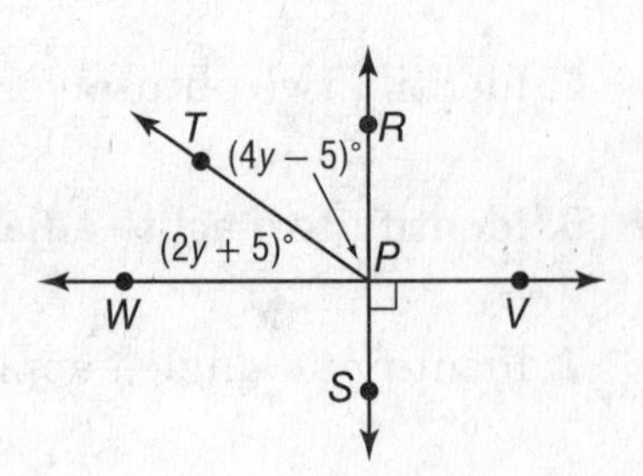

1-6 Study Guide and Intervention

Two-Dimensional Figures

Polygons A **polygon** is a closed figure formed by a finite number of coplanar segments called **sides.** The sides have a common endpoint, are noncollinear, and each side intersects exactly two other sides, but only at their endpoints. In general, a polygon is classified by its number of sides. The vertex of each angle is a **vertex of the polygon.** A polygon is named by the letters of its vertices, written in order of consecutive vertices. Polygons can be **concave** or **convex.** A convex polygon that is both **equilateral** (or has all sides congruent) and **equiangular** (or all angles congruent) is called a regular polygon.

Example **Name each polygon by its number of sides. Then classify it as *convex* or *concave* and *regular* or *irregular*.**

a.

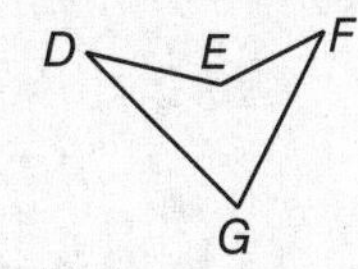

The polygon has four sides, so it is a quadrilateral.

Two of the lines containing the sides of the polygon will pass through the interior of the quadrilateral, so it is concave.

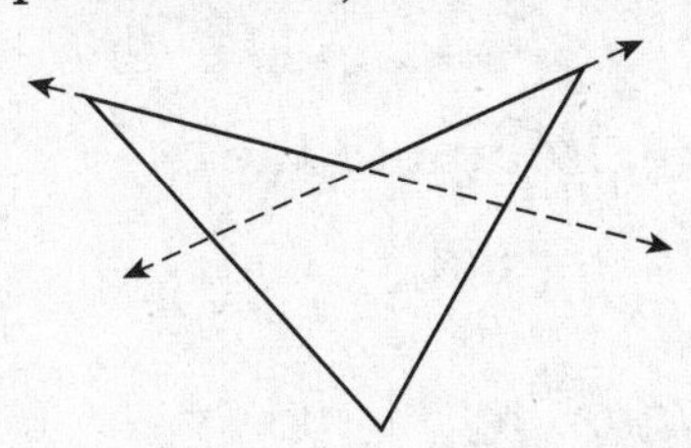

Only convex polygons can be regular, so this is an irregular quadrilateral.

b.

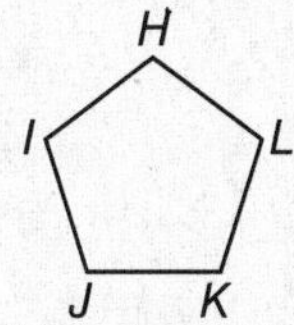

The polygon has five sides, so it is a pentagon.

No line containing any of the sides will pass through the interior of the pentagon, so it is convex.

All of the sides are congruent, so it is equilateral. All of the angles are congruent, so it is equiangular.

Since the polygon is convex, equilateral, and equiangular, it is regular. So this is a regular pentagon.

Exercises

Name each polygon by its number of sides. Then classify it as *convex* or *concave* and *regular* or *irregular*.

1.

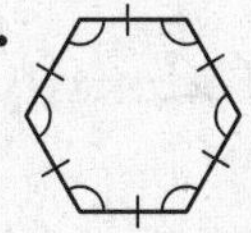

2.

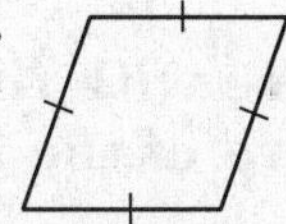

3.

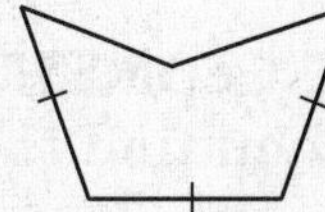

4.

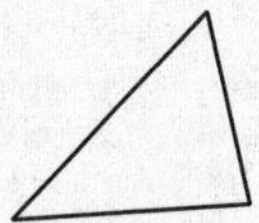

5.

6.

NAME ______________________________ DATE __________ PERIOD __________

1-6 Study Guide and Intervention *(continued)*

Two-Dimensional Figures

Perimeter, Circumference, and Area The **perimeter** of a polygon is the sum of the lengths of all the sides of the polygon. The **circumference** of a circle is the distance around the circle. The **area** of a figure is the number of square units needed to cover a surface.

Example **Write an expression or formula for the perimeter and area of each. Find the perimeter and area. Round to the nearest tenth.**

a.

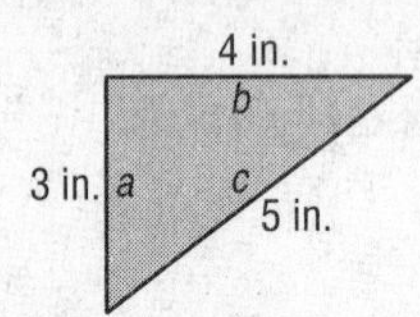

$P = a + b + c$

$= 3 + 4 + 5$

$= 12$ in.

$A = \frac{1}{2}bh$

$= \frac{1}{2}(4)(3)$

$= 6$ in^2

b.

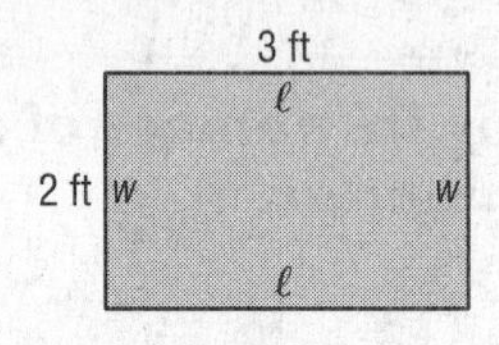

$P = 2\ell + 2w$

$= 2(3) + 2(2)$

$= 10$ ft

$A = lw$

$= (3)(2)$

$= 6$ ft^2

c.

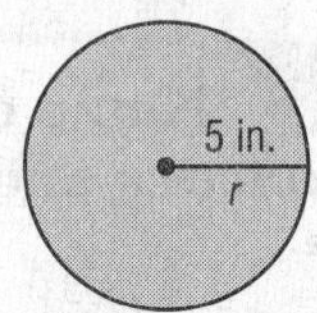

$C = 2\pi r$

$= 2\pi(5)$

$= 10\pi$ or about 31.4 in.

$A = \pi r^2$

$= \pi(5)^2$

$= 25\pi$ or about 78.5 in^2

Exercises

Find the perimeter or circumference and area of each figure. Round to the nearest tenth.

1.

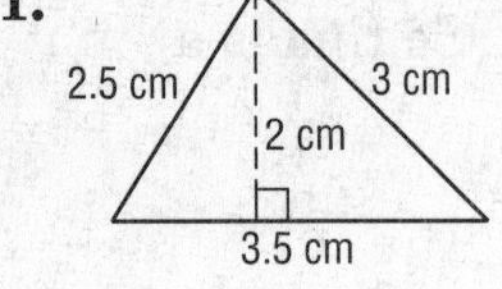

2.

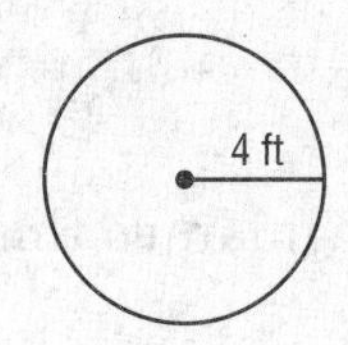

3.

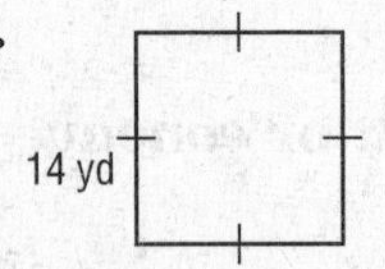

4.

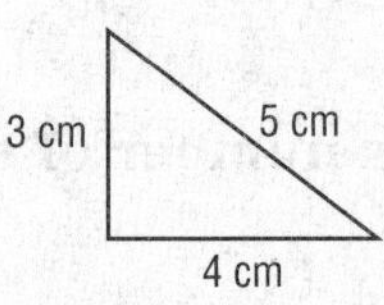

COORDINATE GEOMETRY **Graph each figure with the given vertices and identify the figure. Then find the perimeter and area of the figure.**

5. $A(-2, -4)$, $B(1, 3)$, $C(4, -4)$

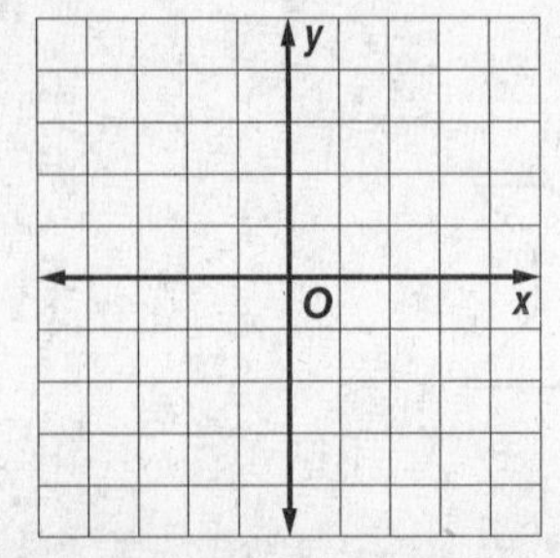

6. $X(-3, -1)$, $Y(-3, 3)$, $Z(4, -1)$, $P(4, 2)$

NAME ______________________________ DATE ____________ PERIOD ____________

1-7 Study Guide and Intervention

Three-Dimensional Figures

Identify Three-Dimensional Figures A solid with all flat surfaces that enclose a single region of space is called a **polyhedron.** Each flat surface, or **face,** is a polygon. The line segments where the faces intersect are called **edges.** The point where three or more edges meet is called a **vertex.** Polyhedrons can be classified as **prisms** or **pyramids.** A prism has two congruent faces called **bases** connected by parallelogram faces. A pyramid has a polygonal base and three or more triangular faces that meet at a common vertex. Polyhedrons or **polyhedra** are named by the shape of their bases. Other solids are a **cylinder**, which has parallel circular bases connected by a curved surface, a **cone** which has a circular base connected by a curved surface to a single vertex, or a **sphere**.

pentagonal prism

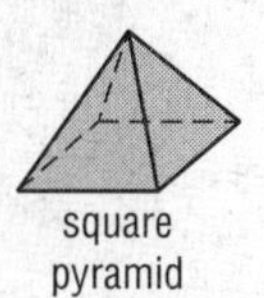

square pyramid

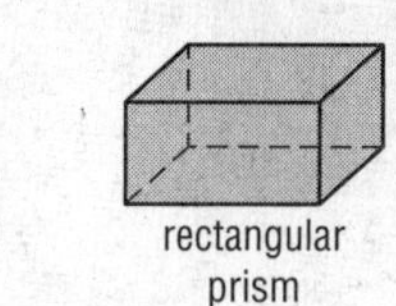

pentagonal pyramid

rectangular prism

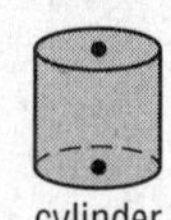

cylinder

cone

sphere

Example **Determine whether each solid is a polyhedron. Then identify the solid. If it is a polyhedron, name the faces, edges, and vertices.**

a.

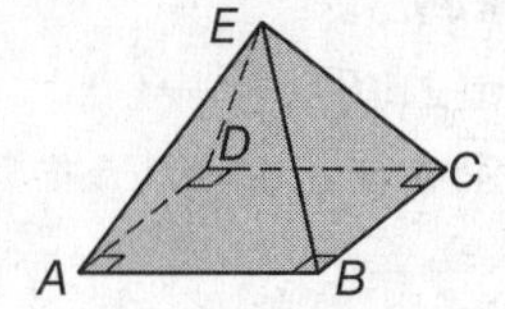

The figure is a rectangular pyramid. The base is rectangle $ABCD$, and the four faces $\triangle ABE$, $\triangle BCE$, $\triangle CDE$, and $\triangle ADE$ meet at vertex E. The edges are $\overline{AB}$, $\overline{BC}$, $\overline{CD}$, $\overline{AD}$, $\overline{AE}$, $\overline{BE}$, $\overline{CE}$, and $\overline{DE}$. The vertices are A, B, C, D, and E.

b.

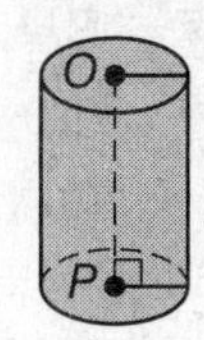

This solid is a cylinder. The two bases are $\odot O$ and $\odot P$.
The solid has a curved surface, so it is not a polyhedron. It has two congruent circular bases, so it is a cylinder.

Exercises

Determine whether each solid is a polyhedron. Then identify the solid. If it is a polyhedron, name the faces, edges, and vertices.

1.

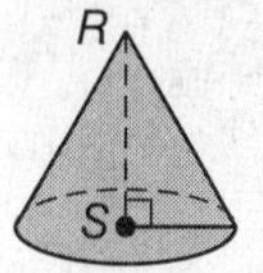

2.

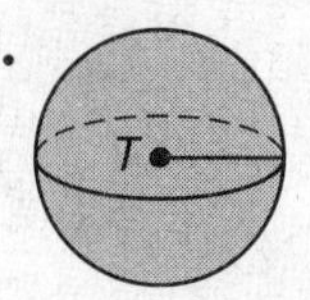

3.

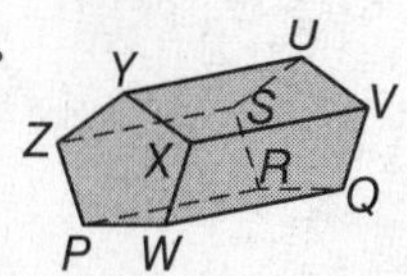

4.

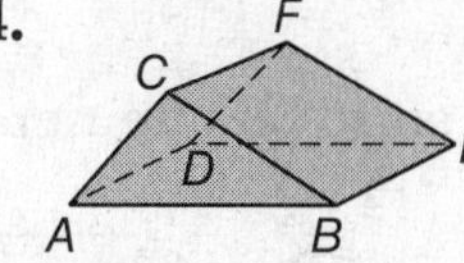

NAME ______________________ DATE __________ PERIOD __________

1-7 Study Guide and Intervention *(continued)*

Three-Dimensional Figures

SURFACE AREA AND VOLUME Surface area is the sum of the areas of each face of a solid. Volume is the measure of the amount of space the solid encloses.

Example

Write an expression or formula for the surface area and volume of each solid. Find the surface area and volume. Round to the nearest tenth.

a.

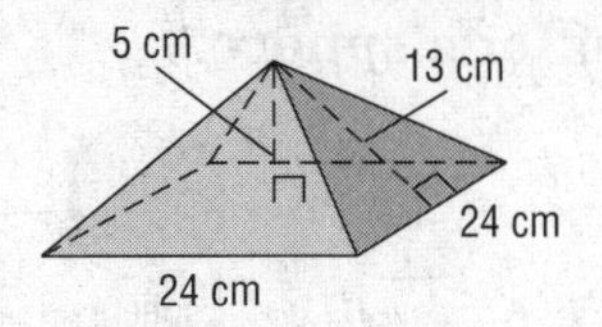

$T = \frac{1}{2}Pl + B$

$= \frac{1}{2}(96)(13) + 576$

$= 1200 \text{ cm}^2$

$V = \frac{1}{3}Bh$

$= \frac{1}{3}(576)(5)$

$= 960 \text{ cm}^3$

b.

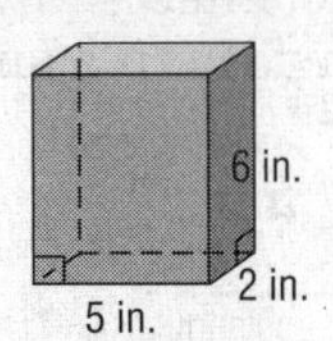

$T = Ph + 2B$

$= (14)(6) + 2(10)$

$= 104 \text{ in}^2$

$V = Bh$

$= (10)(6)$

$= 60 \text{ in}^3$

c.

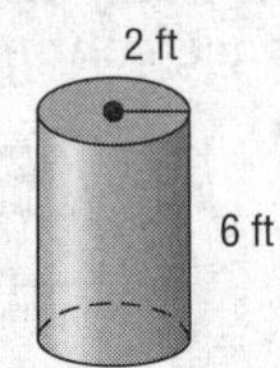

$T = 2\pi rh + 2\pi r^2$

$= 2\pi(2)(6) + 2\pi(2)^2$

$= 32\pi$ or about 100.5 ft^2

$V = \pi r^2 h$

$= \pi(2)^2(6)$

$= 24\pi$ or about 75.4 ft^3

Exercises

Find the surface area of each solid to the nearest tenth.

1.

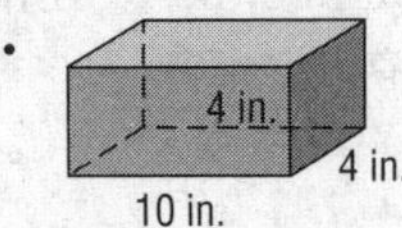

2.

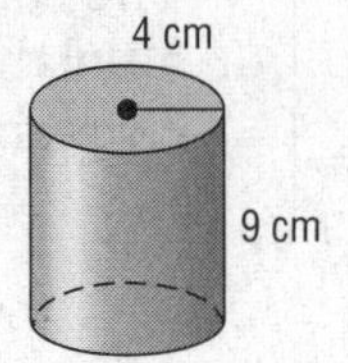

3.

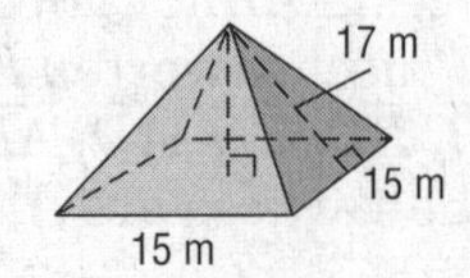

4.

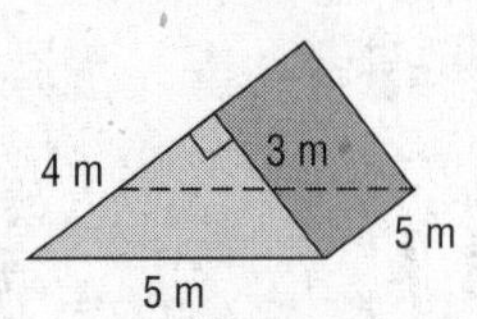

5.

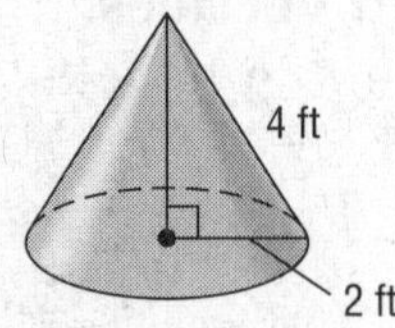

6.

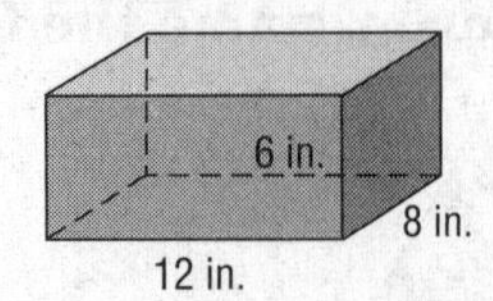

Find the volume of each solid to the nearest tenth.

7.

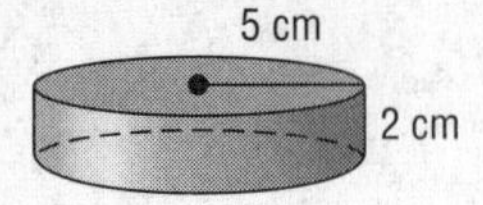

8.

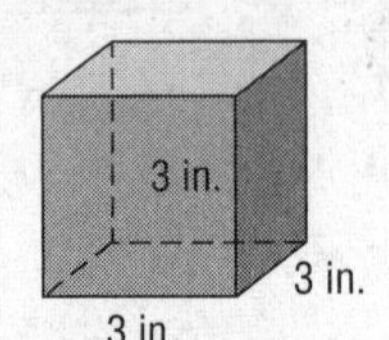

9.

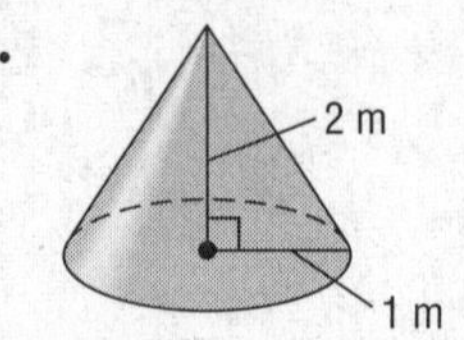

NAME ______________________________ DATE __________ PERIOD __________

2-1 Study Guide and Intervention

Inductive Reasoning and Conjecture

Making Conjectures **Inductive reasoning** is reasoning that uses information from different examples to form a conclusion or statement called a **conjecture**.

Example 1 **Write a conjecture about the next number in the sequence 1, 3, 9, 27, 81.**

Look for a pattern:

Each number is a power of 3.

1	3	9	27	81
3^0	3^1	3^2	3^3	3^4

Conjecture: The next number will be 3^5 or 243.

Example 2 **Write a conjecture about the number of small squares in the next figure.**

Look for a pattern: The sides of the squares have measures 1, 2, and 3 units.

Conjecture: For the next figure, the side of the square will be 4 units, so the figure will have 16 small squares.

Exercises

Write a conjecture that describes the pattern in each sequence. Then use your conjecture to find the next item in the sequence.

1. $-5, 10, -20, 40$

2. 1, 10, 100, 1000

3. $1, \frac{6}{5}, \frac{7}{5}, \frac{8}{5}$

Write a conjecture about each value or geometric relationship.

4. $A(-1, -1)$, $B(2, 2)$, $C(4, 4)$

5. $\angle 1$ and $\angle 2$ form a right angle.

6. $\angle ABC$ and $\angle DBE$ are vertical angles.

7. $\angle E$ and $\angle F$ are right angles.

NAME ______________________ DATE __________ PERIOD __________

2-1 Study Guide and Intervention *(continued)*

Inductive Reasoning and Conjecture

Find Counterexamples A conjecture is false if there is even one situation in which the conjecture is not true. The false example is called a **counterexample**.

Example **Find a counterexample to show the conjecture is false.**

If $\overline{AB} \cong \overline{BC}$, then B is the midpoint of $\overline{AC}$.

Is it possible to draw a diagram with $\overline{AB} \cong \overline{BC}$ such that B is not the midpoint? This diagram is a counterexample because point B is not on $\overline{AC}$. The conjecture is false.

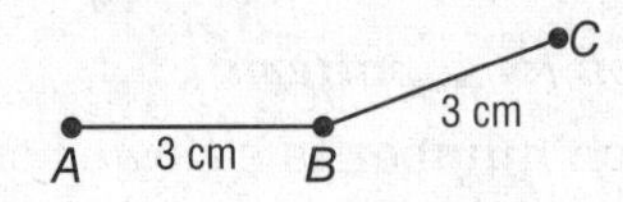

Exercises

Determine whether each conjecture is *true* or *false*. Give a counterexample for any false conjecture.

1. If points A, B, and C are collinear, then $AB + BC = AC$.

2. If $\angle R$ and $\angle S$ are supplementary, and $\angle R$ and $\angle T$ are supplementary, then $\angle T$ and $\angle S$ are congruent.

3. If $\angle ABC$ and $\angle DEF$ are supplementary, then $\angle ABC$ and $\angle DEF$ form a linear pair.

4. If $\overline{DE} \perp \overline{EF}$, then $\angle DEF$ is a right angle.

NAME ______________________ DATE __________ PERIOD __________

2-2 Study Guide and Intervention

Logic

Determine Truth Values A **statement** is any sentence that is either true or false. The **truth value** of a statement is either true (T) or false (F). A statement can be represented by using a letter. For example,

Statement p: Chicago is a city in Illinois. The truth value of statement p is true.

Several statements can be joined in a **compound statement**.

Negation: not p is the negation of the statement p.	Statement p and statement q joined by the word *and* is a **conjunction**.	Statement p and statement q joined by the word *or* is a **disjunction**.
Symbols: $\sim p$ (Read: not p)	Symbols: $p \wedge q$ (Read: p *and* q)	Symbols: $p \vee q$ (Read: p *or* q)
The statements p and $\sim p$ have opposite truth values.	The conjunction $p \wedge q$ is true only when both p and q are true.	The disjunction $p \vee q$ is true if p is true, if q is true, or if both are true.

Example 1 **Write a compound statement for each conjunction. Then find its truth value.**

p: An elephant is a mammal.

q: A square has four right angles.

a. $p \wedge q$

Join the statements with *and*: An elephant is a mammal and a square has four right angles. Both parts of the statement are true so the compound statement is true.

b. $\sim p \wedge q$

$\sim p$ is the statement "An elephant is not a mammal." Join $\sim p$ and q with the word *and*: An elephant is not a mammal and a square has four right angles. The first part of the compound statement, $\sim p$, is false. Therefore the compound statement is false.

Example 2 **Write a compound statement for each disjunction. Then find its truth value.**

p: A diameter of a circle is twice the radius.

q: A rectangle has four equal sides.

a. $p \vee q$

Join the statements p and q with the word *or*: A diameter of a circle is twice the radius or a rectangle has four equal sides. The first part of the compound statement, p, is true, so the compound statement is true.

b. $\sim p \vee q$

Join $\sim p$ and q with the word *or*: A diameter of a circle is not twice the radius or a rectangle has four equal sides. Neither part of the disjunction is true, so the compound statement is false.

Exercises

Use the following statements to write a compound statement for each conjunction or disjunction. Then find its truth value.

p: $10 + 8 = 18$ q: September has 30 days. r: A rectangle has four sides.

1. p and q

2. $p \vee r$

3. q or r

4. $q \wedge \sim r$

2-2 Study Guide and Intervention *(continued)*

Logic

Truth Tables One way to organize the truth values of statements is in a **truth table**. The truth tables for negation, conjunction, and disjunction are shown at the right.

Negation

p	$\sim p$
T	F
F	T

Conjunction

p	q	$p \wedge q$
T	T	T
T	F	F
F	T	F
F	F	F

Disjunction

p	q	$p \vee q$
T	T	T
T	F	T
F	T	T
F	F	F

Example 1 **Construct a truth table for the compound statement *q* or *r*.**
Use the disjunction table.

q	r	q or r
T	T	T
T	F	T
F	T	T
F	F	F

Example 2 **Construct a truth table for the compound statement *p* and (*q* or *r*).**
Use the disjunction table for (q or r). Then use the conjunction table for p and (q or r).

p	q	r	q or r	p and (q or r)
T	T	T	T	T
T	T	F	T	T
T	F	T	T	T
T	F	F	F	F
F	T	T	T	F
F	T	F	T	F
F	F	T	T	F
F	F	F	F	F

Exercises

Construct a truth table for each compound statement.

1. p or r

2. $\sim p \vee q$

3. $q \wedge \sim r$

4. $\sim p \wedge \sim r$

5. (p and r) or q

2-3 Study Guide and Intervention

Conditional Statements

If-Then Statements An if-then statement is a statement such as "If you are reading this page, then you are studying math." A statement that can be written in if-then form is called a **conditional statement**. The phrase immediately following the word *if* is the **hypothesis**. The phrase immediately following the word *then* is the **conclusion**.

A conditional statement can be represented in symbols as $p \rightarrow q$, which is read "p implies q" or "if p, then q."

Example 1 **Identify the hypothesis and conclusion of the conditional statement.**

If $\angle X \cong \angle R$ and $\angle R \cong \angle S$, then $\angle X \cong \angle S$.

hypothesis: $\angle X \cong \angle R$ and $\angle R \cong \angle S$

conclusion: $\angle X \cong \angle S$

Example 2 **Identify the hypothesis and conclusion. Write the statement in if-then form.**

You receive a free pizza with 12 coupons.

If you have 12 coupons, then you receive a free pizza.

hypothesis: you have 12 coupons

conclusion: you receive a free pizza

Exercises

Identify the hypothesis and conclusion of each conditional statement.

1. If it is Saturday, then there is no school.

2. If $x - 8 = 32$, then $x = 40$.

3. If a polygon has four right angles, then the polygon is a rectangle.

Write each statement in if-then form.

4. All apes love bananas.

5. The sum of the measures of complementary angles is 90.

6. Collinear points lie on the same line.

Determine the truth value of each conditional statement. If *true*, explain your reasoning. If *false*, give a counterexample.

7. If today is Wednesday, then yesterday was Friday.

8. If a is positive, then $10a$ is greater than a.

2-3 Study Guide and Intervention *(continued)*

Conditional Statements

Converse, Inverse, and Contrapositive If you change the hypothesis or conclusion of a conditional statement, you form **related conditionals**. This chart shows the three related conditionals, *converse*, *inverse*, and *contrapositive*, and how they are related to a conditional statement.

	Symbols	Formed by	Example
Conditional	$p \to q$	using the given hypothesis and conclusion	If two angles are vertical angles, then they are congruent.
Converse	$q \to p$	exchanging the hypothesis and conclusion	If two angles are congruent, then they are vertical angles.
Inverse	$\sim p \to \sim q$	replacing the hypothesis with its negation and replacing the conclusion with its negation	If two angles are not vertical angles, then they are not congruent.
Contrapositive	$\sim q \to \sim p$	negating the hypothesis, negating the conclusion, and switching them	If two angles are not congruent, then they are not vertical angles.

Just as a conditional statement can be true or false, the related conditionals also can be true or false. A conditional statement always has the same truth value as its contrapositive, and the converse and inverse always have the same truth value.

Exercises

Write the converse, inverse, and contrapositive of each true conditional statement. Determine whether each related conditional is *true* or *false*. If a statement is false, find a counterexample.

1. If you live in San Diego, then you live in California.

2. If a polygon is a rectangle, then it is a square.

3. If two angles are complementary, then the sum of their measures is 90.

NAME ______________________________ DATE ____________ PERIOD ____________

2-4 Study Guide and Intervention

Deductive Reasoning

Law of Detachment **Deductive reasoning** is the process of using facts, rules, definitions, or properties to reach conclusions. One form of deductive reasoning that draws conclusions from a true conditional $p \rightarrow q$ and a true statement p is called the **Law of Detachment.**

Law of Detachment	If $p \rightarrow q$ is true and p is true, then q is true.

Example **Determine whether each conclusion is valid based on the given information. If not, write *invalid*. Explain your reasoning.**

a. Given: Two angles supplementary to the same angle are congruent. $\angle A$ and $\angle C$ are supplementary to $\angle B$.

Conclusion: $\angle A$ is congruent to $\angle C$.

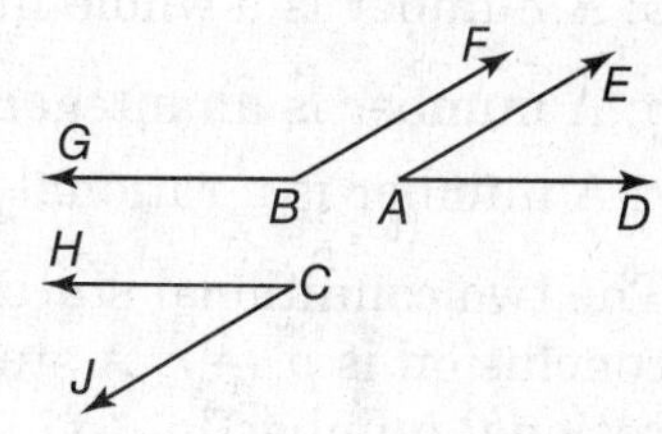

The statement $\angle A$ and $\angle C$ are *supplementary* to $\angle B$ is the hypothesis of the conditional. Therefore, by the Law of Detachment, the conclusion is true.

b. Given: If Helen is going to work, then she is wearing pearls. Helen is wearing pearls.

Conclusion: Helen is going to work.

The given statement *Helen is going to work* satisfies the conclusion of the true conditional. However, knowing that a conditional statement and its conclusion are true does not make the hypothesis true. Helen could be wearing pearls on a date. The conclusion is invalid.

Exercises

Determine whether the stated conclusion is valid based on the given information. If not, write *invalid*. Explain your reasoning.

1. Given: If a number is divisible by 6, then the number is divisible by 3. 18 is divisible by 6.

Conclusion: 18 is divisible by 3.

2. Given: If a pet is a rabbit, then it eats carrots. Jennie's pet eats carrots.

Conclusion: Jennie's pet is a rabbit.

3. Given: If a hen is a Plymouth Rock, then her eggs are brown. Berta is a Plymouth Rock hen.

Conclusion: Berta's eggs are brown.

2-4 Study Guide and Intervention *(continued)*

Deductive Reasoning

Law of Syllogism Another way to make a valid conclusion is to use the **Law of Syllogism**. It allows you to draw conclusions from two true statements when the conclusion of one statement is the hypothesis of another.

Law of Syllogism	If $p \rightarrow q$ is true and $q \rightarrow r$ is true, then $p \rightarrow r$ is also true.

Example **The two conditional statements below are true. Use the Law of Syllogism to find a valid conclusion. State the conclusion.**

(1) If a number is a whole number, then the number is an integer.
(2) If a number is an integer, then it is a rational number.

p: A number is a whole number.

q: A number is an integer.

r: A number is a rational number.

The two conditional statements are $p \rightarrow q$ and $q \rightarrow r$. Using the Law of Syllogism, a valid conclusion is $p \rightarrow r$. A statement of $p \rightarrow r$ is "if a number is a whole number, then it is a rational number."

Exercises

Use the Law of Syllogism to draw a valid conclusion from each set of statements, if possible. If no valid conclusion is possible, write *no valid conclusion*.

1. If a dog eats Superdog Dog Food, he will be happy.
Rover is happy.

2. If an angle is supplementary to an obtuse angle, then it is acute.
If an angle is acute, then its measure is less than 90.

3. If the measure of $\angle A$ is less than 90, then $\angle A$ is acute.
If $\angle A$ is acute, then $\angle A \cong \angle B$.

4. If an angle is a right angle, then the measure of the angle is 90.
If two lines are perpendicular, then they form a right angle.

5. If you study for the test, then you will receive a high grade.
Your grade on the test is high.

NAME ______________________________ DATE ____________ PERIOD ____________

2-5 Study Guide and Intervention

Postulates and Paragraph Proofs

Points, Lines, and Planes In geometry, a **postulate** is a statement that is accepted as true. Postulates describe fundamental relationships in geometry.

Postulate 2.1:	Through any two points, there is exactly one line.
Postulate 2.2:	Through any three noncollinear points, there is exactly one plane.
Postulate 2.3:	A line contains at least two points.
Postulate 2.4:	A plane contains at least three noncollinear points.
Postulate 2.5:	If two points lie in a plane, then the entire line containing those points lies in the plane.
Postulate 2.6:	If two lines intersect, then their intersection is exactly one point.
Postulate 2.7:	If two planes intersect, then their intersection is a line.

Example **Determine whether each statement is *always, sometimes, or never true.***

a. **There is exactly one plane that contains points *A*, *B*, and *C*.**
Sometimes; if *A*, *B*, and *C* are collinear, they are contained in many planes. If they are noncollinear, then they are contained in exactly one plane.

b. **Points *E* and *F* are contained in exactly one line.**
Always; the first postulate states that there is exactly one line through any two points.

c. **Two lines intersect in two distinct points *M* and *N*.**
Never; the intersection of two lines is one point.

Exercises

Determine whether each statement is *always, sometimes,* or *never* true.

1. A line contains exactly one point.

2. Noncollinear points *R*, *S*, and *T* are contained in exactly one plane.

3. Any two lines ℓ and m intersect.

4. If points *G* and *H* are contained in plane $\mathcal{M}$, then $\overline{GH}$ is perpendicular to plane $\mathcal{M}$.

5. Planes $\mathcal{R}$ and $\mathcal{S}$ intersect in point *T*.

6. If points *A*, *B*, and *C* are noncollinear, then segments $\overline{AB}$, $\overline{BC}$, and $\overline{CA}$ are contained in exactly one plane.

In the figure, $\overline{AC}$ and $\overline{DE}$ are in plane *Q* and $\overline{AC} \parallel \overline{DE}$. State the postulate that can be used to show each statement is true.

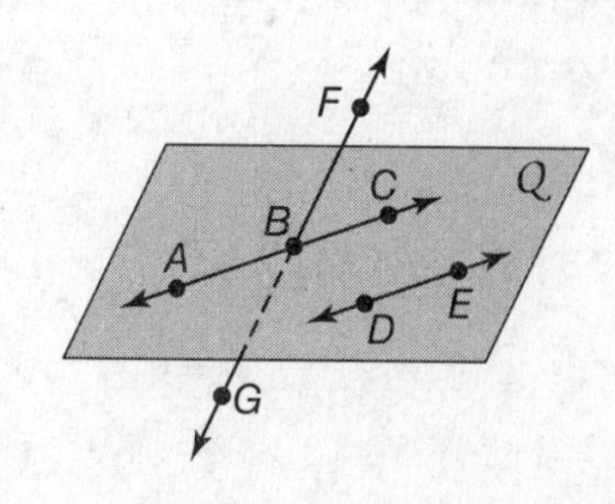

7. Exactly one plane contains points *F*, *B*, and *E*.

8. $\overleftrightarrow{BE}$ lies in plane *Q*.

2-5 Study Guide and Intervention *(continued)*

Postulates and Paragraph Proofs

Paragraph Proofs A logical argument that uses deductive reasoning to reach a valid conclusion is called a **proof**. In one type of proof, a **paragraph proof**, you write a paragraph to explain why a statement is true.

A statement that can be proved true is called a **theorem**. You can use undefined terms, definitions, postulates, and already-proved theorems to prove other statements true.

Example **In $\triangle ABC$, $\overline{BD}$ is an angle bisector. Write a paragraph proof to show that $\angle ABD \cong \angle CBD$.**

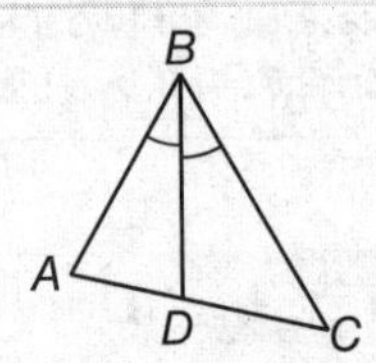

By definition, an angle bisector divides an angle into two congruent angles. Since $\overline{BD}$ is an angle bisector, $\angle ABC$ is divided into two congruent angles. Thus, $\angle ABD \cong \angle CBD$.

Exercises

1. Given that $\angle A \cong \angle D$ and $\angle D \cong \angle E$, write a paragraph proof to show that $\angle A \cong \angle E$.

2. It is given that $\overline{BC} \cong \overline{EF}$, M is the midpoint of $\overline{BC}$, and N is the midpoint of $\overline{EF}$. Write a paragraph proof to show that $BM = EN$.

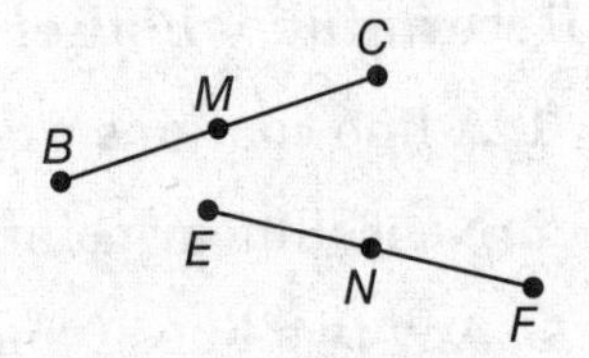

3. Given that S is the midpoint of $\overline{QP}$, T is the midpoint of $\overline{PR}$, and P is the midpoint of $\overline{ST}$, write a paragraph proof to show that $QS = TR$.

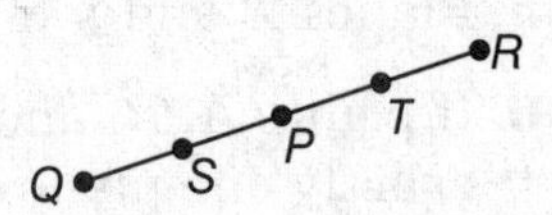

NAME ______________________ DATE ____________ PERIOD ____________

2-6 Study Guide and Intervention

Algebraic Proof

Algebraic Proof A list of algebraic steps to solve problems where each step is justified is called an **algebraic proof**, The table shows properties you have studied in algebra.

The following properties are true for any real numbers *a*, *b*, and *c*.

Addition Property of Equality	If $a = b$, then $a + c = b + c$.
Subtraction Property of Equality	If $a = b$, the $a - c = b - c$.
Multiplication Property of Equality	If $a = b$, then $a \cdot c = b \cdot c$.
Division Property of Equality	If $a = b$ and $c \neq 0$, then, $\frac{a}{c} = \frac{b}{c}$.
Reflexive Property of Equality	$a = a$
Symmetric Property of Equality	If $a = b$ and $b = a$.
Transitive Property of Equality	If $a = b$ and $b = c$, then $a = c$.
Substitution Property of Equality	If $a = b$,then a may be replaced by b in any equation or expression.
Distributive Property	$a(b + c) = ab + ac$

Example **Solve $6x + 2(x - 1) = 30$. Write a justification for each step.**

Algebraic Steps	**Properties**
$6x + 2(x - 1) = 30$	Original equation or Given
$6x + 2x - 2 = 30$	Distributive Property
$8x - 2 = 30$	Substitution Property of Equality
$8x - 2 + 2 = 30 + 2$	Addition Property of Equality
$8x = 32$	Substitution Property of Equality
$\frac{8x}{8} = \frac{32}{8}$	Division Property of Equality
$x = 4$	Substitution Property of Equality

Exercises

Complete each proof.

1. Given: $\frac{4x + 6}{2} = 9$
Prove: $x = 3$
Proof:

Statements	Reasons
a. $\frac{4x + 6}{2} = 9$	**a.** ________
b. _$\left(\frac{4x + 6}{2}\right) = 2(9)$	**b.** Mult. Prop.
c. $4x + 6 = 18$	**c.** ________
d. $4x + 6 - 6 = 18 - 6$	**d.** ________
e. $4x =$ ________	**e.** Substitution
f. $\frac{4x}{4} =$ ________	**f.** Div. Prop.
g. ________	**g.** Substitution

2. Given: $4x + 8 = x + 2$
Prove: $x = -2$
Proof:

Statements	Reasons
a. $4x + 8 = x + 2$	**a.** ________
b. $4x + 8 - x = x + 2 - x$	**b.** ________
c. $3x + 8 = 2$	**c.** Substitution
d. ________	**d.** Subtr. Prop.
e. ________	**e.** Substitution
f. $\frac{3x}{3} = \frac{-6}{3}$	**f.** ________
g. ________	**g.** Substitution

NAME ______________________ DATE __________ PERIOD __________

2-6 Study Guide and Intervention *(continued)*

Algebraic Proof

Geometric Proof Geometry deals with numbers as measures, so geometric proofs use properties of numbers. Here are some of the algebraic properties used in proofs.

Property	Segments	Angles
Reflexive	$AB = AB$	$m\angle 1 = m\angle 1$
Symmetric	If $AB = CD$, then $CD = AB$.	If $m\angle 1 = m\angle 2$, then $m\angle 2 = m\angle 1$.
Transitive	If $AB = CD$ and $CD = EF$, then $AB = EF$.	If $m\angle 1 = m\angle 2$ and $m\angle 2 = m\angle 3$, then $m\angle 1 = m\angle 3$.

Example **Write a two-column proof to verify this conjecture.**

Given: $m\angle 1 = m\angle 2$, $m\angle 2 = m\angle 3$
Prove: $m\angle 1 = m\angle 3$
Proof:

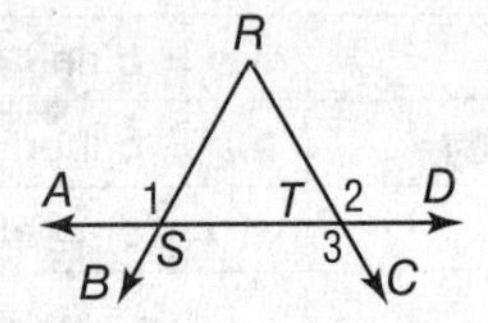

Statements	Reasons
1. $m\angle 1 = m\angle 2$	**1.** Given
2. $m\angle 2 = m\angle 3$	**2.** Given
3. $m\angle 1 = m\angle 3$	**3.** Transitive Property of Equality

Exercises

State the property that justifies each statement.

1. If $m\angle 1 = m\angle 2$, then $m\angle 2 = m\angle 1$.

2. If $m\angle 1 = 90$ and $m\angle 2 = m\angle 1$, then $m\angle 2 = 90$.

3. If $AB = RS$ and $RS = WY$, then $AB = WY$.

4. If $AB = CD$, then $\frac{1}{2}AB = \frac{1}{2}CD$.

5. If $m\angle 1 + m\angle 2 = 110$ and $m\angle 2 = m\angle 3$, then $m\angle 1 + m\angle 3 = 110$.

6. $RS = RS$

7. If $AB = RS$ and $TU = WY$, then $AB + TU = RS + WY$.

8. If $m\angle 1 = m\angle 2$ and $m\angle 2 = m\angle 3$, then $m\angle 1 = m\angle 3$.

9. If the formula for the area of a triangle is $A = \frac{1}{2}bh$, then bh is equal to 2 times the area of the triangle. Write a two-column proof to verify this conjecture.

NAME ______________________ DATE ____________ PERIOD ____________

2-7 Study Guide and Intervention

Proving Segment Relationships

Segment Addition Two basic postulates for working with segments and lengths are the Ruler Postulate, which establishes number lines, and the Segment Addition Postulate, which describes what it means for one point to be between two other points.

Ruler Postulate	The points on any line or line segment can be put into one-to-one correspondence with real numbers.
Segment Addition Postulate	If *A*, *B*, and *C* are collinear, then point *B* is between *A* and *C* if and only if $AB + BC = AC$.

Example **Write a two-column proof.**

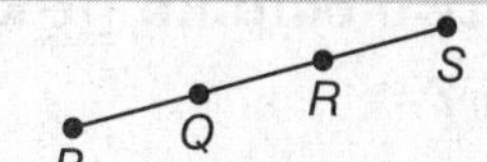

Given: Q is the midpoint of $\overline{PR}$.
R is the midpoint of $\overline{QS}$.

Prove: $PR = QS$

Proof:

Statements	Reasons
1. Q is the midpoint of $\overline{PR}$.	**1.** Given
2. $PQ = QR$	**2.** Definition of midpoint
3. R is the midpoint of $\overline{QS}$.	**3.** Given
4. $QR = RS$	**4.** Definition of midpoint
5. $PQ + QR = QR + RS$	**5.** Addition Property
6. $PQ = RS$	**6.** Transitive Property
7. $PQ + QR = PR$, $QR + RS = QS$	**7.** Segment Addition Postulate
8. $PR = QS$	**8.** Substitution

Exercises

Complete each proof.

1. Given: $BC = DE$
Prove: $AB + DE = AC$
Proof:

Statements	Reasons
1. $BC = DE$	**1.** ________
2. ________	**2.** Seg. Add. Post.
3. $AB + DE = AC$	**3.** ________

2. Given: Q is between P and R, R is between Q and S, $PR = QS$.

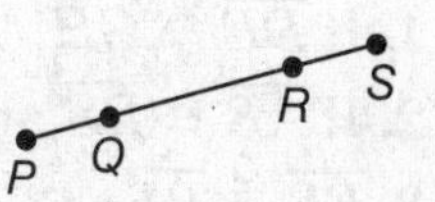

Prove: $PQ = RS$
Proof:

Statements	Reasons
1. Q is between P and R.	**1.** Given
2. $PQ + QR = PR$	**2.** ________
3. R is between Q and S.	**3.** ________
4. ________	**4.** Seg. Add. Post.
5. $PR = QS$	**5.** ________
6. $PQ + QR = QR + RS$	**6.** ________
7. $PQ + QR - QR = QR + RS - QR$	**7.** ________
8. ________	**8.** Substitution

NAME ______________________ DATE ____________ PERIOD ____________

2-7 Study Guide and Intervention *(continued)*

Proving Segment Relationships

Segment Congruence Remember that segment measures are reflexive, symmetric, and transitive. Since segments with the same measure are congruent, congruent segments are also reflexive, symmetric, and transitive.

Reflexive Property	$\overline{AB} \cong \overline{AB}$
Symmetric Property	If $\overline{AB} \cong \overline{CD}$, then $\overline{CD} \cong \overline{AB}$.
Transitive Property	If $\overline{AB} \cong \overline{CD}$ and $\overline{CD} \cong \overline{EF}$, then $\overline{AB} \cong \overline{EF}$.

Example **Write a two-column proof.**

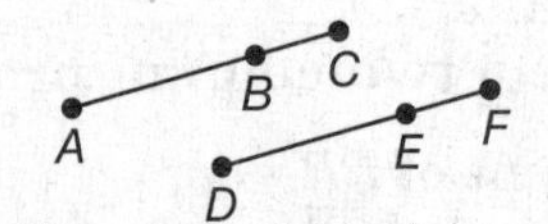

Given: $\overline{AB} \cong \overline{DE}$; $\overline{BC} \cong \overline{EF}$
Prove: $\overline{AC} \cong \overline{DF}$
Proof:

Statements	Reasons
1. $\overline{AB} \cong \overline{DE}$	1. Given
2. $AB = DE$	2. Definition of congruence of segments
3. $\overline{BC} \cong \overline{EF}$	3. Given
4. $BC = EF$	4. Definition of congruence of segments
5. $AB + BC = DE + EF$	5. Addition Property
6. $AB + BC = AC$, $DE + EF = DF$	6. Segment Addition Postulate
7. $AC = DF$	7. Substitution
8. $\overline{AC} \cong \overline{DF}$	8. Definition of congruence of segments

Exercises

Justify each statement with a property of congruence.

1. If $\overline{DE} \cong \overline{GH}$, then $\overline{GH} \cong \overline{DE}$.
2. If $\overline{AB} \cong \overline{RS}$ and $\overline{RS} \cong \overline{WY}$ then $\overline{AB} \cong \overline{WY}$.
3. $\overline{RS} \cong \overline{RS}$
4. Complete the proof.

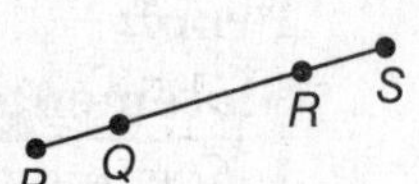

Given: $\overline{PR} \cong \overline{QS}$
Prove: $\overline{PQ} \cong \overline{RS}$
Proof:

Statements	Reasons
a. $\overline{PR} \cong \overline{QS}$	a. ______________
b. $PR = QS$	b. ______________
c. $PQ + QR = PR$	c. ______________
d. ______________	d. Segment Addition Postulate
e. $PQ + QR = QR + RS$	e. ______________
f. ______________	f. Subtraction Property
g. ______________	g. Definition of congruence of segments

NAME _______________ DATE _______ PERIOD _______

2-8 Study Guide and Intervention

Proving Angle Relationships

Supplementary and Complementary Angles There are two basic postulates for working with angles. The Protractor Postulate assigns numbers to angle measures, and the Angle Addition Postulate relates parts of an angle to the whole angle.

Protractor Postulate	Given any angle, the measure can be put into one-to-one correspondance with real numbers between 0 and 180.	
Angle Addition Postulate	R is in the interior of $\angle PQS$ if and only if $m\angle PQR + m\angle RQS = m\angle PQS$.	

The two postulates can be used to prove the following two theorems.

Supplement Theorem	If two angles form a linear pair, then they are supplementary angles. **Example:** If $\angle 1$ and $\angle 2$ form a linear pair, then $m\angle 1 + m\angle 2 = 180$.	
Complement Theorem	If the noncommon sides of two adjacent angles form a right angle, then the angles are complementary angles. **Example:** If $\overleftrightarrow{GF} \perp \overleftrightarrow{GH}$, then $m\angle 3 + m\angle 4 = 90$.	

Example 1 **If $\angle 1$ and $\angle 2$ form a linear pair and $m\angle 2 = 115$, find $m\angle 1$.**

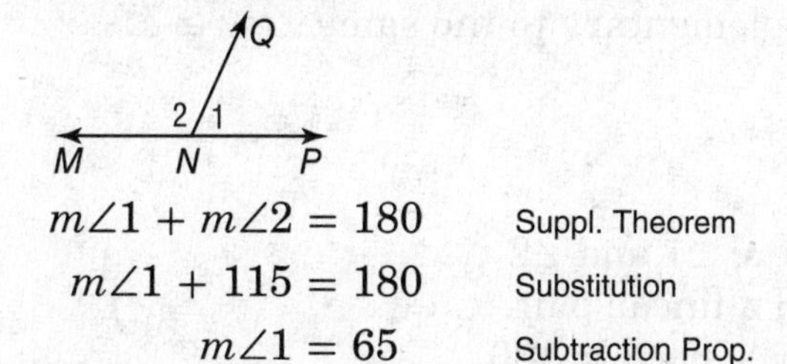

$m\angle 1 + m\angle 2 = 180$	Suppl. Theorem
$m\angle 1 + 115 = 180$	Substitution
$m\angle 1 = 65$	Subtraction Prop.

Example 2 **If $\angle 1$ and $\angle 2$ form a right angle and $m\angle 2 = 20$, find $m\angle 1$.**

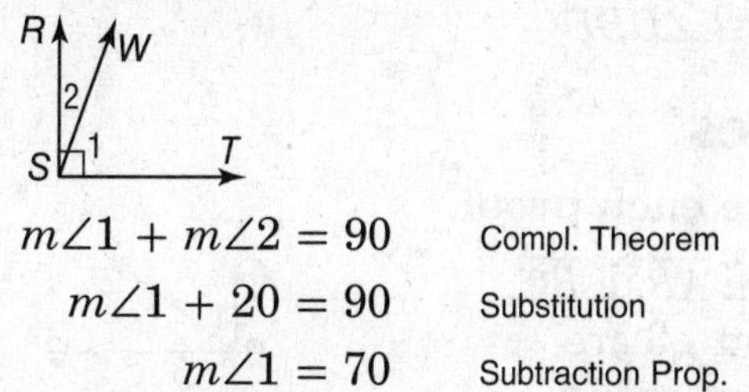

$m\angle 1 + m\angle 2 = 90$	Compl. Theorem
$m\angle 1 + 20 = 90$	Substitution
$m\angle 1 = 70$	Subtraction Prop.

Exercises

Find the measure of each numbered angle and name the theorem that justifies your work.

1.

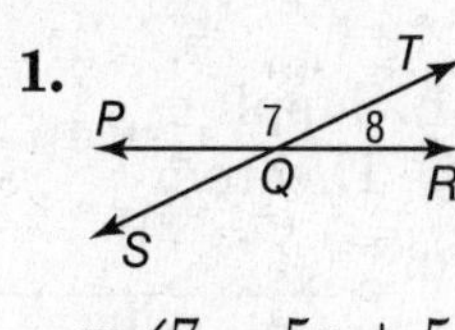

$m\angle 7 = 5x + 5$,
$m\angle 8 = x - 5$

2.

$m\angle 5 = 5x$, $m\angle 6 = 4x + 6$,
$m\angle 7 = 10x$,
$m\angle 8 = 12x - 12$

3.

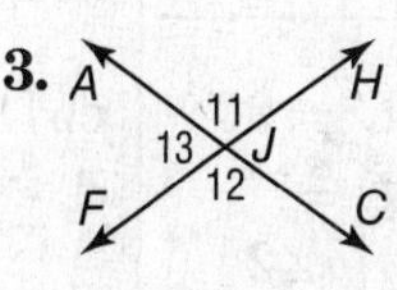

$m\angle 11 = 11x$,
$m\angle 13 = 10x + 12$

2-8 Study Guide and Intervention *(continued)*

Proving Angle Relationships

Congruent and Right Angles The Reflexive Property of Congruence, Symmetric Property of Congruence, and Transitive Property of Congruence all hold true for angles. The following theorems also hold true for angles.

Congruent Supplements Theorem	Angles supplement to the same angle or congruent angles are congruent.
Congruent Compliments Theorem	Angles compliment to the same angle or to congruent angles are congruent.
Vertical Angles Theorem	If two angles are vertical angles, then they are congruent.
Theorem 2.9	Perpendicular lines intersect to form four right angles.
Theorem 2.10	All right angles are congruent.
Theorem 2.11	Perpendicular lines form congruent adjacent angles.
Theorem 2.12	If two angles are congruent and supplementary, then each angle is a right angle.
Theorem 2.13	If two congruent angles form a linear pair, then they are right angles.

Example **Write a two-column proof.**

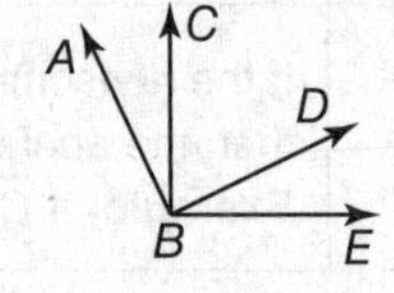

Given: $\angle ABC$ and $\angle CBD$ are complementary.
$\angle DBE$ and $\angle CBD$ form a right angle.

Prove: $\angle ABC \cong \angle DBE$

Statements	**Reasons**
1. $\angle ABC$ and $\angle CBD$ are complementary. $\angle DBE$ and $\angle CBD$ form a right angle.	1. Given
2. $\angle DBE$ and $\angle CBD$ are complementary.	2. Complement Theorm
3. $\angle ABC \cong \angle DBE$	3. $\angle$s complementary to the same $\angle$ or $\cong$ $\angle$s are $\cong$.

Exercises

Complete each proof.

1. Given: $\overline{AB} \perp \overline{BC}$;
$\angle 1$ and $\angle 3$ are complementary.

Prove: $\angle 2 \cong \angle 3$

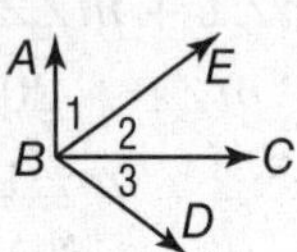

Proof:

Statements	**Reasons**
a. $\overline{AB} \perp \overline{BC}$	**a.** __________
b. __________	**b.** Definition of $\perp$
c. $m\angle ABC = 90$	**c.** Def. of right angle
d. $m\angle ABC = m\angle 1 + m\angle 2$	**d.** __________
e. $90 = m\angle 1 + m\angle 2$	**e.** Substitution
f. $\angle 1$ and $\angle 2$ are compl.	**f.** __________
g. __________	**g.** Given
h. $\angle 2 \cong \angle 3$	**h.** __________

2. Given: $\angle 1$ and $\angle 2$ form a linear pair.
$m\angle 1 + m\angle 3 = 180$

Prove: $\angle 2 \cong \angle 3$

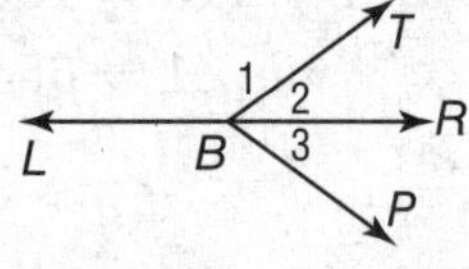

Proof:

Statements	**Reasons**
a. $\angle 1$ and $\angle 2$ form a linear pair. $m\angle 1 + m\angle 3 = 180$	**a.** Given
b. __________	**b.** Suppl. Theorem
c. $\angle 1$ is suppl. to $\angle 3$.	**c.** __________
d. __________	**d.** $\angle$s suppl. to the same $\angle$ or $\cong$ $\angle$s are $\cong$.

NAME _______________ DATE _______ PERIOD _______

3-1 Study Guide and Intervention

Parallel Lines and Transversals

Relationships Between Lines and Planes When two lines lie in the same plane and do not intersect, they are **parallel**. Lines that do not intersect and are not coplanar are **skew lines**. In the figure, ℓ is parallel to m, or $\ell \parallel m$. You can also write $\overline{PQ} \parallel \overline{RS}$. Similarly, if two planes do not intersect, they are **parallel planes**.

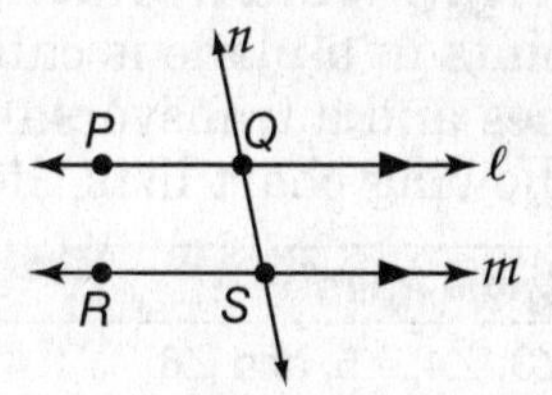

Example **Refer to the figure at the right to identify each of the following.**

a. all planes parallel to plane *ABD*

plane *EFH*

b. all segments parallel to $\overline{CG}$

$\overline{BF}$, $\overline{DH}$, and $\overline{AE}$

c. all segments skew to $\overline{EH}$

$\overline{BF}$, $\overline{CG}$, $\overline{BD}$, $\overline{CD}$, and $\overline{AB}$

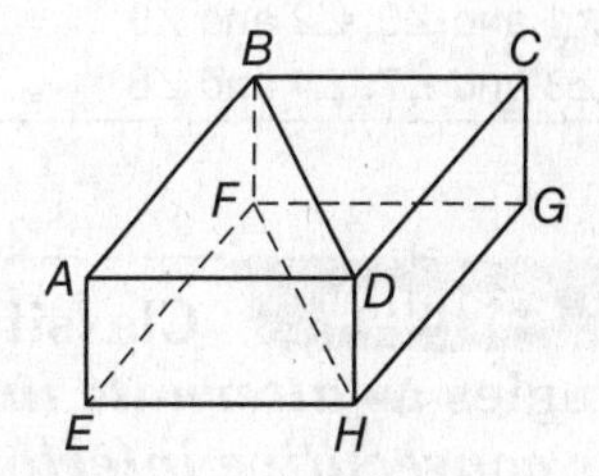

Exercises

Refer to the figure at the right to identify each of the following.

1. all planes that intersect plane *OPT*

2. all segments parallel to $\overline{NU}$

3. all segments that intersect $\overline{MP}$

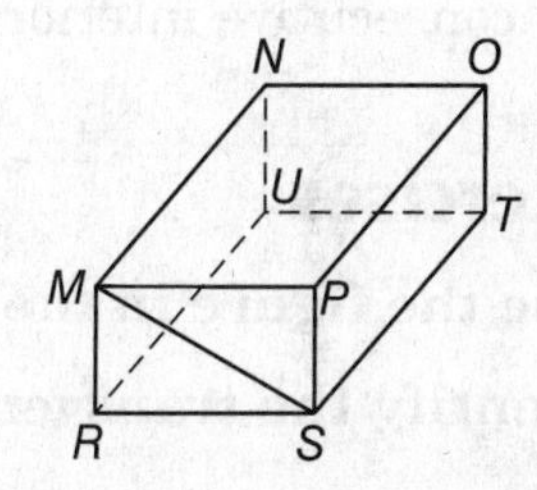

Refer to the figure at the right to identify each of the following.

4. all segments parallel to $\overline{QX}$

5. all planes that intersect plane *MHE*

6. all segments parallel to $\overline{QR}$

7. all segments skew to $\overline{AG}$

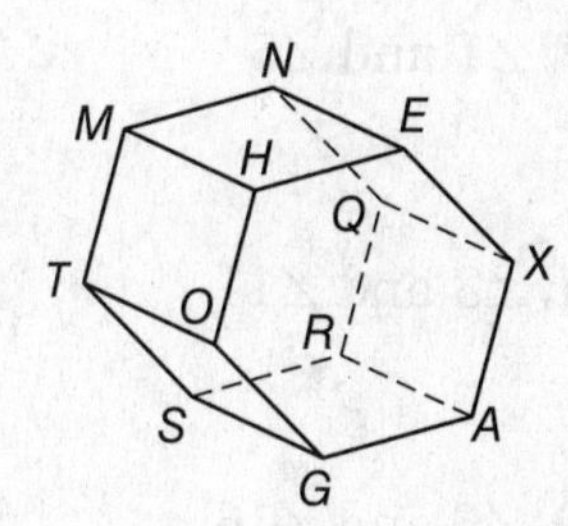

NAME ______________________ DATE __________ PERIOD __________

3-1 Study Guide and Intervention *(continued)*

Parallel Lines and Transversals

Angle Relationships A line that intersects two or more other lines at two different points in a plane is called a **transversal**. In the figure below, line *t* is a transversal. Two lines and a transversal form eight angles. Some pairs of the angles have special names. The following chart lists the pairs of angles and their names.

Angle Pairs	Name
∠3, ∠4, ∠5, and ∠6	interior angles
∠3 and ∠5; ∠4 and ∠6	alternate interior angles
∠3 and ∠6; ∠4 and ∠5	consecutive interior angles
∠1, ∠2, ∠7, and ∠8	exterior angles
∠1 and ∠7; ∠2 and ∠8;	alternate exterior angles
∠1 and ∠5; ∠2 and ∠6; ∠3 and ∠7; ∠4 and ∠8	corresponding angles

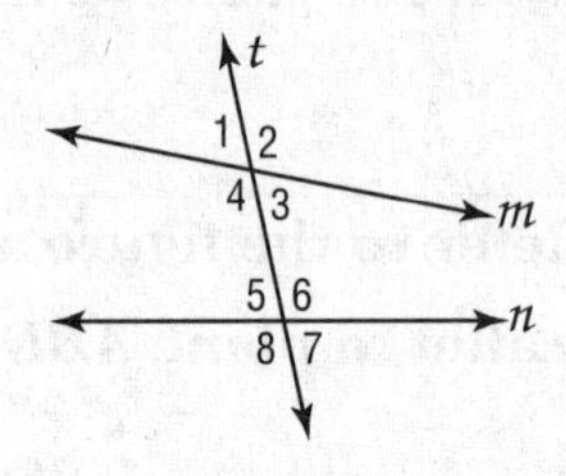

Example **Classify the relationship between each pair of angles as *alternate interior*, *alternate exterior*, *corresponding*, or *consecutive interior* angles.**

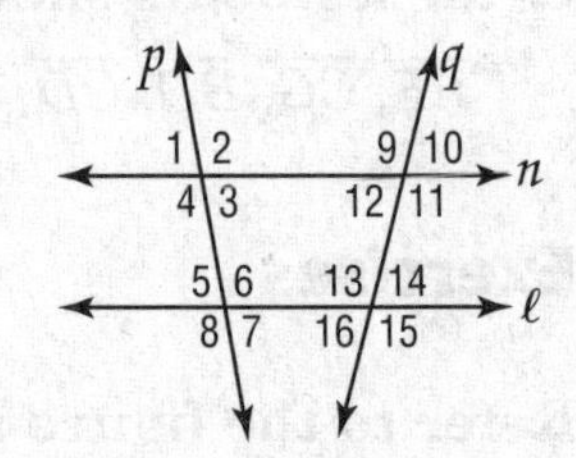

a. ∠10 and ∠16

alternate exterior angles

b. ∠4 and ∠12

corresponding angles

c. ∠12 and ∠13

consecutive interior angles

d. ∠3 and ∠9

alternate interior angles

Exercises

Use the figure in the Example for Exercises 1–12.

Identify the transversal connecting each pair of angles.

1. ∠9 and ∠13

2. ∠5 and ∠14

3. ∠4 and ∠6

Classify the relationship between each pair of angles as *alternate interior*, *alternate exterior*, *corresponding*, or *consecutive interior* angles.

4. ∠1 and ∠5

5. ∠6 and ∠14

6. ∠2 and ∠8

7. ∠3 and ∠11

8. ∠12 and ∠3

9. ∠4 and ∠6

10. ∠6 and ∠16

11. ∠11 and ∠14

12. ∠10 and ∠16

NAME ______________________ DATE ____________ PERIOD ____________

3-2 Study Guide and Intervention

Angles and Parallel Lines

Parallel Lines and Angle Pairs When two parallel lines are cut by a transversal, the following pairs of angles are congruent.

- corresponding angles
- alternate interior angles
- alternate exterior angles

Also, consecutive interior angles are supplementary.

Example **In the figure, $m\angle 2 = 75$. Find the measures of the remaining angles.**

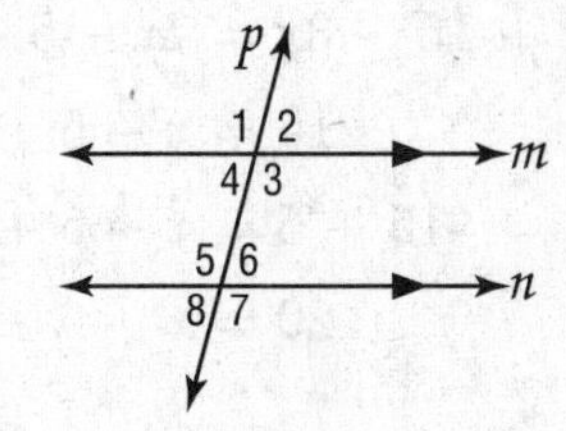

$m\angle 1 = 105$	$\angle 1$ and $\angle 2$ form a linear pair.
$m\angle 3 = 105$	$\angle 3$ and $\angle 2$ form a linear pair.
$m\angle 4 = 75$	$\angle 4$ and $\angle 2$ are vertical angles.
$m\angle 5 = 105$	$\angle 5$ and $\angle 3$ are alternate interior angles.
$m\angle 6 = 75$	$\angle 6$ and $\angle 2$ are corresponding angles.
$m\angle 7 = 105$	$\angle 7$ and $\angle 3$ are corresponding angles.
$m\angle 8 = 75$	$\angle 8$ and $\angle 6$ are vertical angles.

Exercises

In the figure, $m\angle 3 = 102$. Find the measure of each angle. Tell which postulate(s) or theorem(s) you used.

1. $\angle 5$
2. $\angle 6$
3. $\angle 11$
4. $\angle 7$
5. $\angle 15$
6. $\angle 14$

In the figure, $m\angle 9 = 80$ and $m\angle 5 = 68$. Find the measure of each angle. Tell which postulate(s) or theorem(s) you used.

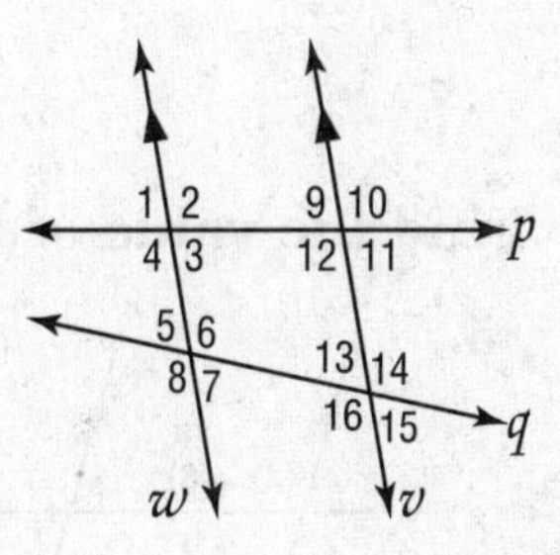

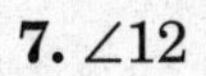

7. $\angle 12$
8. $\angle 1$
9. $\angle 4$
10. $\angle 3$
11. $\angle 7$
12. $\angle 16$

3-2 Study Guide and Intervention *(continued)*

Angles and Parallel Lines

Algebra and Angle Measures Algebra can be used to find unknown values in angles formed by a transversal and parallel lines.

Example **If $m\angle 1 = 3x + 15$, $m\angle 2 = 4x - 5$, and $m\angle 3 = 5y$, find the value of x and y.**

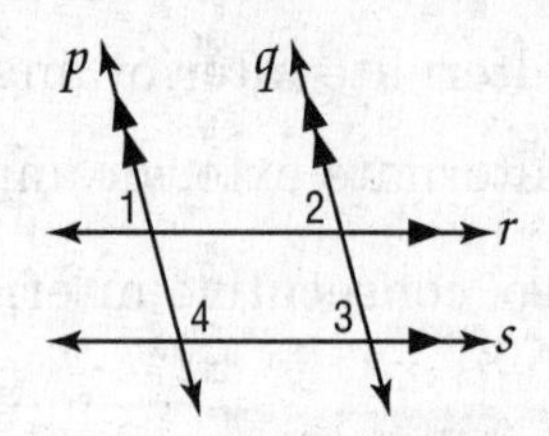

$p \parallel q$, so $m\angle 1 = m\angle 2$ because they are corresponding angles.

$$m\angle 1 = m\angle 2$$
$$3x + 15 = 4x - 5$$
$$3x + 15 - 3x = 4x - 5 - 3x$$
$$15 = x - 5$$
$$15 + 5 = x - 5 + 5$$
$$20 = x$$

$r \parallel s$, so $m\angle 2 = m\angle 3$ because they are corresponding angles.

$$m\angle 2 = m\angle 3$$
$$75 = 5y$$
$$\frac{75}{5} = \frac{5y}{5}$$
$$15 = y$$

Exercises

Find the value of the variable(s) in each figure. Explain your reasoning.

1.

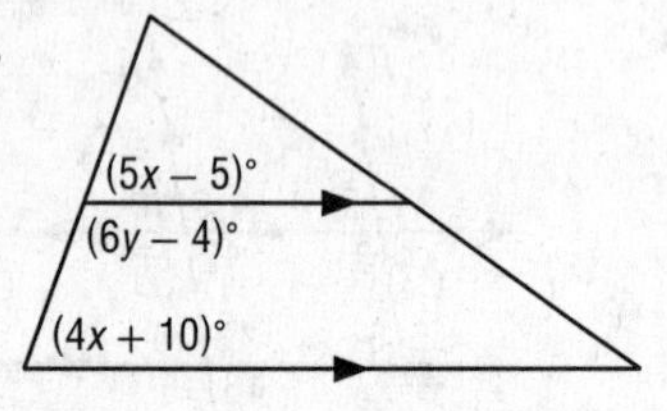

2.

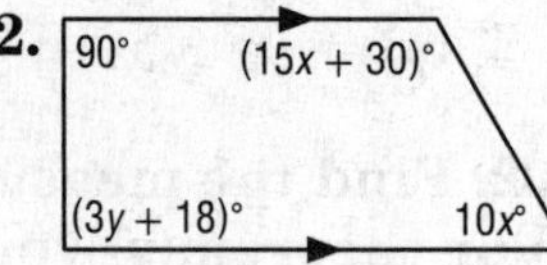

3.

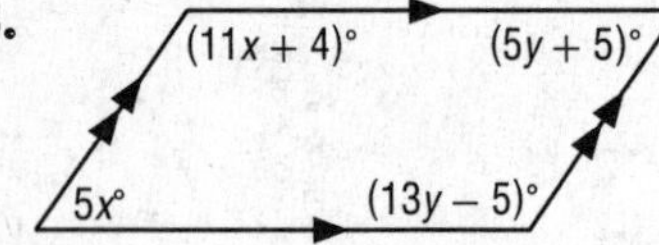

4.

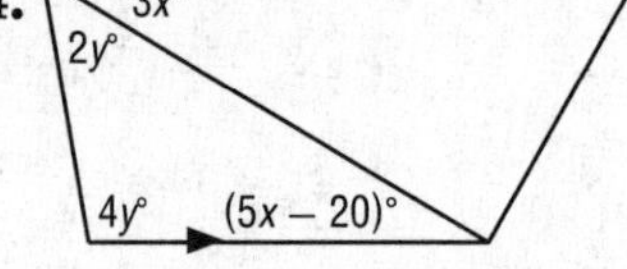

Find the value of the variable(s) in each figure. Explain your reasoning.

5.

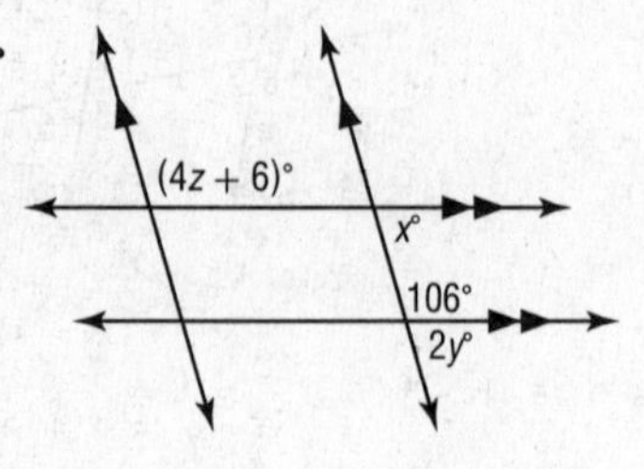

6.

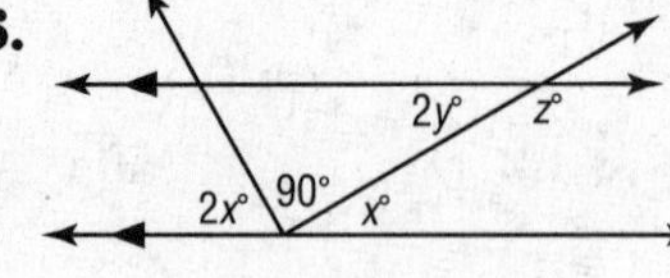

NAME ______________________ DATE ____________ PERIOD ________

3-3 Study Guide and Intervention

Slopes of Lines

Slope of a Line The slope m of a line containing two points with coordinates (x_1, y_1) and (x_2, y_2) is given by the formula $m = \frac{y_2 - y_1}{x_2 - x_1}$, where $x_1 \neq x_2$.

Example **Find the slope of each line.**

For line p, substitute (1, 2) for (x_1, y_1) and (−2, −2) for (x_2, y_2).

$$m = \frac{y_2 - y_1}{x_2 - x_1}$$

$$= \frac{-2 - 2}{-2 - 1} \text{ or } \frac{4}{3}$$

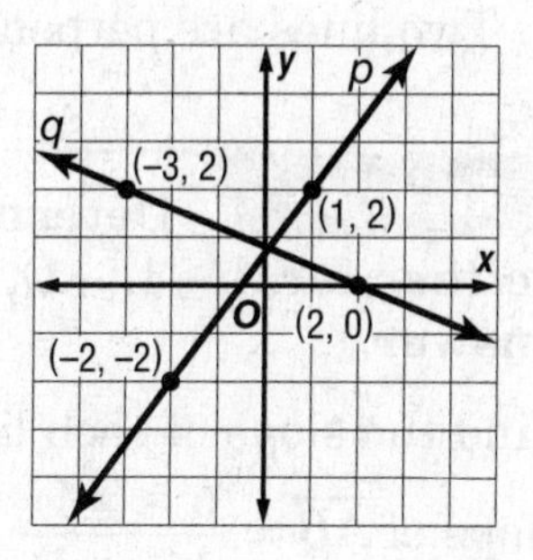

For line q, substitute (2, 0) for (x_1, y_1) and (−3, 2) for (x_2, y_2).

$$m = \frac{y_2 - y_1}{x_2 - x_1}$$

$$= \frac{2 - 0}{-3 - 2} \text{ or } -\frac{2}{5}$$

Exercises

Determine the slope of the line that contains the given points.

1. $J(0, 0)$, $K(-2, 8)$

2. $R(-2, -3)$, $S(3, -5)$

3. $L(1, -2)$, $N(-6, 3)$

4. $P(-1, 2)$, $Q(-9, 6)$

5. $T(1, -2)$, $U(6, -2)$

6. $V(-2, 10)$, $W(-4, -3)$

Find the slope of each line.

7. $\overleftrightarrow{AB}$

8. $\overleftrightarrow{CD}$

9. $\overleftrightarrow{EM}$

10. $\overleftrightarrow{AE}$

11. $\overleftrightarrow{EH}$

12. $\overleftrightarrow{BM}$

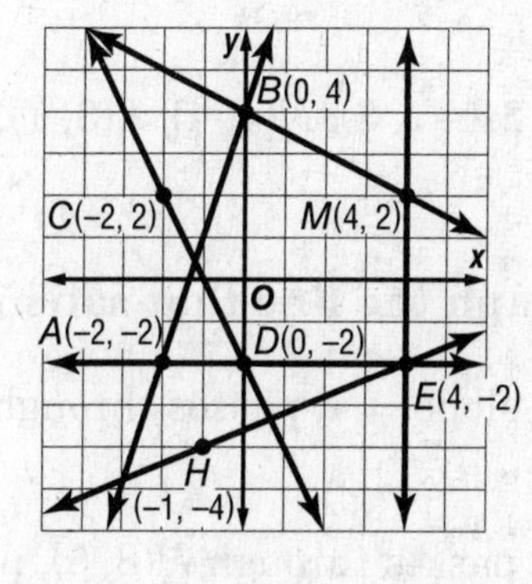

NAME ______________________________ DATE ____________ PERIOD ____________

3-3 Study Guide and Intervention *(continued)*

Slopes of Lines

Parallel and Perpendicular Lines If you examine the slopes of pairs of parallel lines and the slopes of pairs of perpendicular lines, where neither line in each pair is vertical, you will discover the following properties.

Two lines have the same slope if and only if they are parallel.

Two lines are perpendicular if and only if the product of their slopes is −1.

Example **Determine whether $\overleftrightarrow{AB}$ and $\overleftrightarrow{CD}$ are *parallel, perpendicular,* or *neither* for $A(-1, -1)$, $B(1, 5)$, $C(1, 2)$, $D(5, 4)$. Graph each line to verify your answer.**

Find the slope of each line.

slope of $\overleftrightarrow{AB} = \dfrac{5-(-1)}{1-(-1)} = \dfrac{6}{2}$ or 3 slope of $\overleftrightarrow{CD} = \dfrac{4-2}{5-1} = \dfrac{2}{4} = \dfrac{1}{2}$

The two lines do not have the same slope, so they are *not* parallel.
To determine if the lines are perpendicular, find the product of their slopes

$3\left(\dfrac{1}{2}\right) = \dfrac{3}{2}$ or 1.5 Product of slope for $\overleftrightarrow{AB}$ and $\overleftrightarrow{CD}$

Since the product of their slopes is *not* –1, the two lines are *not* perpendicular.
Therefore, there is no relationship between $\overleftrightarrow{AB}$ and $\overleftrightarrow{CD}$.

When graphed, the two lines intersect but not at a right angle.

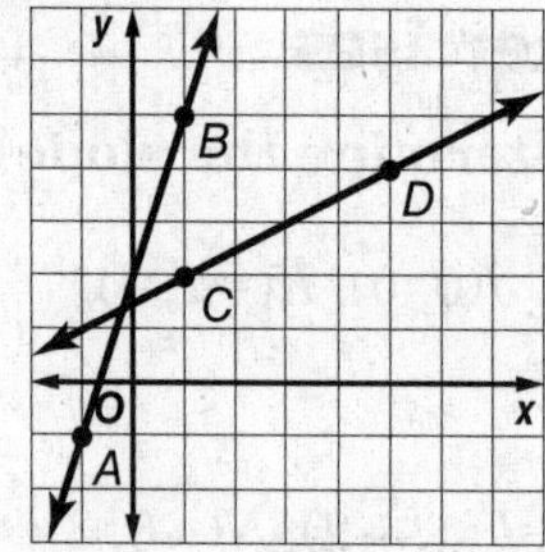

Exercises

Determine whether $\overleftrightarrow{MN}$ and $\overleftrightarrow{RS}$ are *parallel, perpendicular,* or *neither*. Graph each line to verify your answer.

1. $M(0, 3)$, $N(2, 4)$, $R(2, 1)$, $S(8, 4)$

2. $M(-1, 3)$, $N(0, 5)$, $R(2, 1)$, $S(6, -1)$

3. $M(-1, 3)$, $N(4, 4)$, $R(3, 1)$, $S(-2, 2)$

4. $M(0, -3)$, $N(-2, -7)$, $R(2, 1)$, $S(0, -3)$

Graph the line that satisfies each condition.

5. slope = 4, passes through (6, 2)

6. passes through $H(8, 5)$, perpendicular to $\overleftrightarrow{AG}$ with $A(-5, 6)$ and $G(-1, -2)$

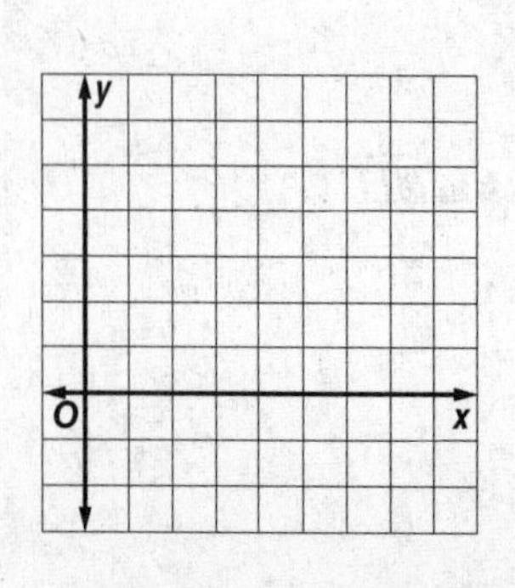

7. passes through $C(-2, 5)$, parallel to $\overleftrightarrow{LB}$ with $L(2, 1)$ and $B(7, 4)$

NAME ______________________ DATE ____________ PERIOD ________

3-4 Study Guide and Intervention

Equations of Lines

Write Equations of Lines You can write an equation of a line if you are given any of the following:

- the slope and the y-intercept,
- the slope and the coordinates of a point on the line, or
- the coordinates of two points on the line.

If m is the slope of a line, b is its y-intercept, and (x_1, y_1) is a point on the line, then:

- the **slope-intercept form** of the equation is $y = mx + b$,
- the **point-slope form** of the equation is $y - y_1 = m(x - x_1)$.

Example 1 **Write an equation in slope-intercept form of the line with slope −2 and y-intercept 4.**

$y = mx + b$ Slope-intercept form

$y = -2x + 4$ $m = -2$, $b = 4$

The slope-intercept form of the equation of the line is $y = -2x + 4$.

Example 2 **Write an equation in point-slope form of the line with slope $-\frac{3}{4}$ that contains (8, 1).**

$y - y_1 = m(x - x_1)$ Point-slope form

$y - 1 = -\frac{3}{4}(x - 8)$ $m = -\frac{3}{4}$, $(x_1, y_1) = (8, 1)$

The point-slope form of the equation of the line is $y - 1 = -\frac{3}{4}(x - 8)$.

Exercises

Write an equation in slope-intercept form of the line having the given slope and y-intercept or given points. Then graph the line.

1. m: 2, b: −3

2. m: $-\frac{1}{2}$, b: 4

3. m: $\frac{1}{4}$, b: 5

4. m: 0, b: −2

5. m: $-\frac{5}{3}$, $(0, \frac{1}{3})$

6. m: −3, (1,−11)

Write an equation in point-slope form of the line having the given slope that contains the given point. Then graph the line.

7. $m = \frac{1}{2}$, (3, −1)

8. $m = -2$, (4, −2)

9. $m = -1$, (−1, 3)

10. $m = \frac{1}{4}$, (−3, −2)

11. $m = -\frac{5}{2}$, (0, −3)

12. $m = 0$, (−2, 5)

NAME ______________________ DATE __________ PERIOD __________

3-4 Study Guide and Intervention *(continued)*

Equations of Lines

Write Equations to Solve Problems Many real-world situations can be modeled using linear equations.

Example **Donna offers computer services to small companies in her city. She charges \$55 per month for maintaining a web site and \$45 per hour for each service call.**

a. Write an equation to represent the total monthly cost, *C*, for maintaining a web site and for *h* hours of service calls.

For each hour, the cost increases \$45. So the rate of change, or slope, is 45. The *y*-intercept is located where there are 0 hours, or \$55.

$C = mh + b$

$= 45h + 55$

b. Donna may change her costs to represent them by the equation $C = 25h + 125$, where \$125 is the fixed monthly fee for a web site and the cost per hour is \$25. Compare her new plan to the old one if a company has $5\frac{1}{2}$ hours of service calls. Under which plan would Donna earn more?

First plan

For $5\frac{1}{2}$ hours of service Donna would earn

$C = 45h + 55 = 45\left(5\frac{1}{2}\right) + 55$

$= 247.5 + 55$ or \$302.50

Second Plan

For $5\frac{1}{2}$ hours of service Donna would earn

$C = 25h + 125 = 25(5.5) + 125$

$= 137.5 + 125$ or \$262.50

Donna would earn more with the first plan.

Exercises

For Exercises 1–4, use the following information.

Jerri's current satellite television service charges a flat rate of \$34.95 per month for the basic channels and an additional \$10 per month for each premium channel. A competing satellite television service charges a flat rate of \$39.99 per month for the basic channels and an additional \$8 per month for each premium channel.

1. Write an equation in slope-intercept form that models the total monthly cost for each satellite service, where *p* is the number of premium channels.

2. If Jerri wants to include three premium channels in her package, which service would be less, her current service or the competing service?

3. A third satellite company charges a flat rate of \$69 for all channels, including the premium channels. If Jerri wants to add a fourth premium channel, which service would be least expensive?

4. Write a description of how the fee for the number of premium channels is reflected in the equation.

NAME ______________________ DATE __________ PERIOD __________

3-5 Study Guide and Intervention

Proving Lines Parallel

Identify Parallel Lines If two lines in a plane are cut by a transversal and certain conditions are met, then the lines must be parallel.

If	then
• corresponding angles are congruent, • alternate exterior angles are congruent, • consecutive interior angles are supplementary, • alternate interior angles are congruent, or • two lines are perpendicular to the same line,	the lines are parallel.

Example 1 **If $m\angle 1 = m\angle 2$, determine which lines, if any, are parallel. State the postulate or theorem that justifies your answer.**

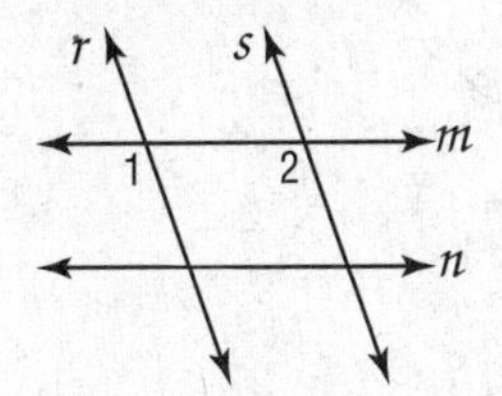

$\angle 1$ and $\angle 2$ are corresponding angles of lines r and s. Since $\angle 1 \cong \angle 2$, $r \parallel s$ by the Converse of the Corresponding Angles Postulate.

Example 2 **Find $m\angle ABC$ so that $m \parallel n$.**

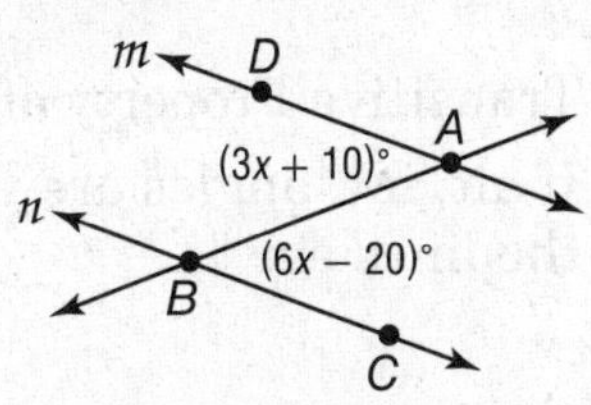

We can conclude that $m \parallel n$ if alternate interior angles are congruent.

$m\angle BAD = m\angle ABC$

$3x + 10 = 6x - 20$

$10 = 3x - 20$

$30 = 3x$

$10 = x$

$m\angle ABC = 6x - 20$

$= 6(10) - 20$ or 40

Exercises

Find x so that $\ell \parallel m$. Identify the postulate or theorem you used.

1.

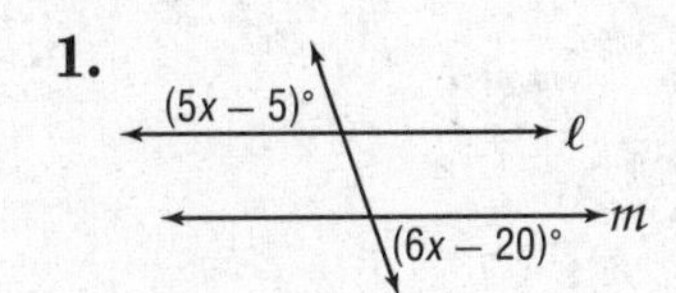

2.

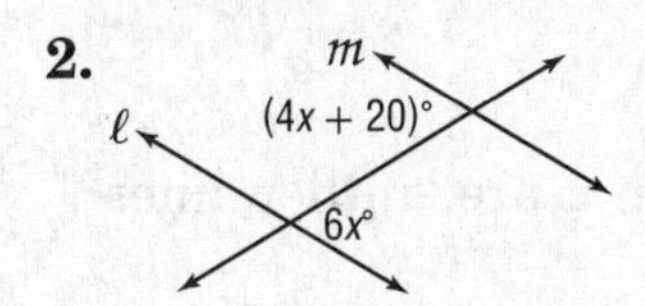

3.

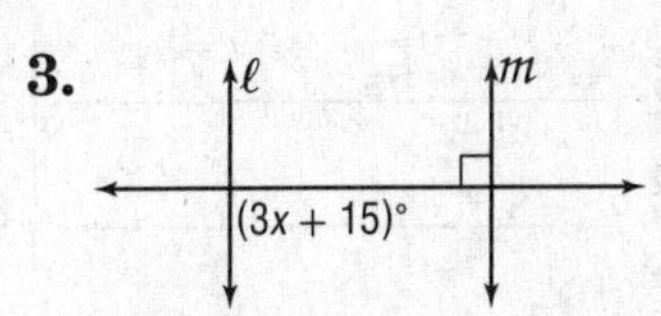

4.

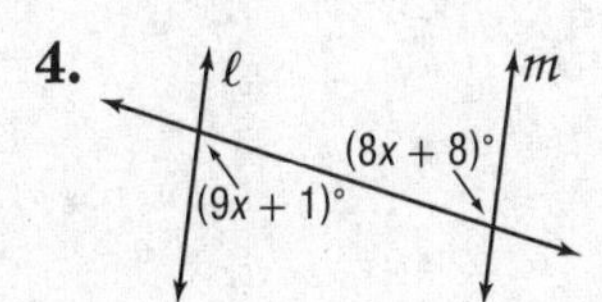

5.

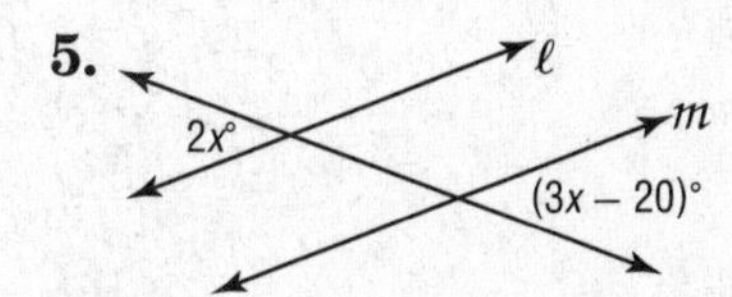

6.

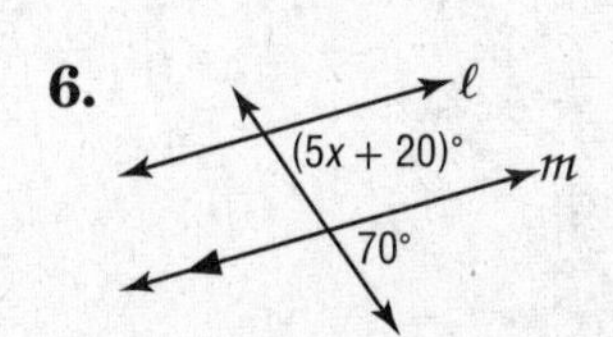

NAME _______________________________ DATE ____________ PERIOD ____________

3-5 Study Guide and Intervention *(continued)*

Proving Lines Parallel

Prove Lines Parallel You can prove that lines are parallel by using postulates and theorems about pairs of angles.

Example

Given: $\angle 1 \cong \angle 2$, $\angle 1 \cong \angle 3$

Prove: $\overline{AB} \parallel \overline{DC}$

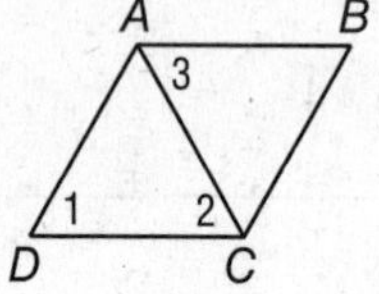

Proof:

Statements	Reasons
1. $\angle 1 \cong \angle 2$ $\angle 1 \cong \angle 3$	1. Given
2. $\angle 2 \cong \angle 3$	2. Transitive Property of $\cong$
3. $\overline{AB} \parallel \overline{DC}$	3. If alt. int. angles are $\cong$, then the lines are $\parallel$.

Exercises

1. Complete the proof.

Given: $\angle 1 \cong \angle 5$, $\angle 15 \cong \angle 5$

Prove: $\ell \parallel m$, $r \parallel s$

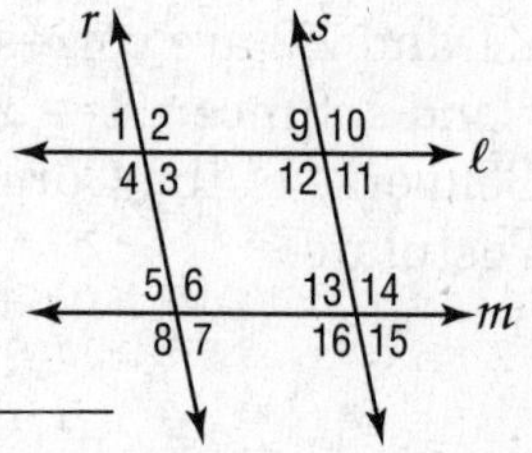

Proof:

Statements	Reasons
1. $\angle 15 \cong \angle 5$	1. ____________________
2. $\angle 13 \cong \angle 15$	2. ____________________
3. $\angle 5 \cong \angle 13$	3. ____________________
4. $r \parallel s$	4. ____________________
5. ____________	5. Given
6. ____________	6. If corr ∠s are $\cong$, then lines $\parallel$.

3-6 Study Guide and Intervention

Perpendiculars and Distance

Distance From a Point to a Line

Distance From a Point to a Line When a point is not on a line, the distance from the point to the line is the length of the segment that contains the point and is perpendicular to the line.

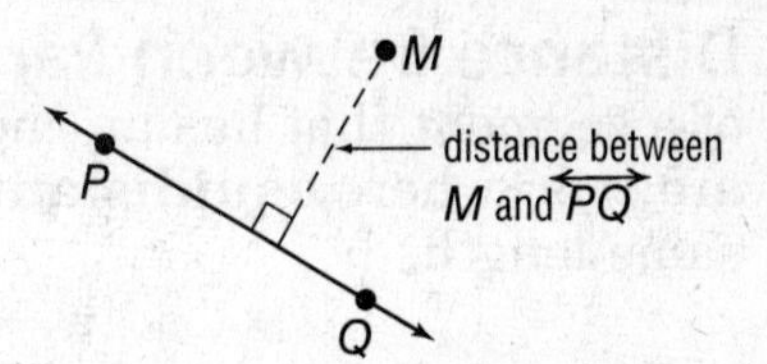

Example **Construct the segment that represents the distance from E to $\overleftrightarrow{AF}$.**

Extend $\overleftrightarrow{AF}$. Draw $\overleftrightarrow{EG} \perp \overleftrightarrow{AF}$.

$\overline{EG}$ represents the distance from E to $\overleftrightarrow{AF}$.

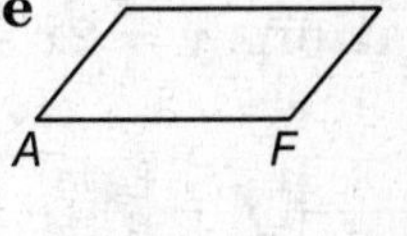

Exercises

Construct the segment that represents the distance indicated.

1. C to $\overleftrightarrow{AB}$

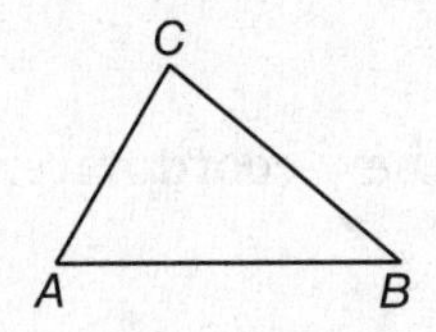

2. D to $\overleftrightarrow{AB}$

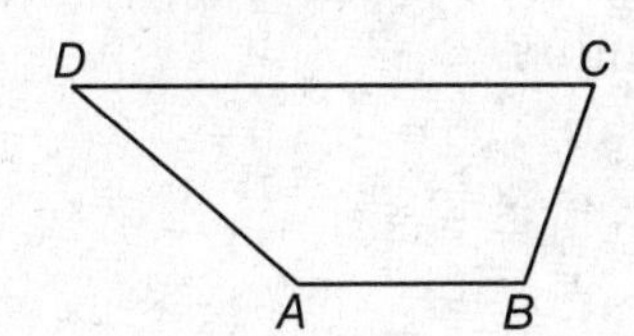

3. T to $\overleftrightarrow{RS}$

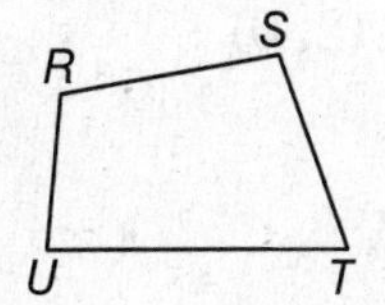

4. S to $\overleftrightarrow{PQ}$

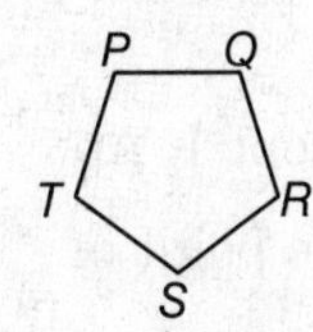

5. S to $\overleftrightarrow{QR}$

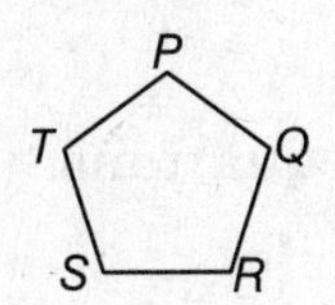

6. S to $\overleftrightarrow{RT}$

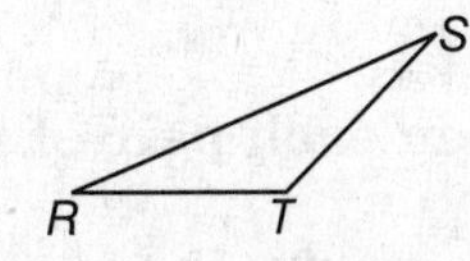

3-6 Study Guide and Intervention *(continued)*

Perpendiculars and Distance

Distance Between Parallel Lines The distance between parallel lines is the length of a segment that has an endpoint on each line and is perpendicular to them. Parallel lines are everywhere **equidistant**, which means that all such perpendicular segments have the same length.

Example **Find the distance between the parallel lines ℓ and m with the equations $y = 2x + 1$ and $y = 2x - 4$, respectively.**

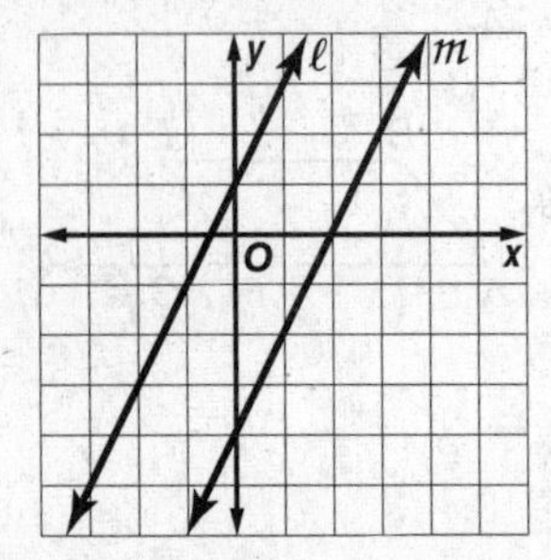

Draw a line p through (0, 1) that is perpendicular to ℓ and m.

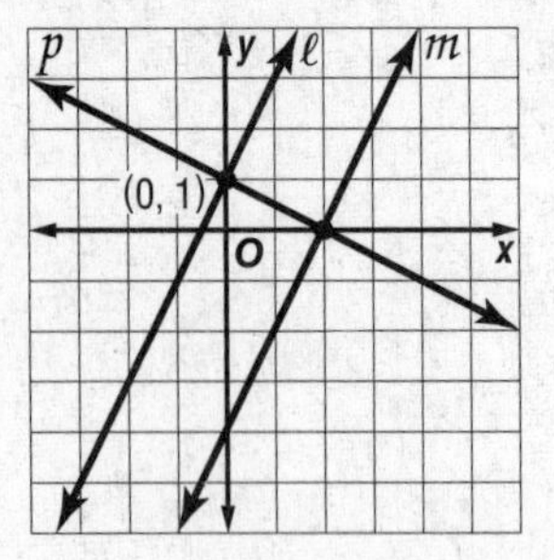

Line p has slope $-\frac{1}{2}$ and y-intercept 1. An equation of p is $y = -\frac{1}{2}x + 1$. The point of intersection for p and ℓ is (0, 1).

To find the point of intersection of p and m, solve a system of equations.

Line m: $y = 2x - 4$

Line p: $y = -\frac{1}{2}x + 1$

Use substitution.

$$2x - 4 = -\frac{1}{2}x + 1$$
$$4x - 8 = -x + 2$$
$$5x = 10$$
$$x = 2$$

Substitute 2 for x to find the y-coordinate.

$$y = -\frac{1}{2}x + 1$$
$$= -\frac{1}{2}(2) + 1 = -1 + 1 = 0$$

The point of intersection of p and m is (2, 0).

Use the Distance Formula to find the distance between (0, 1) and (2, 0).

$$d = \sqrt{(x_2 - x_1)^2 + (y_2 - y_1)^2}$$
$$= \sqrt{(2 - 0)^2 + (0 - 1)^2}$$
$$= \sqrt{5}$$

The distance between ℓ and m is $\sqrt{5}$ units.

Exercises

Find the distance between each pair of parallel lines with the given equations.

1. $y = 8$
$y = -3$

2. $y = x + 3$
$y = x - 1$

3. $y = -2x$
$y = -2x - 5$

NAME ______________________ DATE ____________ PERIOD ____________

4-1 Study Guide and Intervention

Classifying Triangles

Classify Triangles by Angles One way to classify a triangle is by the measures of its angles.

• If *all three* of the angles of a triangle are acute angles, then the triangle is an **acute triangle.**
• If *all three* angles of an acute triangle are congruent, then the triangle is an **equiangular triangle.**
• If *one* of the angles of a triangle is an obtuse angle, then the triangle is an **obtuse triangle.**
• If *one* of the angles of a triangle is a right angle, then the triangle is a **right triangle.**

Example **Classify each triangle.**

a.

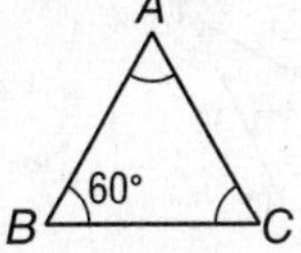

All three angles are congruent, so all three angles have measure 60°. The triangle is an equiangular triangle.

b.

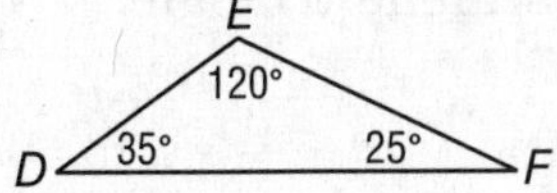

The triangle has one angle that is obtuse. It is an obtuse triangle.

c.

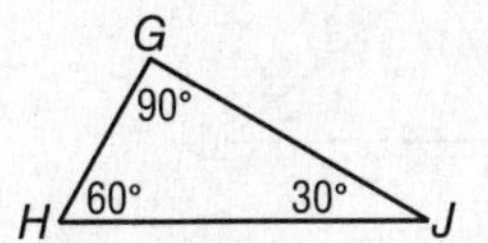

The triangle has one right angle. It is a right triangle.

Exercises

Classify each triangle as *acute*, *equiangular*, *obtuse*, or *right*.

1.

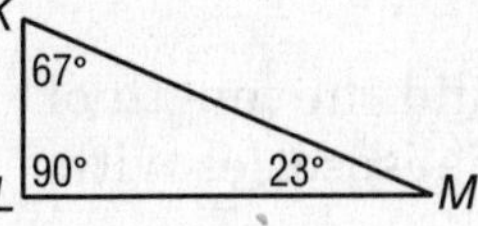

2.

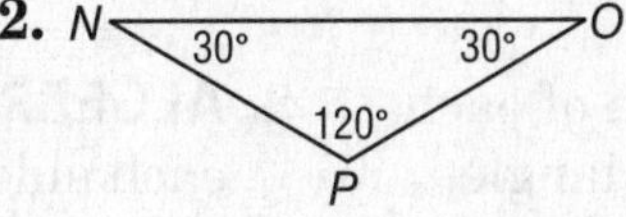

3.

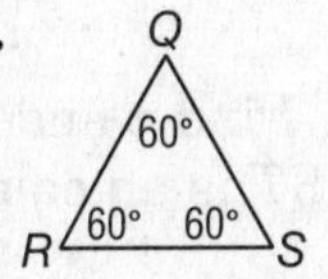

4.

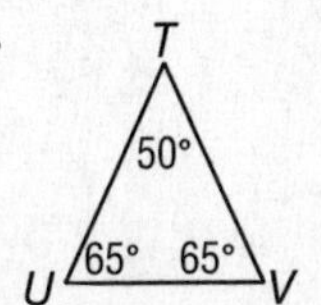

5.

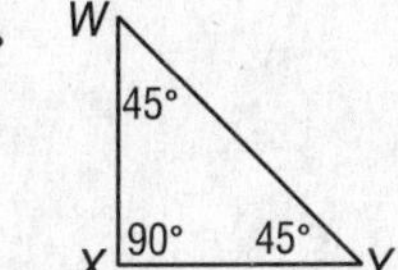

6.

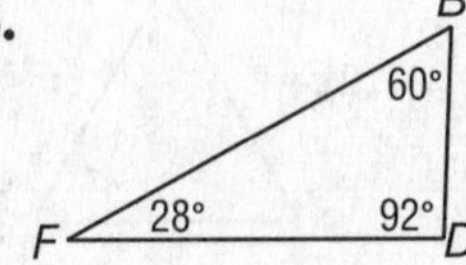

4-1 Study Guide and Intervention *(continued)*

Classifying Triangles

Classify Triangles by Sides You can classify a triangle by the number of congruent sides. Equal numbers of hash marks indicate congruent sides.

• If *all three* sides of a triangle are congruent, then the triangle is an **equilateral triangle**.
• If *at least two* sides of a triangle are congruent, then the triangle is an **isosceles triangle**. Equilateral triangles can also be considered isosceles.
• If *no two* sides of a triangle are congruent, then the triangle is a **scalene triangle**.

Example **Classify each triangle.**

a.

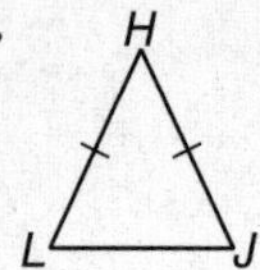

Two sides are congruent. The triangle is an isosceles triangle.

b.

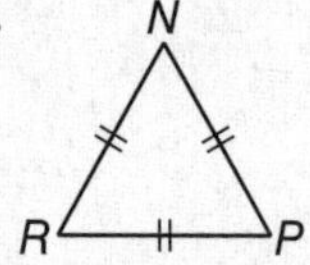

All three sides are congruent. The triangle is an equilateral triangle.

c.

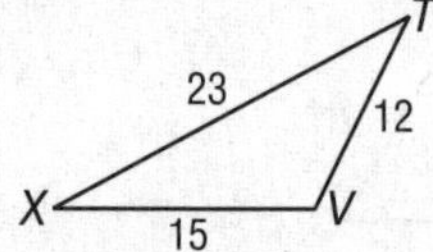

The triangle has no pair of congruent sides. It is a scalene triangle.

Exercises

Classify each triangle as *equilateral, isosceles,* or *scalene.*

1.

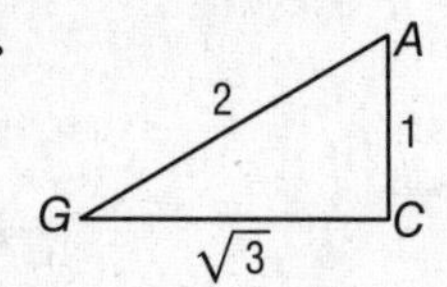

2.

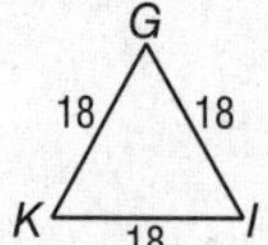

3.

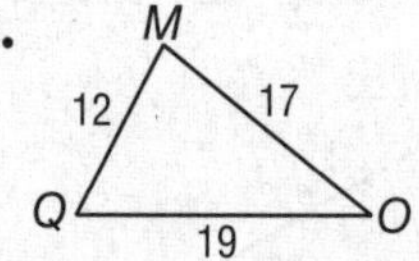

4.

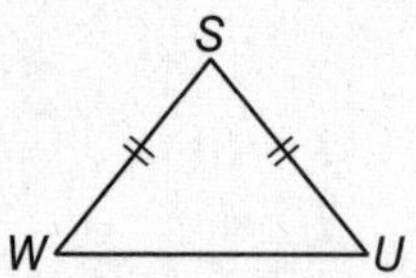

5.

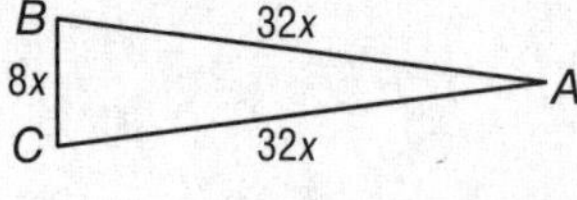

6.

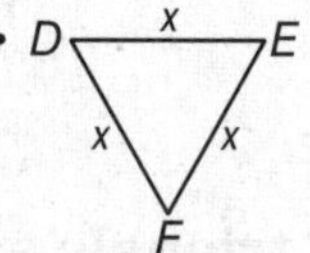

7. ALGEBRA Find x and the length of each side if $\triangle RST$ is an equilateral triangle.

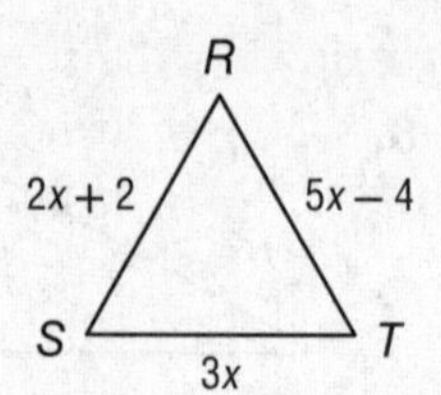

8. ALGEBRA Find x and the length of each side if $\triangle ABC$ is isosceles with $AB = BC$.

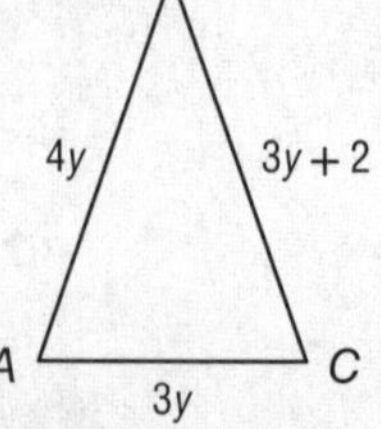

NAME ______________________ DATE ____________ PERIOD ____________

4-2 Study Guide and Intervention

Angles of Triangles

Triangle Angle-Sum Theorem If the measures of two angles of a triangle are known, the measure of the third angle can always be found.

Triangle Angle Sum Theorem	The sum of the measures of the angles of a triangle is 180. In the figure at the right, $m\angle A + m\angle B + m\angle C = 180$. 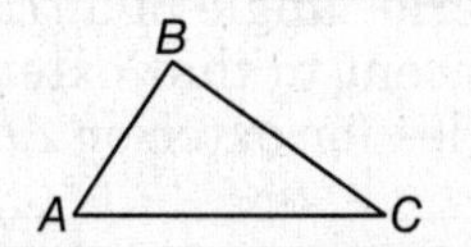

Example 1 **Find $m\angle T$.**

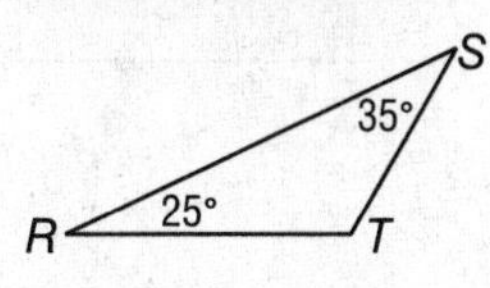

$m\angle R + m\angle S + m\angle T = 180$	Triangle Angle-Sum Theorem
$25 + 35 + m\angle T = 180$	Substitution
$60 + m\angle T = 180$	Simplify.
$m\angle T = 120$	Subtract 60 from each side.

Example 2 **Find the missing angle measures.**

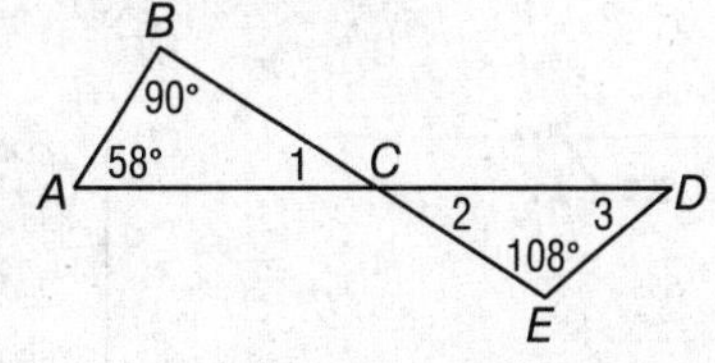

$m\angle 1 + m\angle A + m\angle B = 180$	Triangle Angle-Sum Theorem
$m\angle 1 + 58 + 90 = 180$	Substitution
$m\angle 1 + 148 = 180$	Simplify.
$m\angle 1 = 32$	Subtract 148 from each side.
$m\angle 2 = 32$	Vertical angles are congruent.
$m\angle 3 + m\angle 2 + m\angle E = 180$	Triangle Angle-Sum Theorem
$m\angle 3 + 32 + 108 = 180$	Substitution
$m\angle 3 + 140 = 180$	Simplify.
$m\angle 3 = 40$	Subtract 140 from each side.

Exercises

Find the measure of each numbered angle.

1.

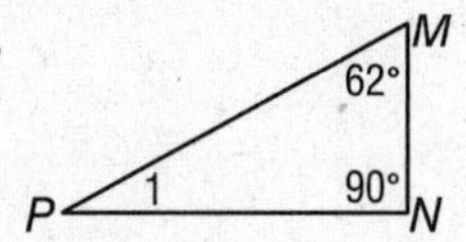

2.

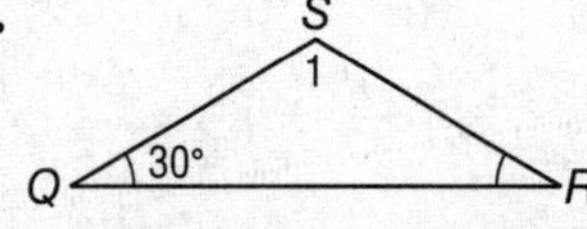

3.

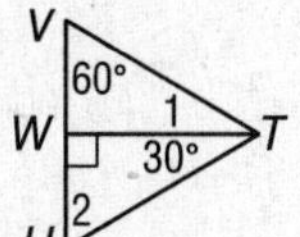

4.

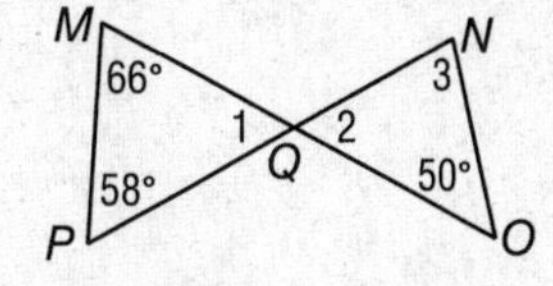

5.

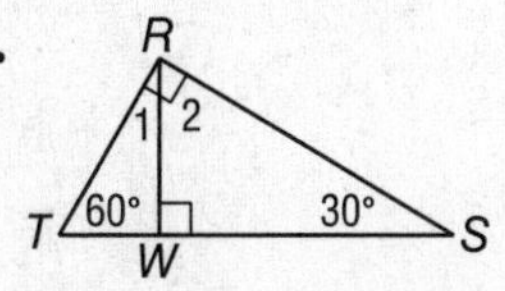

6.

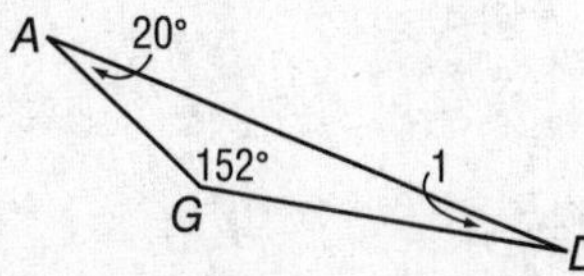

NAME ______________________ DATE __________ PERIOD __________

4-2 Study Guide and Intervention *(continued)*

Angles of Triangles

Exterior Angle Theorem At each vertex of a triangle, the angle formed by one side and an extension of the other side is called an **exterior angle** of the triangle. For each exterior angle of a triangle, the **remote interior angles** are the interior angles that are not adjacent to that exterior angle. In the diagram below, $\angle B$ and $\angle A$ are the remote interior angles for exterior $\angle DCB$.

Exterior Angle Theorem	The measure of an exterior angle of a triangle is equal to the sum of the measures of the two remote interior angles. $m\angle 1 = m\angle A + m\angle B$ 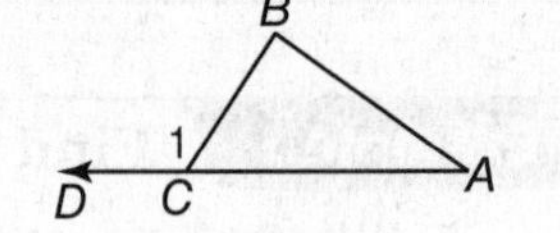

Example 1 **Find $m\angle 1$.**

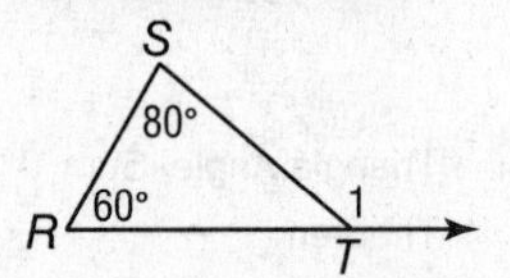

$m\angle 1 = m\angle R + m\angle S$ Exterior Angle Theorem

$= 60 + 80$ Substitution

$= 140$ Simplify.

Example 2 **Find x.**

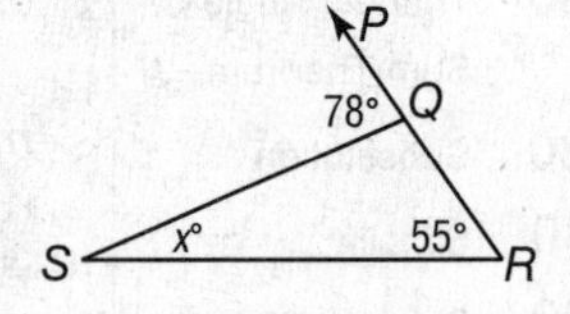

$m\angle PQS = m\angle R + m\angle S$ Exterior Angle Theorem

$78 = 55 + x$ Substitution

$23 = x$ Subtract 55 from each side.

Exercises

Find the measures of each numbered angle.

1.

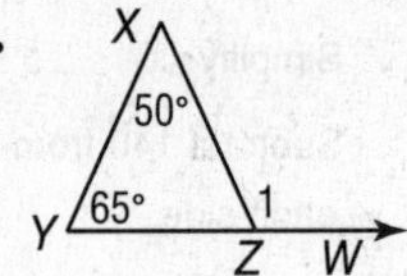

2.

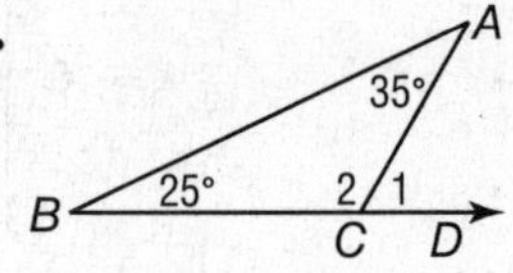

3.

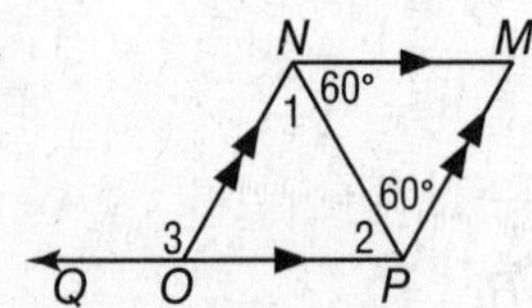

4.

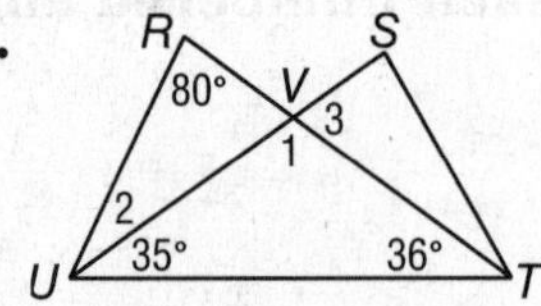

Find each measure.

5. $m\angle ABC$

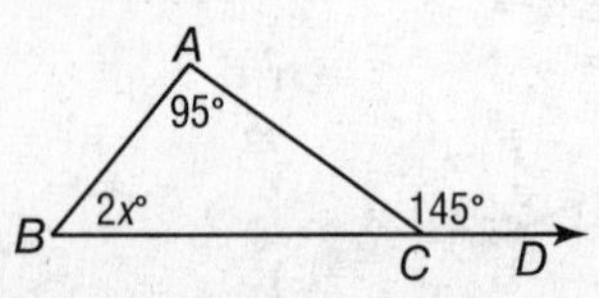

6. $m\angle F$

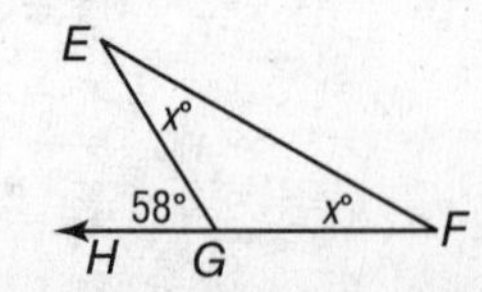

NAME ________________ DATE ________ PERIOD ________

4-3 Study Guide and Intervention

Congruent Triangles

Congruence and Corresponding Parts

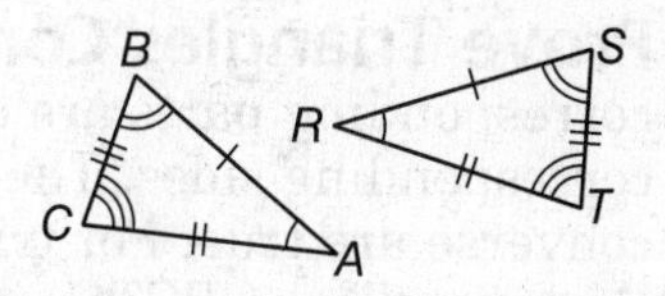

Triangles that have the same size and same shape are **congruent triangles**. Two triangles are congruent if and only if all three pairs of corresponding angles are congruent and all three pairs of corresponding sides are congruent. In the figure, $\triangle ABC \cong \triangle RST$.

Third Angles Theorem	If two angles of one triangle are congruent to two angles of a second triangle, then the third angles of the triangles are congruent.

Example **If $\triangle XYZ \cong \triangle RST$, name the pairs of congruent angles and congruent sides.**

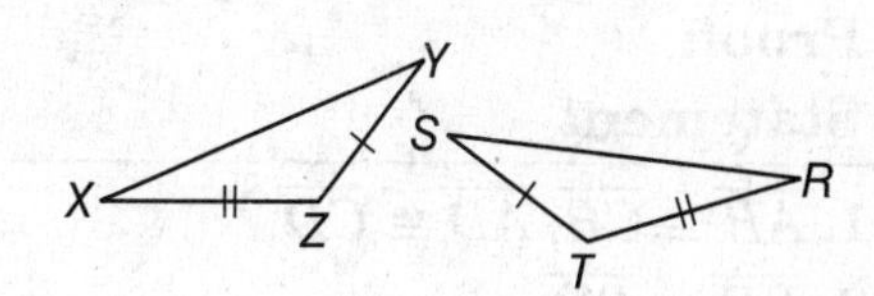

$\angle X \cong \angle R$, $\angle Y \cong \angle S$, $\angle Z \cong \angle T$

$\overline{XY} \cong \overline{RS}$, $\overline{XZ} \cong \overline{RT}$, $\overline{YZ} \cong \overline{ST}$

Exercises

Show that the polygons are congruent by identifying all congruent corresponding parts. Then write a congruence statement.

1.

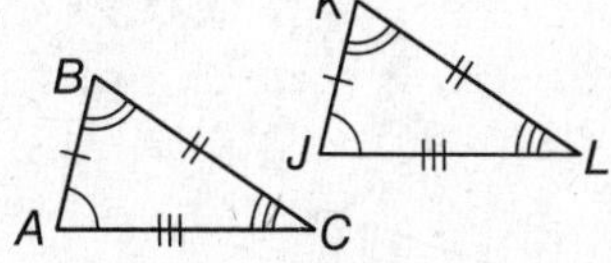

2.

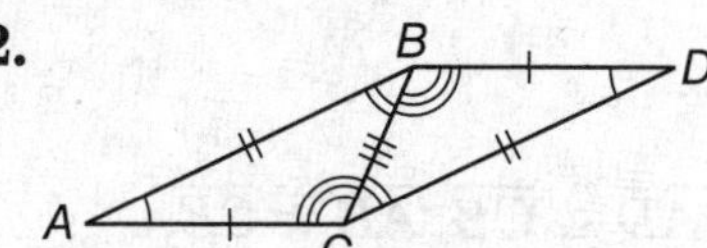

3.

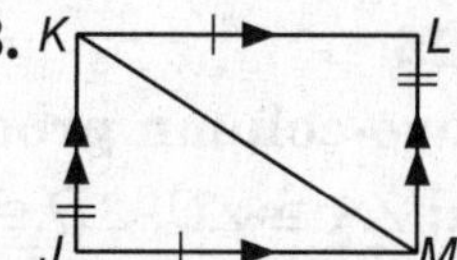

4. F G L K E J

5.

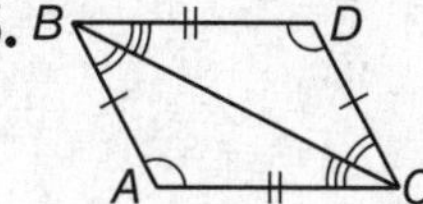

6.

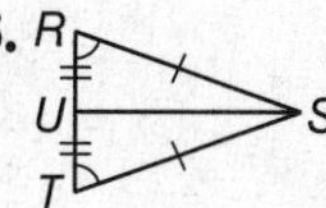

$\triangle ABC \cong \triangle DEF$.

7. Find the value of x.

8. Find the value of y.

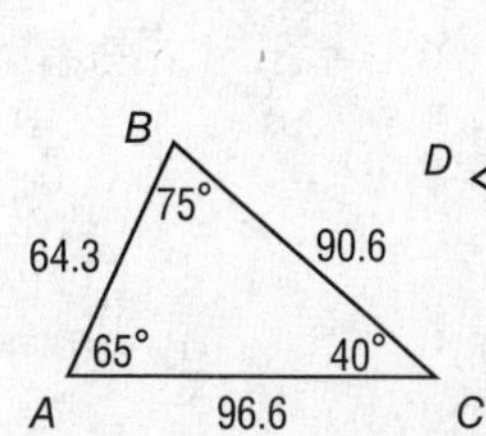

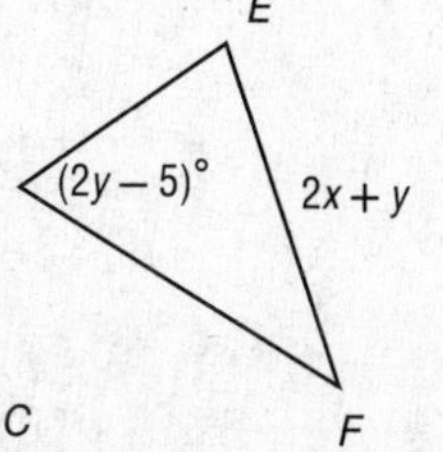

NAME ______________________ DATE __________ PERIOD __________

4-3 Study Guide and Intervention *(continued)*

Congruent Triangles

Prove Triangles Congruent Two triangles are congruent if and only if their corresponding parts are congruent. Corresponding parts include corresponding angles and corresponding sides. The phrase "if and only if" means that both the conditional and its converse are true. For triangles, we say, "Corresponding parts of congruent triangles are congruent," or CPCTC.

Example **Write a two-column proof.**

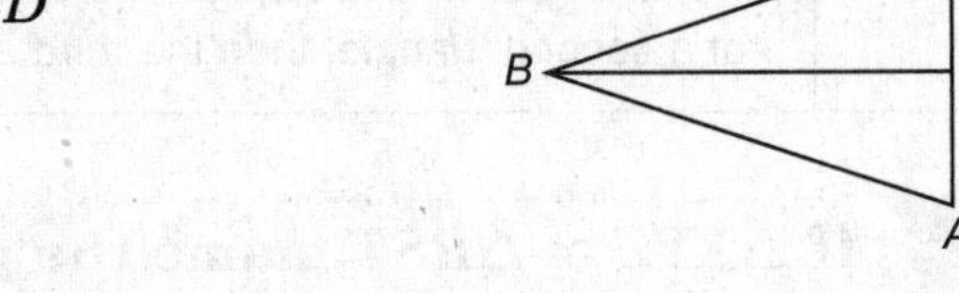

Given: $\overline{AB} \cong \overline{CB}$, $\overline{AD} \cong \overline{CD}$, $\angle BAD \cong \angle BCD$
$\overline{BD}$ bisects $\angle ABC$

Prove: $\triangle ABD \cong \triangle CBD$

Proof:

Statement	**Reason**
1. $\overline{AB} \cong \overline{CB}$, $\overline{AD} \cong \overline{CD}$	**1.** Given
2. $\overline{BD} \cong \overline{BD}$	**2.** Reflexive Property of congruence
3. $\angle BAD \cong \angle BCD$	**3.** Given
4. $\angle ABD \cong \angle CBD$	**4.** Definition of angle bisector
5. $\angle BDA \cong \angle BDC$	**5.** Third Angles Theorem
6. $\triangle ABD \cong \triangle CBD$	**6.** CPCTC

Exercises

Write a two-column proof.

1. Given: $\angle A \cong \angle C$, $\angle D \cong \angle B$, $\overline{AD} \cong \overline{CB}$, $\overline{AE} \cong \overline{CE}$,
$\overline{AC}$ bisects $\overline{BD}$

Prove: $\triangle AED \cong \triangle CEB$

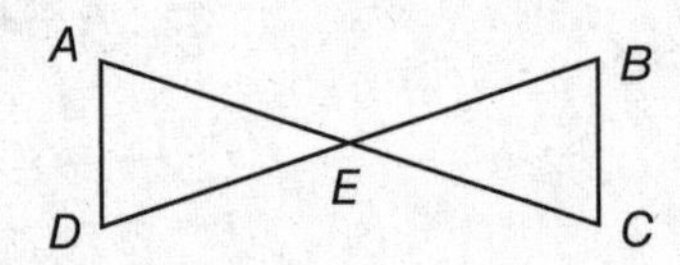

Write a paragraph proof.

2. Given: $\overline{BD}$ bisects $\angle ABC$ and $\angle ADC$,
$\overline{AB} \cong \overline{CB}$, $\overline{AB} \cong \overline{AD}$, $\overline{CB} \cong \overline{DC}$

Prove: $\triangle ABD \cong \triangle CBD$

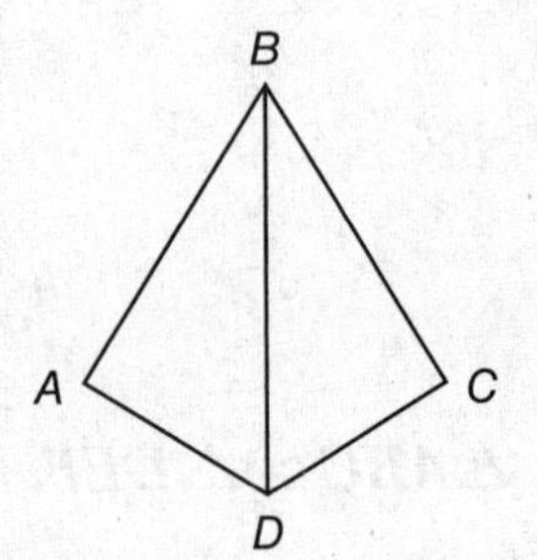

NAME ______________________________ DATE ____________ PERIOD ____________

4-4 Study Guide and Intervention

Proving Triangles Congruent—SSS, SAS

SSS Postulate You know that two triangles are congruent if corresponding sides are congruent and corresponding angles are congruent. The Side-Side-Side (SSS) Postulate lets you show that two triangles are congruent if you know only that the sides of one triangle are congruent to the sides of the second triangle.

SSS Postulate	If three sides of one triangle are congruent to three sides of a second triangle, then the triangles are congruent.

Example **Write a two-column proof.**

Given: $\overline{AB} \cong \overline{DB}$ and C is the midpoint of $\overline{AD}$.

Prove: $\triangle ABC \cong \triangle DBC$

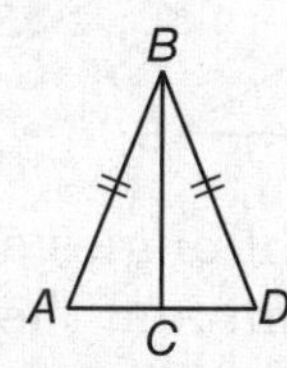

Statements	Reasons
1. $\overline{AB} \cong \overline{DB}$	1. Given
2. C is the midpoint of $\overline{AD}$.	2. Given
3. $\overline{AC} \cong \overline{DC}$	3. Midpoint Theorem
4. $\overline{BC} \cong \overline{BC}$	4. Reflexive Property of ≅
5. $\triangle ABC \cong \triangle DBC$	5. SSS Postulate

Exercises

Write a two-column proof.

1.

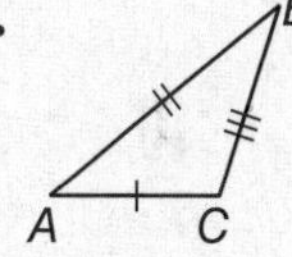

Y Z X

Given: $AB \cong \overline{XY}$, $\overline{AC} \cong \overline{XZ}$, $\overline{BC} \cong \overline{YZ}$

Prove: $\triangle ABC \cong \triangle XYZ$

2.

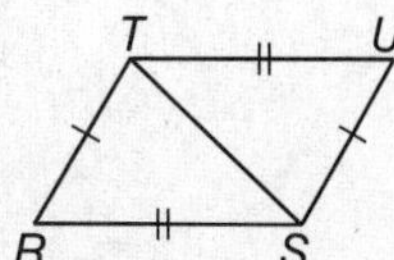

Given: $\overline{RS} \cong \overline{UT}$, $\overline{RT} \cong \overline{US}$

Prove: $\triangle RST \cong \triangle UTS$

NAME ______________________________ DATE ______________ PERIOD ____________

4-4 Study Guide and Intervention *(continued)*

Proving Triangles Congruent—SSS, SAS

SAS Postulate Another way to show that two triangles are congruent is to use the Side-Angle-Side (SAS) Postulate.

SAS Postulate	If two sides and the included angle of one triangle are congruent to two sides and the included angle of another triangle, then the triangles are congruent.

Example **For each diagram, determine which pairs of triangles can be proved congruent by the SAS Postulate.**

a.

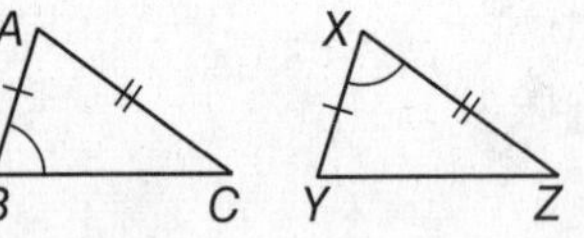

In $\triangle ABC$, the angle is not "included" by the sides $\overline{AB}$ and $\overline{AC}$. So the triangles cannot be proved congruent by the SAS Postulate.

b.

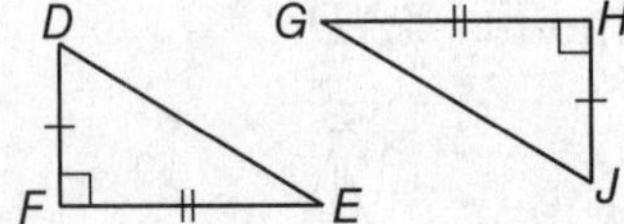

The right angles are congruent and they are the included angles for the congruent sides. $\triangle DEF \cong \triangle JGH$ by the SAS Postulate.

c.

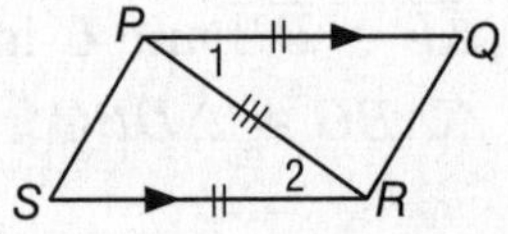

The included angles, $\angle 1$ and $\angle 2$, are congruent because they are alternate interior angles for two parallel lines. $\triangle PSR \cong \triangle RQP$ by the SAS Postulate.

Exercises

Write the specified type of proof.

1. Write a two column proof.
Given: $NP = PM$, $\overline{NP} \perp \overline{PL}$
Prove: $\triangle NPL \cong \triangle MPL$

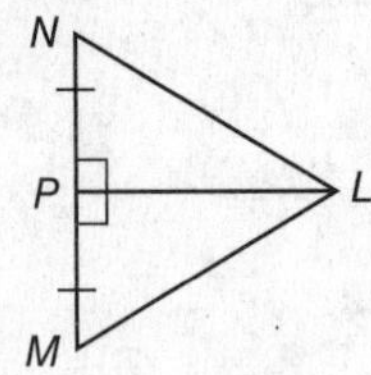

2. Write a two-column proof.
Given: $AB = CD$, $\overline{AB} \parallel \overline{CD}$
Prove: $\triangle ACD \cong \triangle CAB$

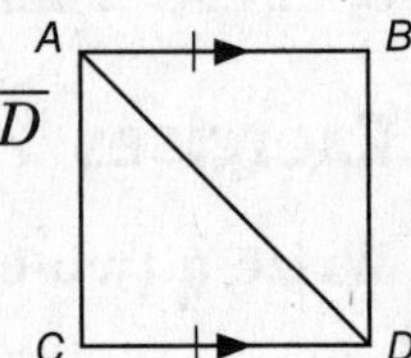

3. Write a paragraph proof.
Given: V is the midpoint of $\overline{YZ}$.
V is the midpoint of $\overline{WX}$.
Prove: $\triangle XVZ \cong \triangle WVY$

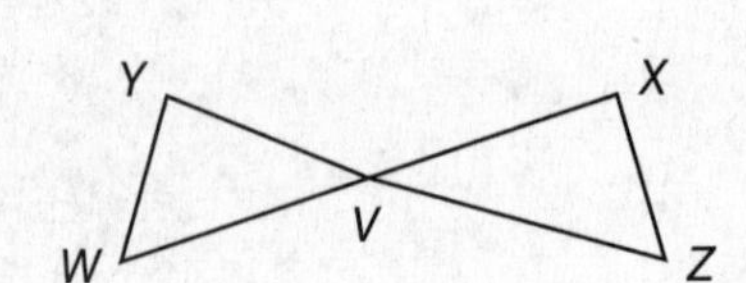

NAME ______________________ DATE ____________ PERIOD ____________

4-5 Study Guide and Intervention

Proving Triangles Congruent—ASA, AAS

ASA Postulate The Angle-Side-Angle (ASA) Postulate lets you show that two triangles are congruent.

ASA Postulate	If two angles and the included side of one triangle are congruent to two angles and the included side of another triangle, then the triangles are congruent.

Example **Write a two column proof.**

Given: $\overline{AB} \parallel \overline{CD}$

$\angle CBD \cong \angle ADB$

Prove: $\triangle ABD \cong \triangle CDB$

Statements	**Reasons**
1. $\overline{AB} \parallel \overline{CD}$	**1.** Given
2. $\angle CBD \cong \angle ADB$	**2.** Given
3. $\angle ABD \cong \angle BDC$	**3.** Alternate Interior Angles Theorem
4. $\overline{BD} \cong \overline{BD}$	**4.** Reflexive Property of Congruence
5. $\triangle ABD \cong \triangle CDB$	**5.** ASA

Exercises

PROOF Write the specified type of proof.

1. Write a two column proof.

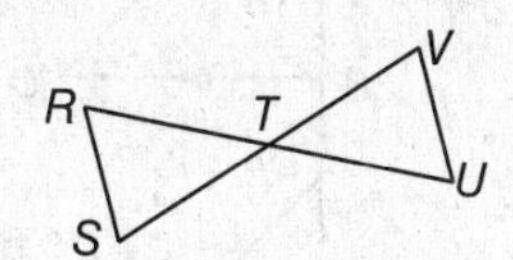

Given: $\angle S \cong \angle V$,

T is the midpoint *of* $\overline{SV}$.

Prove: $\triangle RTS \cong \triangle UTV$

2. Write a paragraph proof.

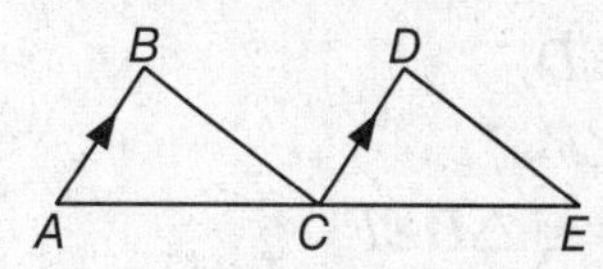

Given: $\overline{CD}$ bisects $\overline{AE}$, $\overline{AB} \parallel \overline{CD}$

$\angle E \cong \angle BCA$

Prove: $\triangle ABC \cong \triangle CDE$

4-5 Study Guide and Intervention *(continued)*

Proving Triangles Congruent—ASA, AAS

AAS Theorem Another way to show that two triangles are congruent is the Angle-Angle-Side (AAS) Theorem.

AAS Theorem	If two angles and a nonincluded side of one triangle are congruent to the corresponding two angles and side of a second triangle, then the two triangles are congruent.

You now have five ways to show that two triangles are congruent.

- definition of triangle congruence
- SSS Postulate
- SAS Postulate
- ASA Postulate
- AAS Theorem

Example **In the diagram, $\angle BCA \cong \angle DCA$. Which sides are congruent? Which additional pair of corresponding parts needs to be congruent for the triangles to be congruent by the AAS Theorem?**

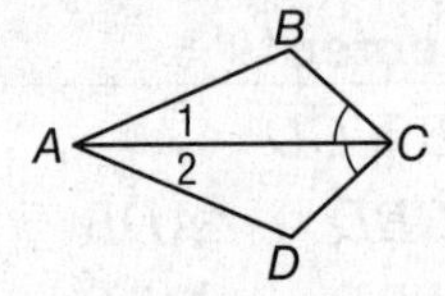

$\overline{AC} \cong \overline{AC}$ by the Reflexive Property of congruence. The congruent angles cannot be $\angle 1$ and $\angle 2$, because $\overline{AC}$ would be the included side. If $\angle B \cong \angle D$, then $\triangle ABC \cong \triangle ADC$ by the AAS Theorem.

Exercises

PROOF Write the specified type of proof.

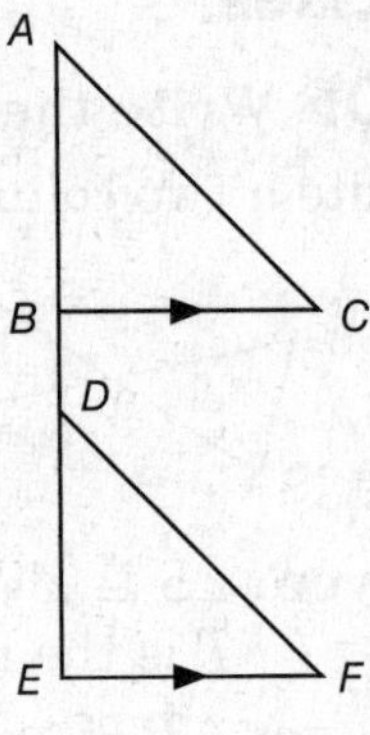

1. Write a two column proof.

Given: $\overline{BC} \parallel \overline{EF}$

$AB = ED$

$\angle C \cong \angle F$

Prove: $\triangle ABD \cong \triangle DEF$

2. Write a flow proof.

Given: $\angle S \cong \angle U$; $\overline{TR}$ bisects $\angle STU$.

Prove: $\angle SRT \cong \angle URT$

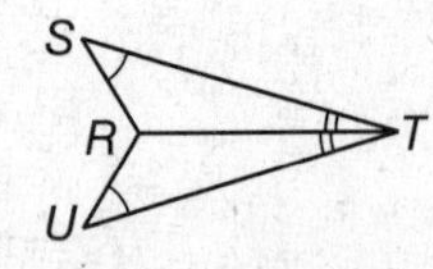

NAME ______________________ DATE ____________ PERIOD ____________

4-6 Study Guide and Intervention

Isosceles and Equilateral Triangles

Properties of Isosceles Triangles An **isosceles triangle** has two congruent sides called the legs. The angle formed by the legs is called the **vertex angle**. The other two angles are called **base angles**. You can prove a theorem and its converse about isosceles triangles.

- If two sides of a triangle are congruent, then the angles opposite those sides are congruent. **(Isosceles Triangle Theorem)**
- If two angles of a triangle are congruent, then the sides opposite those angles are congruent. **(Converse of Isosceles Triangle Theorem)**

If $\overline{AB} \cong \overline{CB}$, then $\angle A \cong \angle C$.

If $\angle A \cong \angle C$, then $\overline{AB} \cong \overline{CB}$.

Example 1 **Find x, given $\overline{BC} \cong \overline{BA}$.**

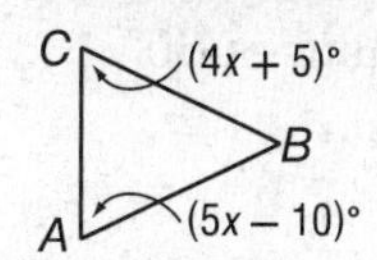

$BC = BA$, so

$m\angle A = m\angle C$	Isos. Triangle Theorem
$5x - 10 = 4x + 5$	Substitution
$x - 10 = 5$	subtract $4x$ from each side.
$x = 15$	Add 10 to each side.

Example 2 **Find x.**

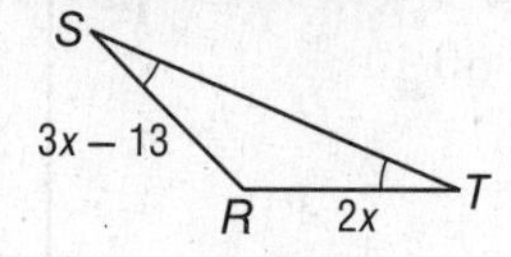

$m\angle S = m\angle T$, so

$SR = TR$	Converse of Isos. △ Thm.
$3x - 13 = 2x$	Substitution
$3x = 2x + 13$	Add 13 to each side.
$x = 13$	Subtract $2x$ from each side.

Exercises

ALGEBRA Find the value of each variable.

1.

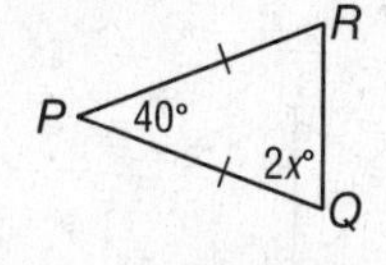

2.

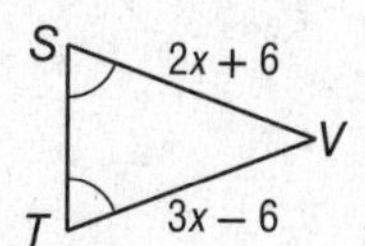

3.

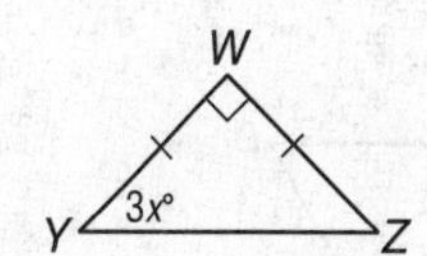

4.

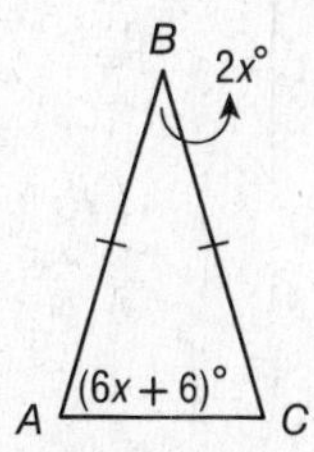

5.

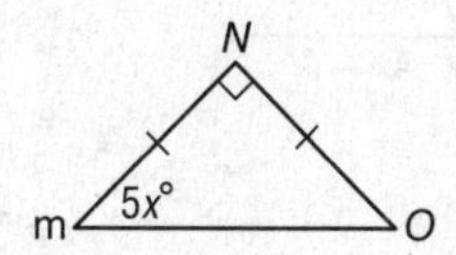

6.

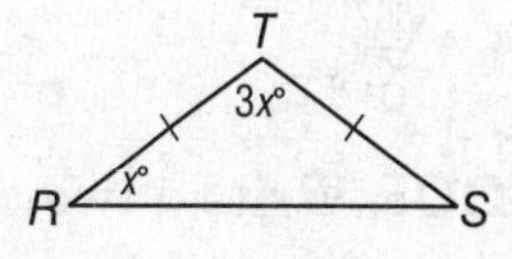

7. PROOF Write a two-column proof.

Given: $\angle 1 \cong \angle 2$

Prove: $\overline{AB} \cong \overline{CB}$

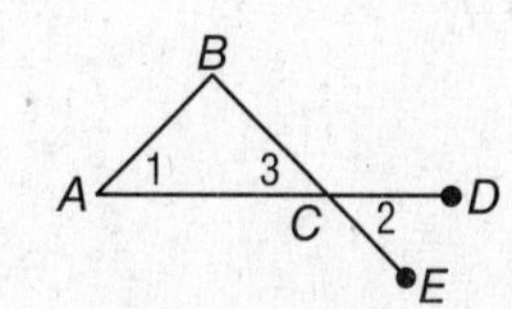

4-6 Study Guide and Intervention *(continued)*

Isosceles and Equilateral Triangles

Properties of Equilateral Triangles An **equilateral triangle** has three congruent sides. The Isosceles Triangle Theorem leads to two corollaries about equilateral triangles.

> 1. A triangle is equilateral if and only if it is equiangular.
> 2. Each angle of an equilateral triangle measures 60°.

Example **Prove that if a line is parallel to one side of an equilateral triangle, then it forms another equilateral triangle.**

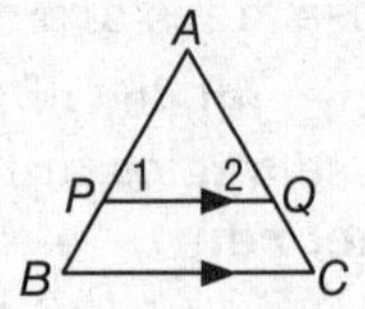

Proof:

Statements	Reasons
1. $\triangle ABC$ is equilateral; $\overline{PQ} \parallel \overline{BC}$.	**1.** Given
2. $m\angle A = m\angle B = m\angle C = 60$	**2.** Each $\angle$ of an equilateral $\triangle$ measures 60°.
3. $\angle 1 \cong \angle B$, $\angle 2 \cong \angle C$	**3.** If $\parallel$ lines, then corres. $\angle$s are $\cong$.
4. $m\angle 1 = 60$, $m\angle 2 = 60$	**4.** Substitution
5. $\triangle APQ$ is equilateral.	**5.** If a $\triangle$ is equiangular, then it is equilateral.

Exercises

ALGEBRA Find the value of each variable.

1.

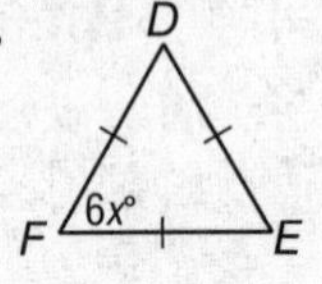

2.

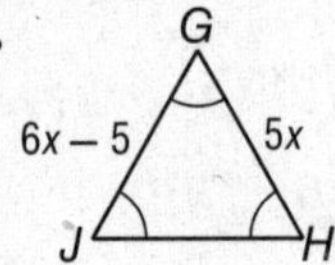

3.

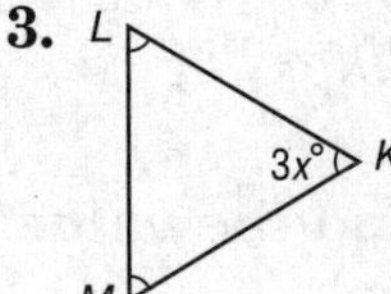

4.

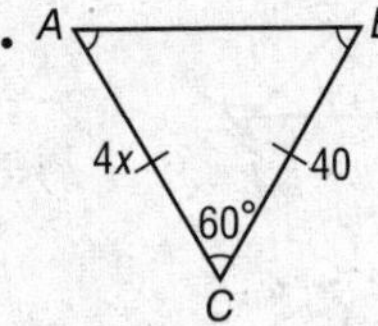

5.

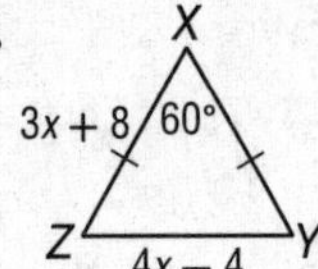

6.

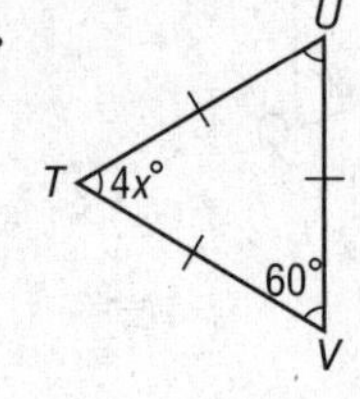

7. PROOF Write a two-column proof.

Given: $\triangle ABC$ is equilateral; $\angle 1 \cong \angle 2$.

Prove: $\angle ADB \cong \angle CDB$

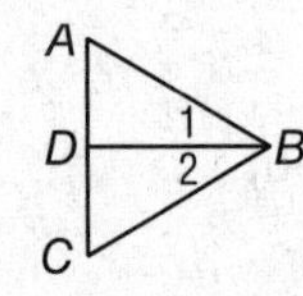

NAME ______________________ DATE ____________ PERIOD ____________

4-7 Study Guide and Intervention

Congruence Transformations

Identify Congruence Transformations A **congruence transformation** is a transformation where the original figure, or preimage, and the transformed figure, or image, figure are still congruent. The three types of congruence transformations are **reflection** (or flip), **translation** (or slide), and **rotation** (or turn).

Example **Identify the type of congruence transformation shown as a *reflection, translation,* or *rotation.***

a.	**b.**	**c.**
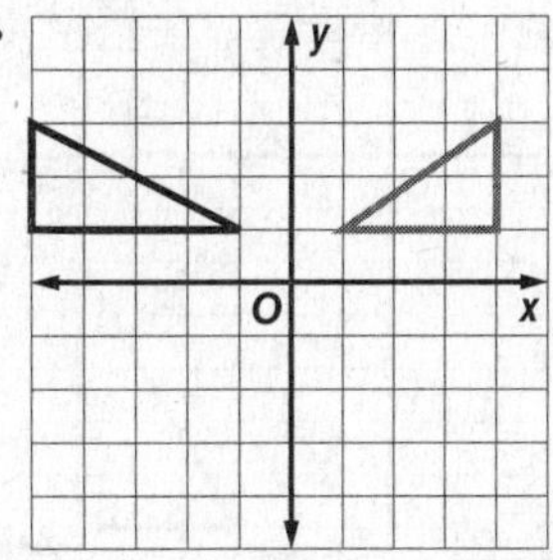	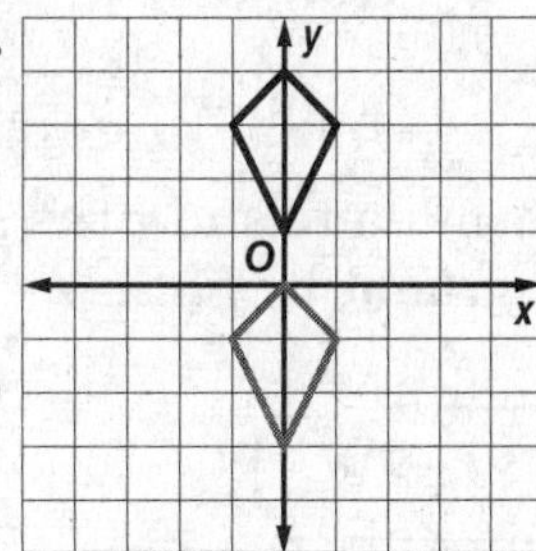	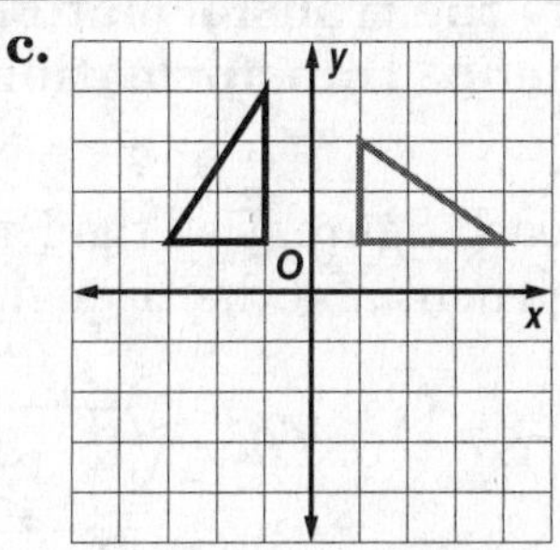
Each vertex and its image are the same distance from the *y*-axis. This is a reflection.	Each vertex and its image are in the same position, just two units down. This is a translation.	Each vertex and its image are the same distance from the origin. The angles formed by each pair of corresponding points and the origin are congruent. This is a rotation.

Exercises

Identify the type of congruence transformation shown as a *reflection, translation,* or *rotation.*

1.

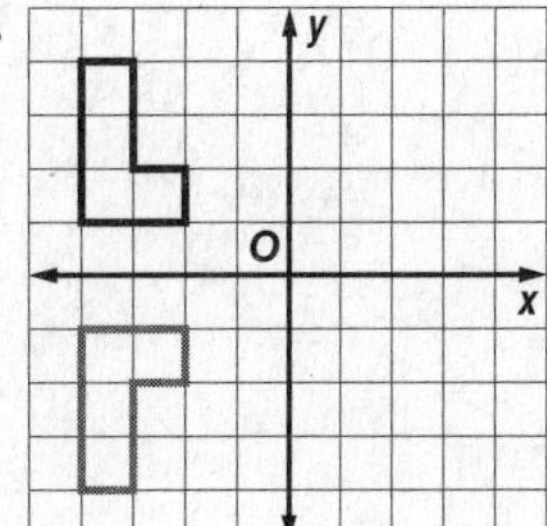

2.

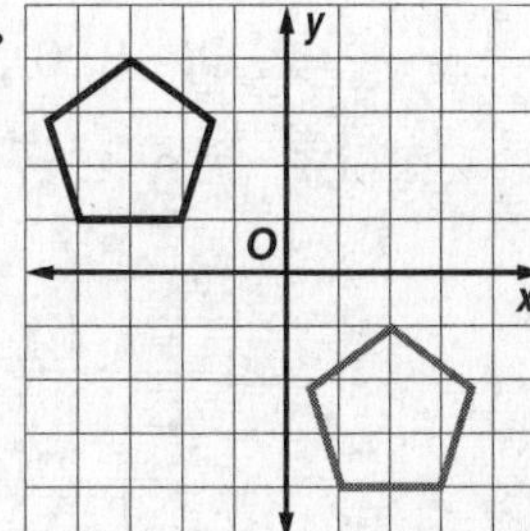

3.

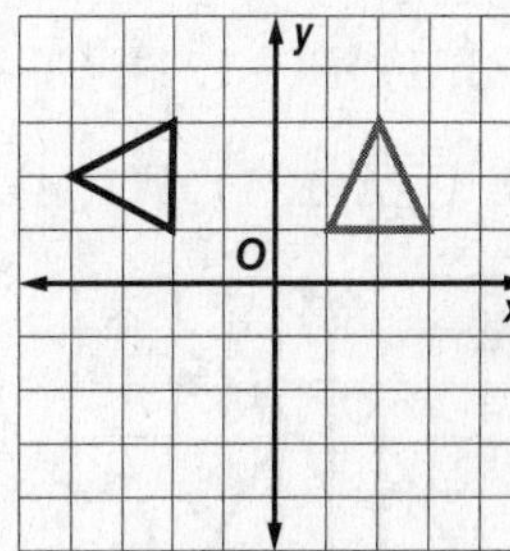

4.

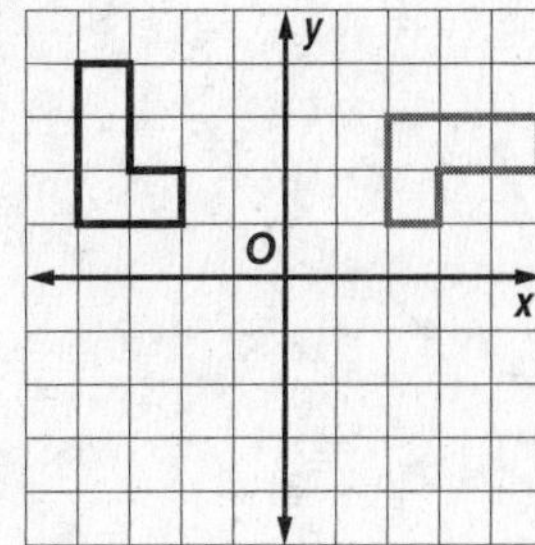

5.

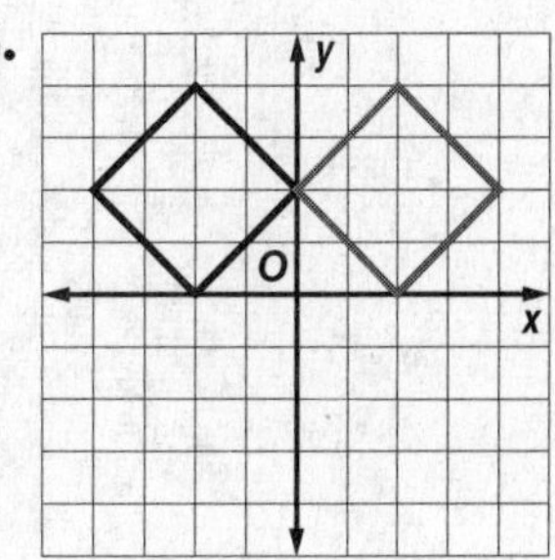

6.

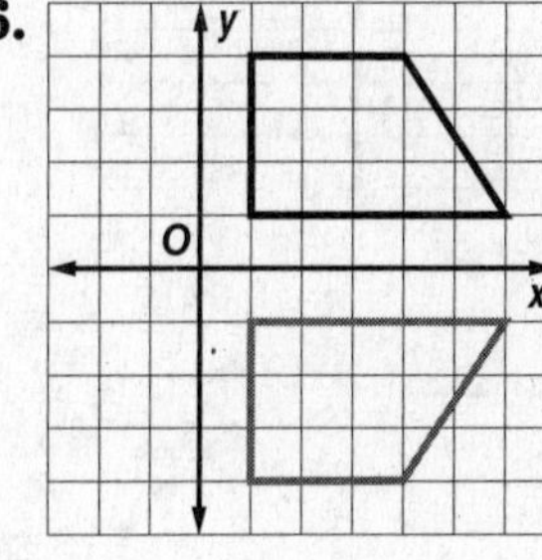

4-7 Study Guide and Intervention *(continued)*

Congruence Transformations

Verify Congruence You can verify that reflections, translations, and rotations of triangles produce congruent triangles using SSS.

Example **Verify congruence after a transformation.**

$\triangle WXY$ with vertices $W(3, -7)$, $X(6, -7)$, $Y(6, -2)$ is a transformation of $\triangle RST$ with vertices $R(2, 0)$, $S(5, 0)$, and $T(5, 5)$. Graph the original figure and its image. Identify the transformation and verify that it is a congruence transformation.

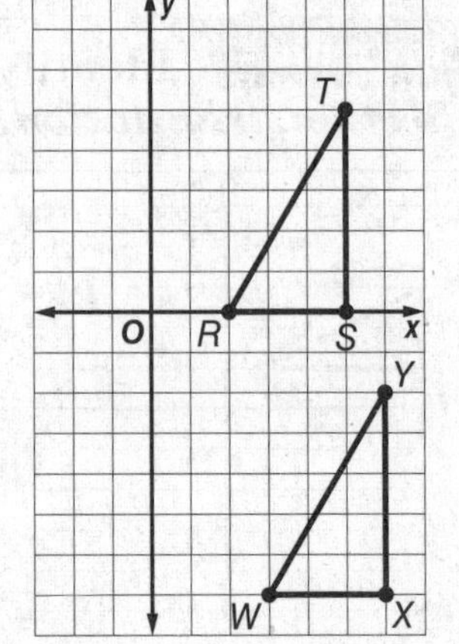

Graph each figure. Use the Distance Formula to show the sides are congruent and the triangles are congruent by SSS.

$RS = 3$, $ST = 5$, $TR = \sqrt{(5-2)^2 + (5-0)^2} = \sqrt{34}$

$WX = 3$, $XY = 5$, $YW = \sqrt{(6-3)^2 + (-2-(-7))^2} = \sqrt{34}$

$\overline{RS} \cong \overline{WX}$, $\overline{ST}$, $\cong \overline{XY}$, $\overline{TR} \cong \overline{YW}$

By SSS, $\triangle RST \cong \triangle WXY$.

Exercises

COORDINATE GEOMETRY Graph each pair of triangles with the given vertices. Then identify the transformation, and verify that it is a congruence transformation.

1. $A(-3, 1)$, $B(-1, 1)$, $C(-1, 4)$;

$D(3, 1)$, $E(1, 1)$, $F(1, 4)$

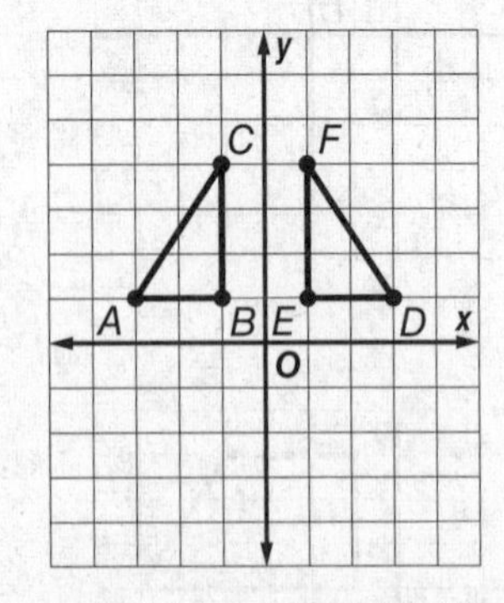

2. $Q(-3, 0)$, $R(-2, 2)$, $S(-1, 0)$;

$T(2, -4)$, $U(3, -2)$, $V(4, -4)$

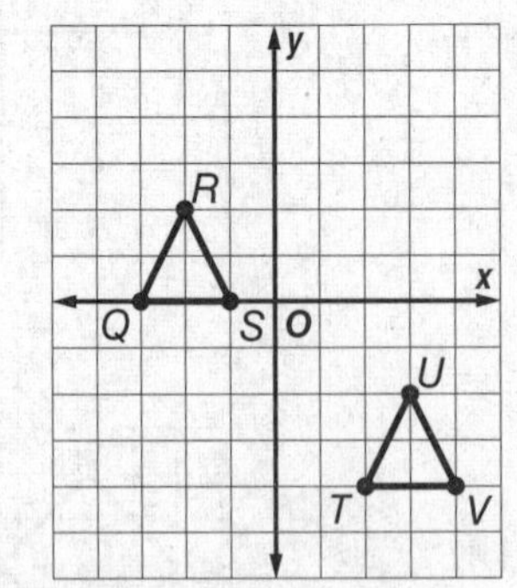

4-8 Study Guide and Intervention

Triangles and Coordinate Proof

Position and Label Triangles A coordinate proof uses points, distances, and slopes to prove geometric properties. The first step in writing a coordinate proof is to place a figure on the coordinate plane and label the vertices. Use the following guidelines.

1. Use the origin as a vertex or center of the figure.
2. Place at least one side of the polygon on an axis.
3. Keep the figure in the first quadrant if possible.
4. Use coordinates that make computations as simple as possible.

Example **Position an isosceles triangle on the coordinate plane so that its sides are a units long and one side is on the positive x-axis.**

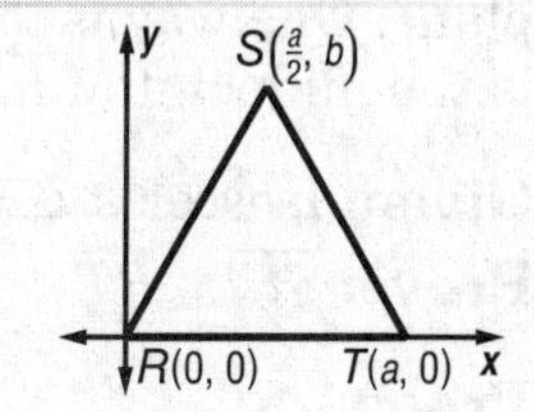

Start with $R(0, 0)$. If RT is a, then another vertex is $T(a, 0)$.
For vertex S, the x-coordinate is $\frac{a}{2}$. Use b for the y-coordinate, so the vertex is $S\left(\frac{a}{2}, b\right)$.

Exercises

Name the missing coordinates of each triangle.

1.

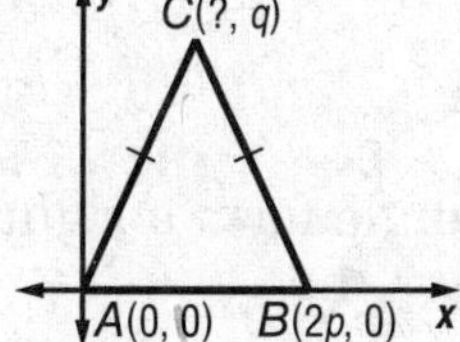

2.

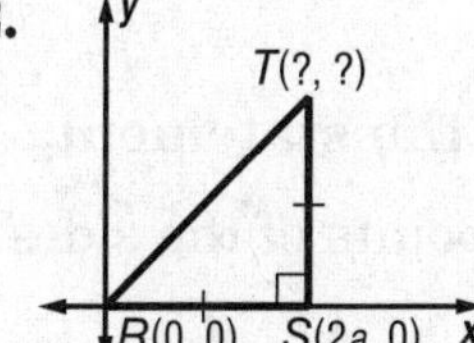

3.

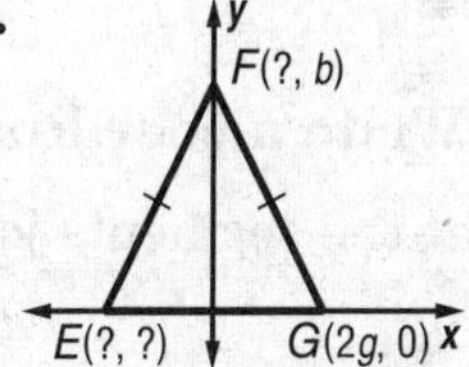

Position and label each triangle on the coordinate plane.

4. isosceles triangle $\triangle RST$ with base $\overline{RS}$ $4a$ units long

5. isosceles right $\triangle DEF$ with legs e units long

6. equilateral triangle $\triangle EQI$ with vertex $Q(0, \sqrt{3}b)$ and sides $2b$ units long

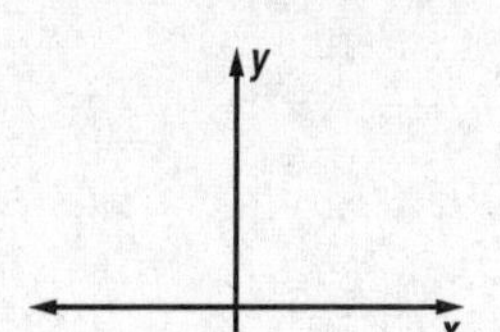

NAME ______________________ DATE ____________ PERIOD ____________

4-8 Study Guide and Intervention *(continued)*

Triangles and Coordinate Proof

Write Coordinate Proofs Coordinate proofs can be used to prove theorems and to verify properties. Many coordinate proofs use the Distance Formula, Slope Formula, or Midpoint Theorem.

Example **Prove that a segment from the vertex angle of an isosceles triangle to the midpoint of the base is perpendicular to the base.**

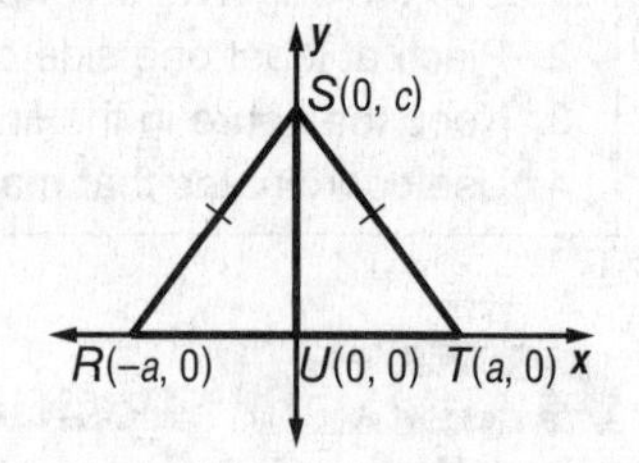

First, position and label an isosceles triangle on the coordinate plane. One way is to use $T(a, 0)$, $R(-a, 0)$, and $S(0, c)$. Then $U(0, 0)$ is the midpoint of $\overline{RT}$.

Given: Isosceles $\triangle RST$; U is the midpoint of base $\overline{RT}$.

Prove: $\overline{SU} \perp \overline{RT}$

Proof:

U is the midpoint of $\overline{RT}$ so the coordinates of U are $\left(\frac{-a + a}{2}, \frac{0 + 0}{2}\right) = (0, 0)$. Thus $\overline{SU}$ lies on the y-axis, and $\triangle RST$ was placed so $\overline{RT}$ lies on the x-axis. The axes are perpendicular, so $\overline{SU} \perp \overline{RT}$.

Exercise

PROOF Write a coordinate proof for the statement.

Prove that the segments joining the midpoints of the sides of a right triangle form a right triangle.

NAME ______________________ DATE __________ PERIOD __________

5-1 Study Guide and Intervention

Bisectors of Triangles

Perpendicular Bisector A perpendicular bisector is a line, segment, or ray that is perpendicular to the given segment and passes through its midpoint. Some theorems deal with perpendicular bisectors.

Perpendicular Bisector Theorem	If a point is on the perpendicular bisector of a segment, then it is equidistant from the endpoints of the segment.
Converse of Perpendicular Bisector Theorem	If a point is equidistant from the endpoints of a segment, then it is on the perpendicular bisector of the segment.
Circumcenter Theorem	The perpendicular bisectors of the sides of a triangle intersect at a point called the circumcenter that is equidistant from the vertices of the triangle.

Example 1 **Find the measure of $\overline{FM}$.**

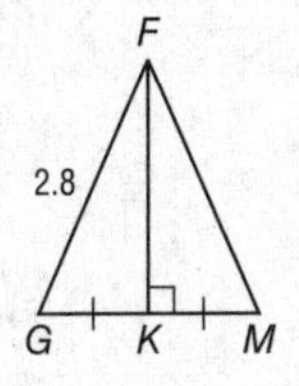

$\overline{FK}$ is the perpendicular bisector of $\overline{GM}$.

$FG = FM$

$2.8 = FM$

Example 2 **$\overline{BD}$ is the perpendicular bisector of $\overline{AC}$. Find x.**

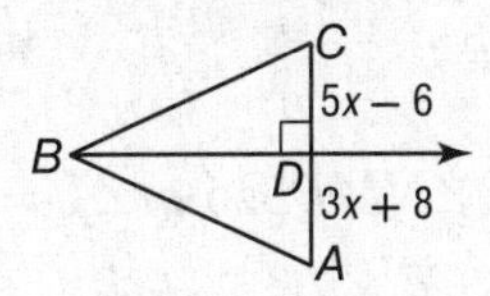

$$AD = DC$$
$$3x + 8 = 5x - 6$$
$$14 = 2x$$
$$7 = x$$

Exercises

Find each measure.

1. XW

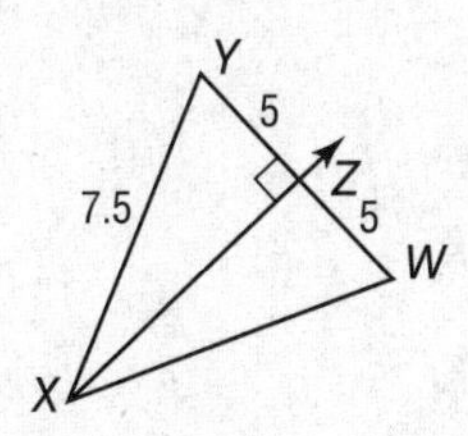

2. BF

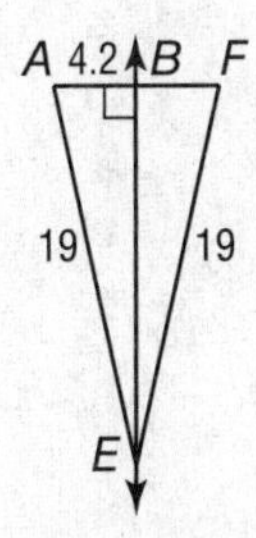

Point P is the circumcenter of $\triangle EMK$. List any segment(s) congruent to each segment below.

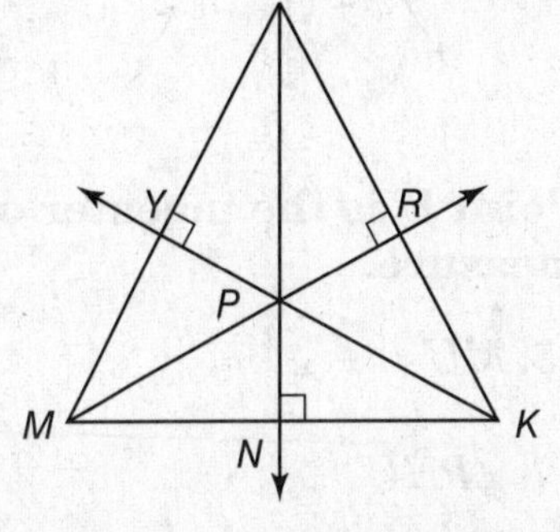

3. $\overline{MY}$

4. $\overline{KP}$

5. $\overline{MN}$

6. $\overline{ER}$

5-1 Study Guide and Intervention *(continued)*

Bisectors of Triangles

Angle Bisectors Another special segment, ray, or line is an angle bisector, which divides an angle into two congruent angles.

Angle Bisector Theorem	If a point is on the bisector of an angle, then it is equidistant from the sides of the angle.
Converse of Angle Bisector Theorem	If a point in the interior of an angle if equidistant from the sides of the angle, then it is on the bisector of the angle.
Incenter Theorem	The angle bisectors of a triangle intersect at a point called the incenter that is equidistant from the sides of the triangle.

Example **$\overrightarrow{MR}$ is the angle bisector of $\angle NMP$. Find x if $m\angle 1 = 5x + 8$ and $m\angle 2 = 8x - 16$.**

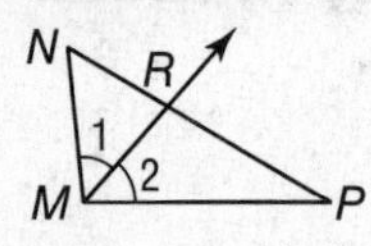

$\overrightarrow{MR}$ is the angle bisector of $\angle NMP$, so $m\angle 1 = m\angle 2$.

$5x + 8 = 8x - 16$

$24 = 3x$

$8 = x$

Exercises

Find each measure.

1. $\angle ABE$

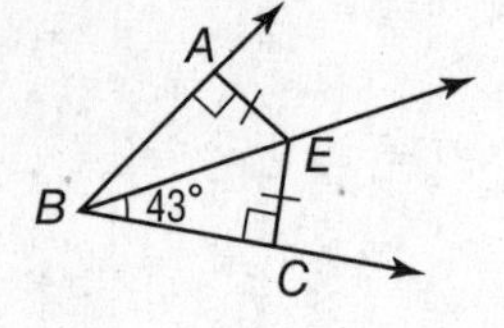

2. $\angle YBA$

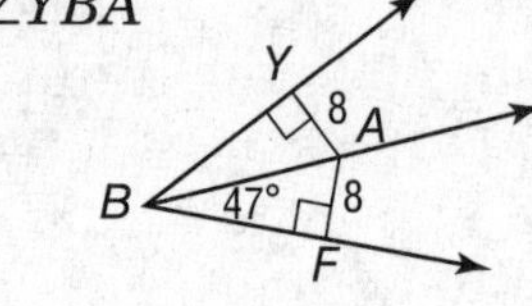

3. MK

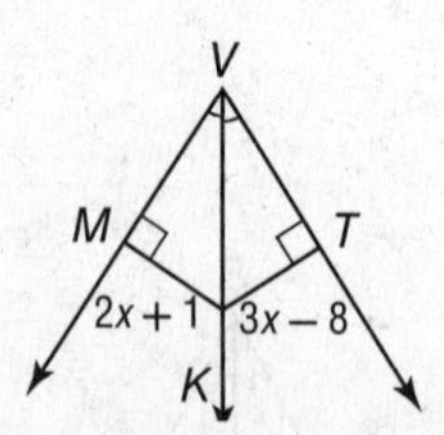

4. $\angle EWL$

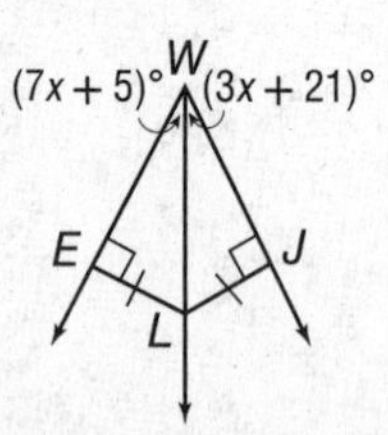

Point U is the incenter of $\triangle GHY$. Find each measure.

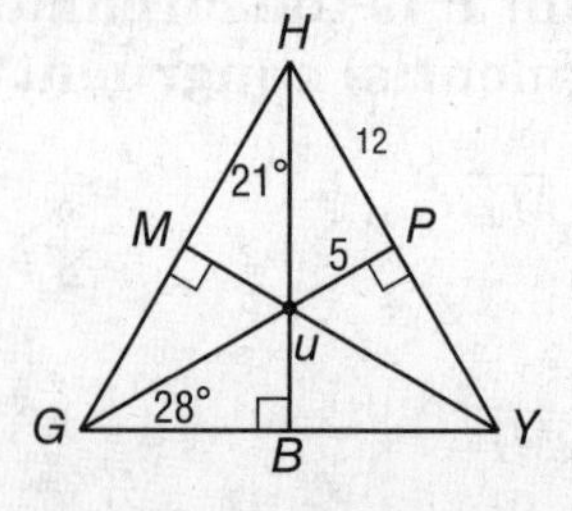

5. MU

6. $\angle UGM$

7. $\angle PHU$

8. HU

NAME ______________________ DATE __________ PERIOD __________

5-2 Study Guide and Intervention

Medians and Altitudes of Triangles

Medians A **median** is a line segment that connects a vertex of a triangle to the midpoint of the opposite side. The three medians of a triangle intersect at the **centroid** of the triangle. The centroid is located two thirds of the distance from a vertex to the midpoint of the side opposite the vertex on a median.

Example **In $\triangle ABC$, U is the centroid and $BU = 16$. Find UK and BK.**

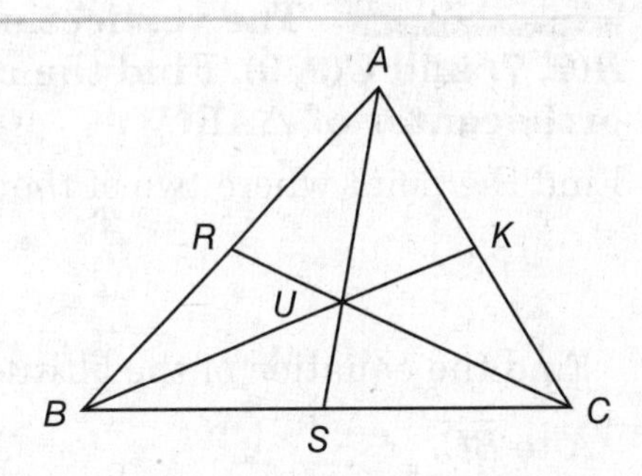

$BU = \frac{2}{3}BK$

$16 = \frac{2}{3}BK$

$24 = BK$

$BU + UK = BK$

$16 + UK = 24$

$UK = 8$

Exercises

In $\triangle ABC$, $AU = 16$, $BU = 12$, and $CF = 18$. Find each measure.

1. UD
2. EU
3. CU
4. AD
5. UF
6. BE

In $\triangle CDE$, U is the centroid, $UK = 12$, $EM = 21$, and $UD = 9$. Find each measure.

7. CU
8. MU
9. CK
10. JU
11. EU
12. JD

5-2 Study Guide and Intervention *(continued)*

Medians and Altitudes of Triangles

Altitudes An **altitude** of a triangle is a segment from a vertex to the line containing the opposite side meeting at a right angle. Every triangle has three altitudes which meet at a point called the **orthocenter.**

Example **The vertices of $\triangle ABC$ are $A(1, 3)$, $B(7, 7)$ and $C(9, 3)$. Find the coordinates of the orthocenter of $\triangle ABC$.**

Find the point where two of the three altitudes intersect.

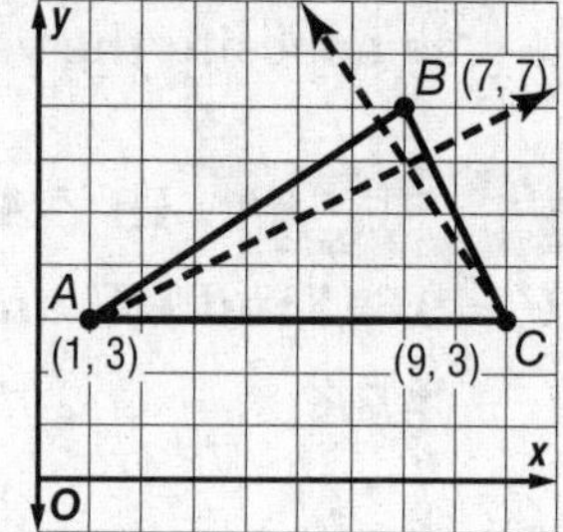

Find the equation of the altitude from A to $\overline{BC}$.

If $\overline{BC}$ has a slope of -2, then the altitude has a slope of $\frac{1}{2}$.

$y - y_1 = m(x - x_1)$ Point-slope form

$y - 3 = \frac{1}{2}(x - 1)$ $m = \frac{1}{2}$, $(x_1, y_1) = A(1, 3)$

$y - 3 = \frac{1}{2}x - \frac{1}{2}$ Distributive Property

$y = \frac{1}{2}x + \frac{5}{2}$ Simplify.

Find the equation of the altitude from C to $\overline{AB}$.

If $\overline{AB}$ has a slope of $\frac{2}{3}$, then the altitude has a slope of $-\frac{3}{2}$.

$y - y_1 = m(x - x_1)$ Point-slope form

$y - 3 = -\frac{3}{2}(x - 9)$ $m = -\frac{3}{2}$, $(x_1, y_1) = C(9, 3)$

$y - 3 = -\frac{3}{2}x + \frac{27}{2}$ Distributive Property

$y = -\frac{3}{2}x + \frac{33}{2}$ Simplify.

Solve the system of equations and find where the altitudes meet.

$y = \frac{1}{2}x + \frac{5}{2}$ $\quad$ $y = -\frac{3}{2}x + \frac{33}{2}$

$\frac{1}{2}x + \frac{5}{2} = -\frac{3}{2}x + \frac{33}{2}$ Subtract $\frac{1}{2}x$ from each side.

$\frac{5}{2} = -2x + \frac{33}{2}$ Subtract $\frac{33}{2}$ from each side.

$-14 = -2x$ Divide both sides by -2.

$7 = x$

$y = \frac{1}{2}x + \frac{5}{2} = \frac{1}{2}(7) + \frac{5}{2} = \frac{7}{2} + \frac{5}{2} = 6$

The coordinates of the orthocenter of $\triangle ABC$ are (6, 7).

Exercises

COORDINATE GEOMETRY Find the coordinates of the orthocenter of the triangle with the given vertices.

1. $J(1, 0)$, $H(6, 0)$, $I(3, 6)$

2. $S(1, 0)$, $T(4, 7)$, $U(8, -3)$

5-3 Study Guide and Intervention

Inequalities in One Triangle

Angle Inequalities Properties of inequalities, including the Transitive, Addition, and Subtraction Properties of Inequality, can be used with measures of angles and segments. There is also a Comparison Property of Inequality.

For any real numbers a and b, either $a < b$, $a = b$, or $a > b$.

The Exterior Angle Inequality Theorem can be used to prove this inequality involving an exterior angle.

Exterior Angle Inequality Theorem	The measure of an exterior angle of a triangle is greater than the measure of either of its corresponding remote interior angles.	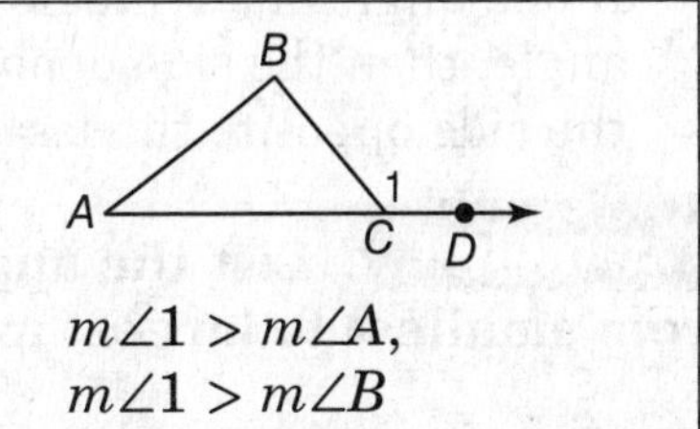 $m\angle 1 > m\angle A$, $m\angle 1 > m\angle B$

Example **List all angles of $\triangle EFG$ with measures that are less than $m\angle 1$.**

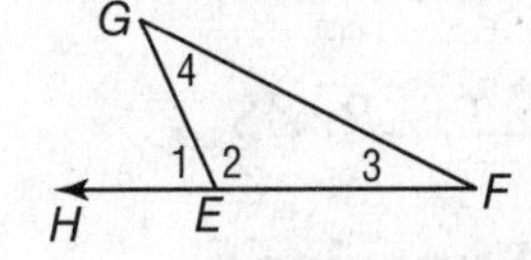

The measure of an exterior angle is greater than the measure of either remote interior angle. So $m\angle 3 < m\angle 1$ and $m\angle 4 < m\angle 1$.

Exercises

Use the Exterior Angle Inequality Theorem to list all of the angles that satisfy the stated condition.

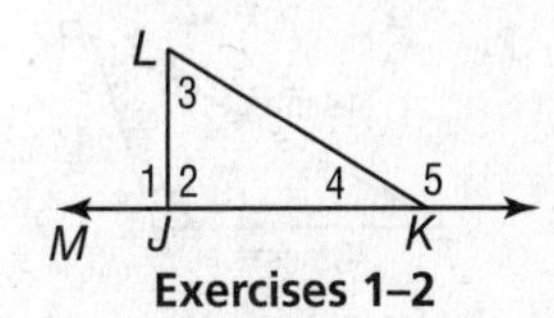

Exercises 1–2

1. measures are less than $m\angle 1$

2. measures are greater than $m\angle 3$

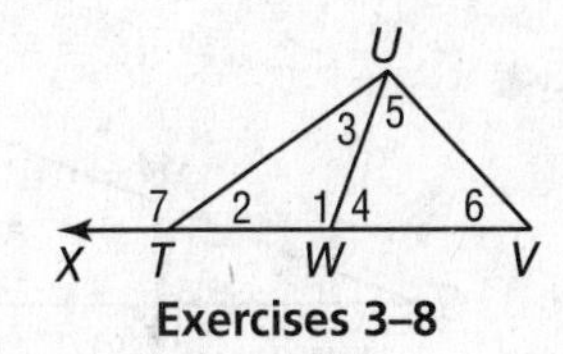

Exercises 3–8

3. measures are less than $m\angle 1$

4. measures are greater than $m\angle 1$

5. measures are less than $m\angle 7$

6. measures are greater than $m\angle 2$

7. measures are greater than $m\angle 5$

8. measures are less than $m\angle 4$

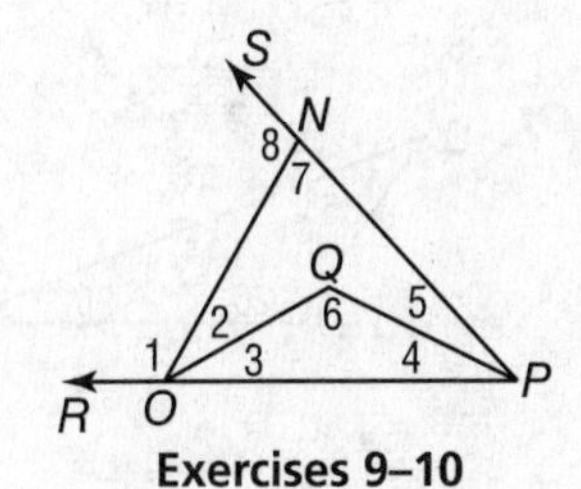

Exercises 9–10

9. measures are less than $m\angle 1$

10. measures are greater than $m\angle 4$

5-3 Study Guide and Intervention *(continued)*

Inequalities in One Triangle

Angle-Side Relationships When the sides of triangles are not congruent, there is a relationship between the sides and angles of the triangles.

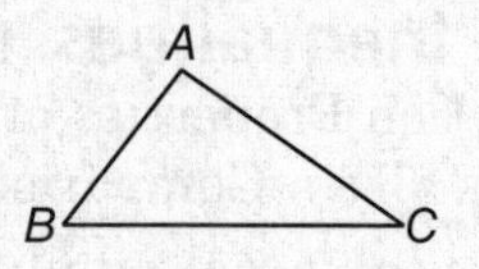

If $AC > AB$, then $m\angle B > m\angle C$.
If $m\angle A > m\angle C$, then $BC > AB$.

- If one side of a triangle is longer than another side, then the angle opposite the longer side has a greater measure than the angle opposite the shorter side.
- If one angle of a triangle has a greater measure than another angle, then the side opposite the greater angle is longer than the side opposite the lesser angle.

Example 1 **List the angles in order from smallest to largest measure.**

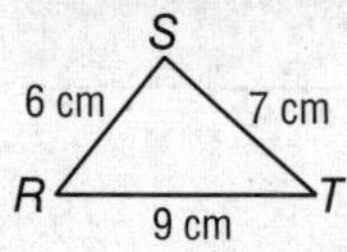

$\angle T, \angle R, \angle S$

Example 2 **List the sides in order from shortest to longest.**

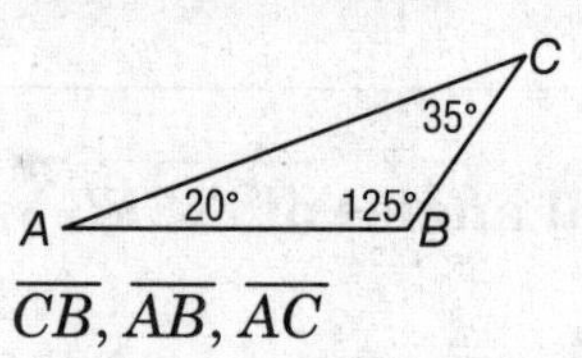

$\overline{CB}, \overline{AB}, \overline{AC}$

Exercises

List the angles and sides in order from smallest to largest.

1.

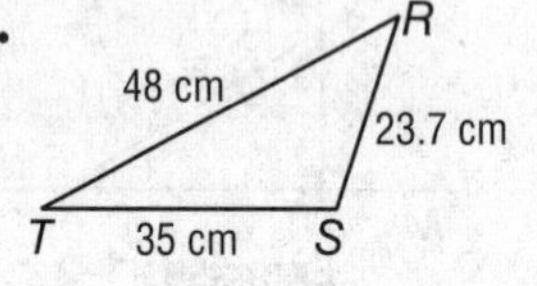

2.

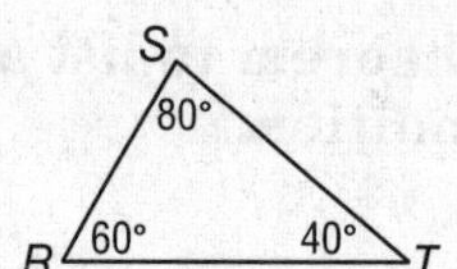

3.

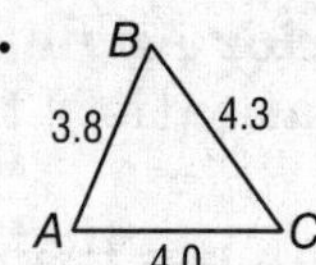

4.

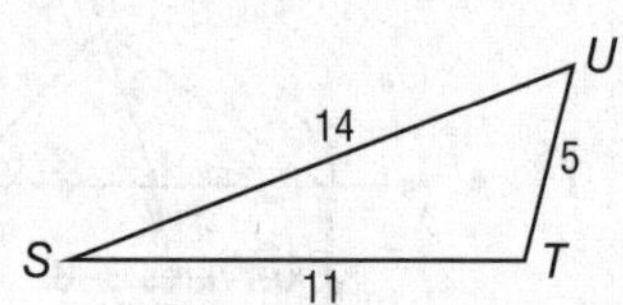

5.

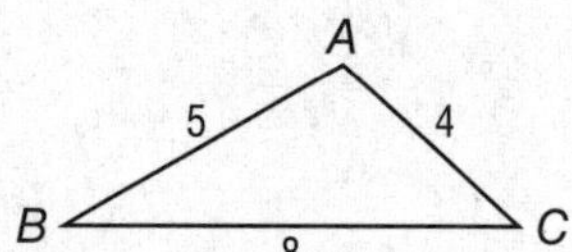

6.

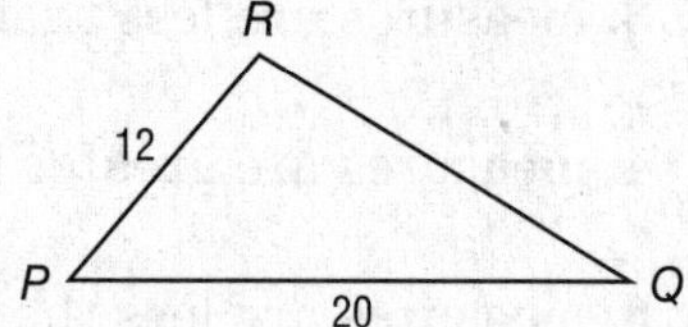

7.

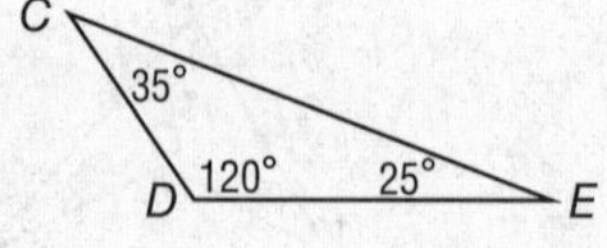

8.

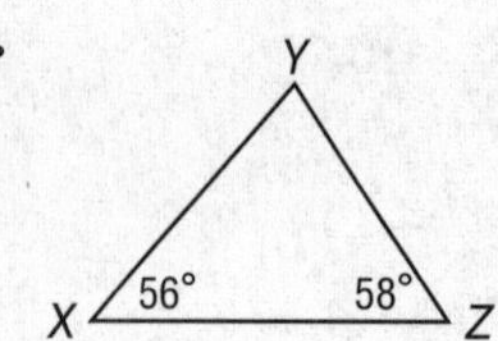

9.

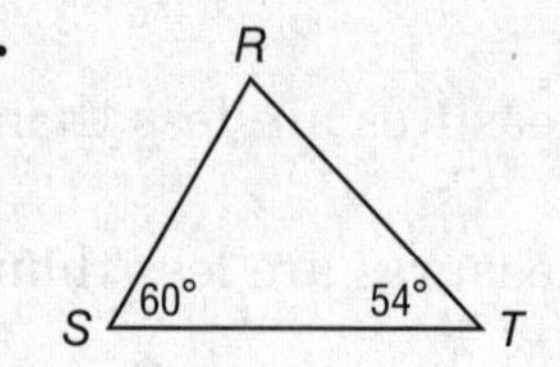

NAME ______________________ DATE ____________ PERIOD ____________

5-4 Study Guide and Intervention

Indirect Proof

Indirect Algebraic Proof One way to prove that a statement is true is to temporarily assume that what you are trying to prove is false. By showing this assumption to be logically impossible, you prove your assumption false and the original conclusion true. This is known as an **indirect proof**.

Steps for Writing an Indirect Proof

1. Assume that the conclusion is false by assuming the oppposite is true.
2. Show that this assumption leads to a contradiction of the hypothesis or some other fact.
3. Point out that the assumption must be false, and therefore, the conclusion must be true.

Example **Given:** $3x + 5 > 8$
Prove: $x > 1$

Step 1 Assume that x is not greater than 1. That is, $x = 1$ or $x < 1$.

Step 2 Make a table for several possibilities for $x = 1$ or $x < 1$. When $x = 1$ or $x < 1$, then $3x + 5$ is not greater than 8.

Step 3 This contradicts the given information that $3x + 5 > 8$. The assumption that x is not greater than 1 must be false, which means that the statement "$x > 1$" must be true.

x	3x + 5
1	8
0	5
−1	2
−2	−1
−3	−4

Exercises

State the assumption you would make to start an indirect proof of each statement.

1. If $2x > 14$, then $x > 7$.

2. For all real numbers, if $a + b > c$, then $a > c - b$.

Complete the indirect proof.

Given: n is an integer and n^2 is even.

Prove: n is even.

3. Assume that ______________________

4. Then n can be expressed as $2a + 1$ by ______________________

5. $n^2 =$ ______________ Substitution

6. $=$ ______________ Multiply.

7. $=$ ______________ Simplify.

8. $= 2(2a^2 + 2a) + 1$ ______________________

9. $2(2a^2 + 2a) + 1$ is an odd number. This contradicts the given that n^2 is even, so the assumption must be ______________________

10. Therefore, ______________________

5-4 Study Guide and Intervention *(continued)*

Indirect Proof

Indirect Proof with Geometry To write an indirect proof in geometry, you assume that the conclusion is false. Then you show that the assumption leads to a contradiction. The contradiction shows that the conclusion cannot be false, so it must be true.

Example

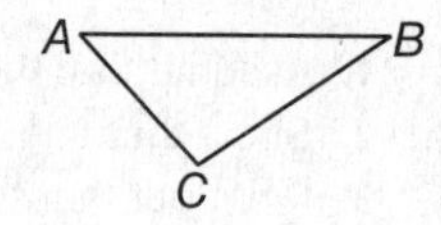

Given: $m\angle C = 100$
Prove: $\angle A$ is not a right angle.

Step 1 Assume that $\angle A$ is a right angle.

Step 2 Show that this leads to a contradiction. If $\angle A$ is a right angle, then $m\angle A = 90$ and $m\angle C + m\angle A = 100 + 90 = 190$. Thus the sum of the measures of the angles of $\triangle ABC$ is greater than 180.

Step 3 The conclusion that the sum of the measures of the angles of $\triangle ABC$ is greater than 180 is a contradiction of a known property. The assumption that $\angle A$ is a right angle must be false, which means that the statement "$\angle A$ is not a right angle" must be true.

Exercises

State the assumption you would make to start an indirect proof of each statement.

1. If $m\angle A = 90$, then $m\angle B = 45$.

2. If $\overline{AV}$ is not congruent to $\overline{VE}$, then $\triangle AVE$ is not isosceles.

Complete the indirect proof.

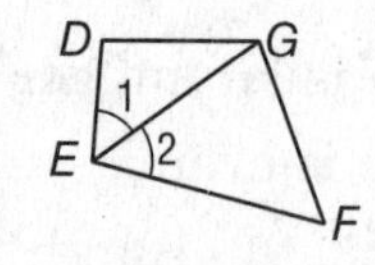

Given: $\angle 1 \cong \angle 2$ and $\overline{DG}$ is not congruent to $\overline{FG}$.
Prove: $\overline{DE}$ is not congruent to $\overline{FE}$.

3. Assume that ____________ Assume the conclusion is false.

4. $\overline{EG} \cong \overline{EG}$ ______________________________

5. $\triangle EDG \cong \triangle EFG$ ______________________________

6. ______________________________

7. This contradicts the given information, so the assumption must be ______________________________

8. Therefore, ______________________________

NAME ______________________ DATE ____________ PERIOD ____________

5-5 Study Guide and Intervention

The Triangle Inequality

The Triangle Inequality If you take three straws of lengths 8 inches, 5 inches, and 1 inch and try to make a triangle with them, you will find that it is not possible. This illustrates the Triangle Inequality Theorem.

Triangle Inequality Theorem	The sum of the lengths of any two sides of a triangle must be greater than the length of the third side.	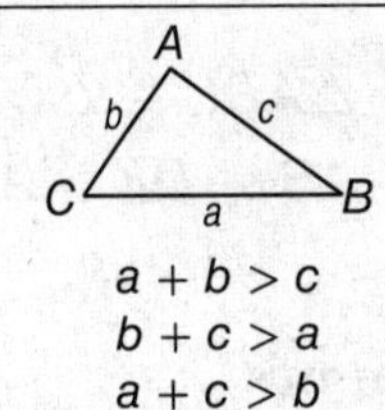 $a + b > c$ $b + c > a$ $a + c > b$

Example **The measures of two sides of a triangle are 5 and 8. Find a range for the length of the third side.**

By the Triangle Inequality Theorem, all three of the following inequalities must be true.

$5 + x > 8$	$8 + x > 5$	$5 + 8 > x$
$x > 3$	$x > -3$	$13 > x$

Therefore x must be between 3 and 13.

Exercises

Is it possible to form a triangle with the given side lengths? If not, explain why not.

1. 3, 4, 6

2. 6, 9, 15

3. 8, 8, 8

4. 2, 4, 5

5. 4, 8, 16

6. 1.5, 2.5, 3

Find the range for the measure of the third side of a triangle given the measures of two sides.

7. 1 cm and 6 cm

8. 12 yd and 18 yd

9. 1.5 ft and 5.5 ft

10. 82 m and 8 m

11. Suppose you have three different positive numbers arranged in order from least to greatest. What single comparison will let you see if the numbers can be the lengths of the sides of a triangle?

5-5 Study Guide and Intervention *(continued)*

The Triangle Inequality

Proofs Using The Triangle Inequality Theorem You can use the Triangle Inequality Theorem as a reason in proofs.

Complete the following proof.

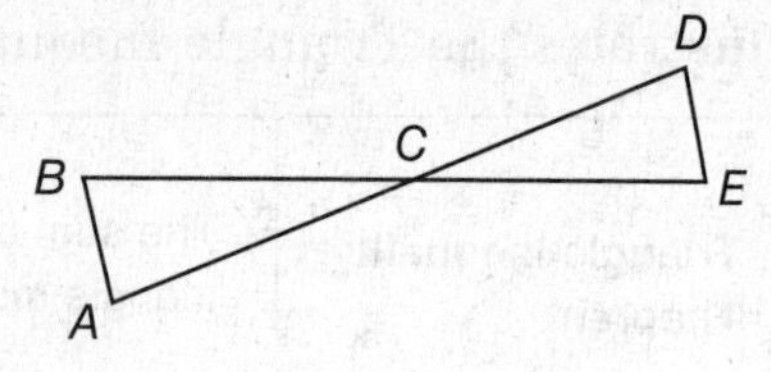

Given: $\triangle ABC \cong \triangle DEC$

Prove: $AB + DE > AD - BE$

Proof:

Statements	**Reasons**
1. $\triangle ABC \cong \triangle DEC$	**1.** Given
2. $AB + BC > AC$ $DE + EC > CD$	**2.** Triangle Inequality Theorem
3. $AB > AC - BC$ $DE > CD - EC$	**3.** Subtraction
4. $AB + DE > AC - BC + CD - EC$	**4.** Addition
5. $AB + DE > AC + CD - BC - EC$	**5.** Commutative
6. $AB + DE > AC + CD - (BC + EC)$	**6.** Distributive
7. $AC + CD = AD$ $BC + EC = BE$	**7.** Segment Addition Postulate
8. $AB + DE > AD - BE$	**8.** Substitution

Exercises

PROOF Write a two column proof.

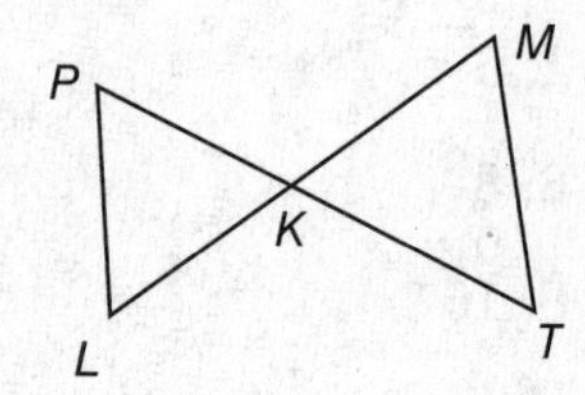

Given: $\overline{PL} \parallel \overline{MT}$

K is the midpoint of $\overline{PT}$.

Prove: $PK + KM > PL$

Proof:

Statements	**Reasons**
1. $\overline{PL} \parallel \overline{MT}$	**1.** ____________
2. $\angle P \cong \angle T$	**2.** ____________
3. K is the midpoint of $\overline{PT}$.	**3.** Given
4. $PK = KT$	**4.** ____________
5. ____________	**5.** Vertical Angles Theorem
6. $\triangle PKL \cong \triangle TKM$	**6.** ____________
7. ____________	**7.** Triangle Inequality Theorem
8. ____________	**8.** CPCTC
9. $PK + KM > PL$	**9.** ____________

NAME ____________________ DATE __________ PERIOD __________

5-6 Study Guide and Intervention

Inequalities in Two Triangles

Hinge Theorem The following theorem and its converse involve the relationship between the sides of two triangles and an angle in each triangle.

Hinge Theorem	If two sides of a triangle are congruent to two sides of another triangle and the included angle of the first is larger than the included angle of the second, then the third side of the first triangle is longer than the third side of the second triangle.	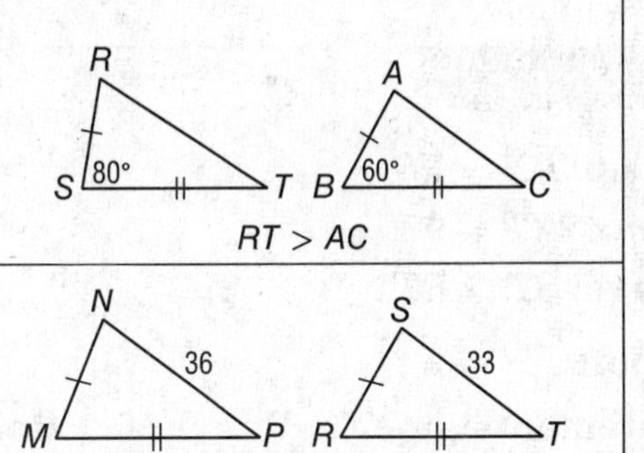 $RT > AC$
Converse of the Hinge Theorem	If two sides of a triangle are congruent to two sides of another triangle, and the third side in the first is longer than the third side in the second, then the included angle in the first triangle is greater than the included angle in the second triangle.	$m\angle M > m\angle R$

Example 1 **Compare the measures of $\overline{GF}$ and $\overline{FE}$.**

Two sides of $\triangle HGF$ are congruent to two sides of $\triangle HEF$, and $m\angle GHF > m\angle EHF$. By the Hinge Theorem, $GF > FE$.

Example 2 **Compare the measures of $\angle ABD$ and $\angle CBD$.**

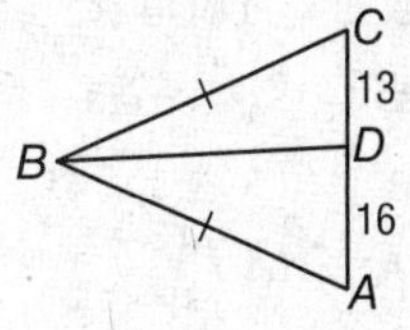

Two sides of $\triangle ABD$ are congruent to two sides of $\triangle CBD$, and $AD > CD$. By the Converse of the Hinge Theorem, $m\angle ABD > m\angle CBD$.

Exercises

Compare the given measures.

1. MR and RP

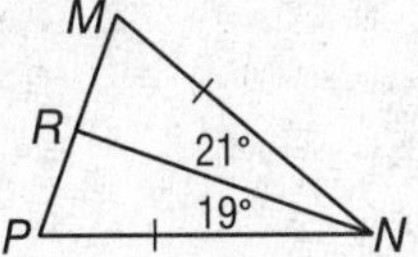

2. AD and CD

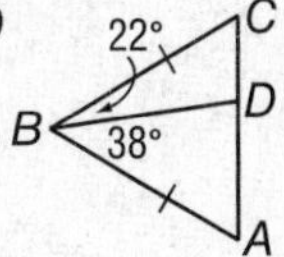

3. $m\angle C$ and $m\angle Z$

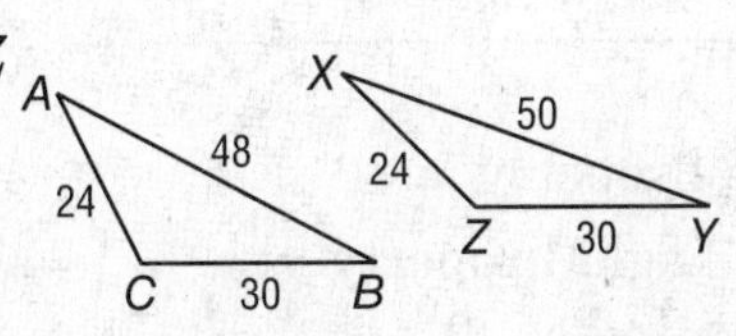

4. $m\angle XYW$ and $m\angle WYZ$

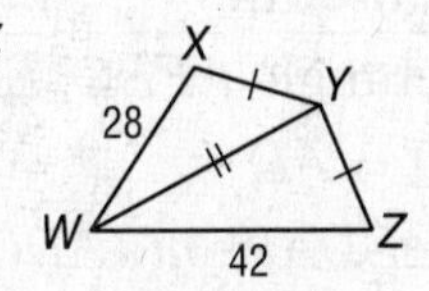

Write an inequality for the range of values of *x*.

5.

6.

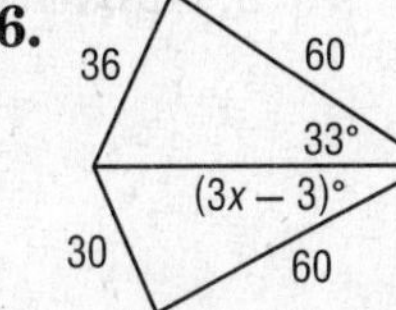

5-6 Study Guide and Intervention *(continued)*

Inequalities Involving Two Triangles

PROVE RELATIONSHIPS IN TWO TRIANGLES You can use the Hinge Theorem and its converse to prove relationships in two triangles.

Example

Given: $RX = XS$

$m\angle SXT = 97$

Prove: $ST > RT$

Proof:

Statements	Reasons
1. $\angle SXT$ and $\angle RXT$ are supplementary.	**1.** Def. of linear pair
2. $m\angle SXT + m\angle RXT = 180$	**2.** Def. of supplementary
3. $m\angle SXT = 97$	**3.** Given
4. $97 + m\angle RXT = 180$	**4.** Substitution
5. $m\angle RXT = 83$	**5.** Subtraction
6. $97 > 83$	**6.** Inequality
7. $m\angle SXT > m\angle RXT$	**7.** Substitution
8. $RX = XS$	**8.** Given
9. $TX = TX$	**9.** Reflexive Property
10. $ST > RT$	**10.** Hinge Theorem

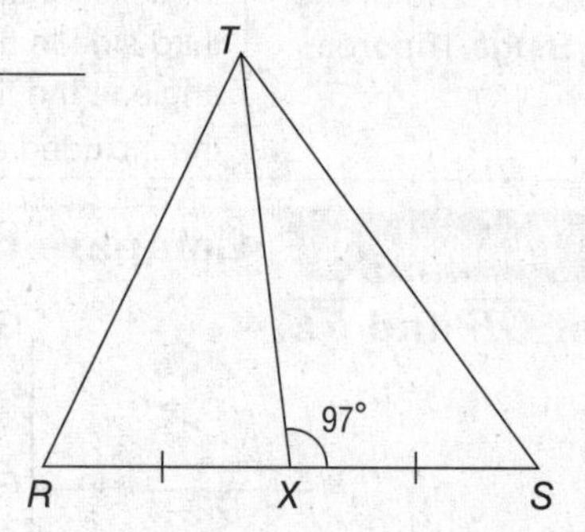

Exercises

Complete the proof.

Given: rectangle $AFBC$

$ED = DC$

Prove: $AE > FB$

Proof:

Statements	Reasons
1. rectangle $AFBC$, $ED = DC$	**1.** Given
2. $AD = AD$	**2.** Reflexive Property
3. $m\angle EDA > m\angle ADC$	**3.** Exterior Angle Property
4.	**4.** Hinge Theorem
5.	**5.** Opp sides in rectangle are ≅.
6. $AE > FB$	**6.** Substitution

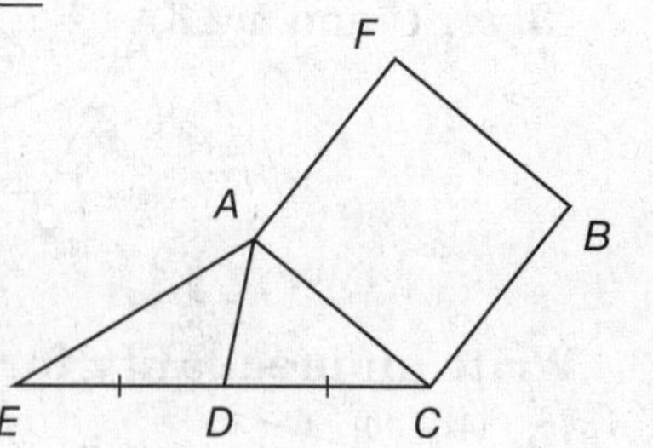

NAME ______________________ DATE ____________ PERIOD ____________

6-1 Study Guide and Intervention

Angles of Polygons

Polygon Interior Angles Sum The segments that connect the nonconsecutive vertices of a polygon are called **diagonals**. Drawing all of the diagonals from one vertex of an ***n*-gon** separates the polygon into $n - 2$ triangles. The sum of the measures of the interior angles of the polygon can be found by adding the measures of the interior angles of those $n - 2$ triangles.

Polygon Interior Angle Sum Theorem	The sum of the interior angle measures of an n-sided convex polygon is $(n - 2) \cdot 180$.

Example 1 **A convex polygon has 13 sides. Find the sum of the measures of the interior angles.**

$$(n - 2) \cdot 180 = (13 - 2) \cdot 180$$
$$= 11 \cdot 180$$
$$= 1980$$

Example 2 **The measure of an interior angle of a regular polygon is 120. Find the number of sides.**

The number of sides is n, so the sum of the measures of the interior angles is $120n$.

$$120n = (n - 2) \cdot 180$$
$$120n = 180n - 360$$
$$-60n = -360$$
$$n = 6$$

Exercises

Find the sum of the measures of the interior angles of each convex polygon.

1. decagon

2. 16-gon

3. 30-gon

4. octagon

5. 12-gon

6. 35-gon

The measure of an interior angle of a regular polygon is given. Find the number of sides in the polygon.

7. 150

8. 160

9. 175

10. 165

11. 144

12. 135

13. Find the value of x.

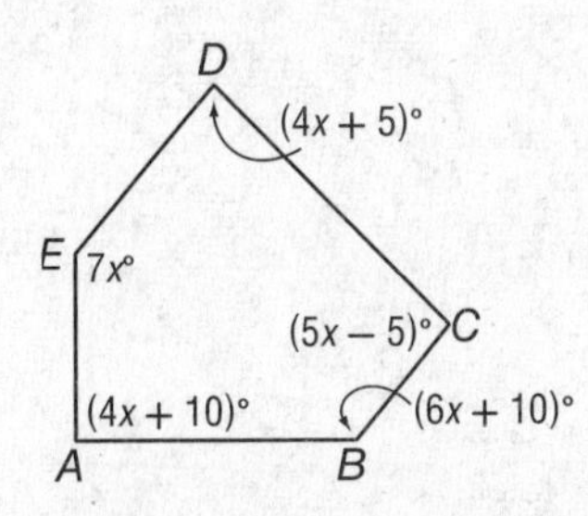

NAME ______________________ DATE ____________ PERIOD ____________

6-1 Study Guide and Intervention *(continued)*

Angles of Polygons

Polygon Exterior Angles Sum There is a simple relationship among the exterior angles of a convex polygon.

Polygon Exterior Angle Sum Theorem	The sum of the exterior angle measures of a convex polygon, one angle at each vertex, is 360.

Example 1 **Find the sum of the measures of the exterior angles, one at each vertex, of a convex 27-gon.**

For *any* convex polygon, the sum of the measures of its exterior angles, one at each vertex, is 360.

Example 2 **Find the measure of each exterior angle of regular hexagon *ABCDEF*.**

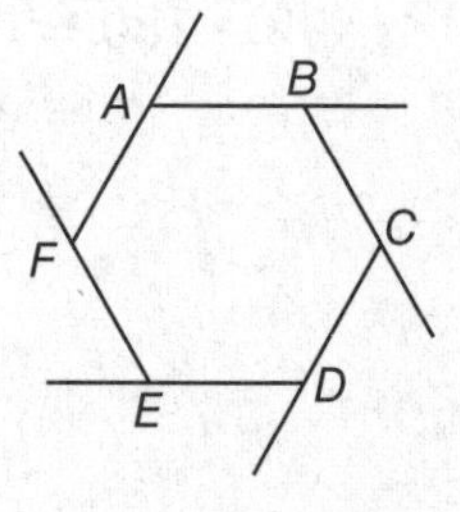

The sum of the measures of the exterior angles is 360 and a hexagon has 6 angles. If n is the measure of each exterior angle, then

$6n = 360$

$n = 60$

The measure of each exterior angle of a regular hexagon is 60.

Exercises

Find the sum of the measures of the exterior angles of each convex polygon.

1. decagon **2.** 16-gon **3.** 36-gon

Find the measure of each exterior angle for each regular polygon.

4. 12-gon **5.** hexagon **6.** 20-gon

7. 40-gon **8.** heptagon **9.** 12-gon

10. 24-gon **11.** dodecagon **12.** octagon

NAME ______________________ DATE ____________ PERIOD ____________

6-2 Study Guide and Intervention

Parallelograms

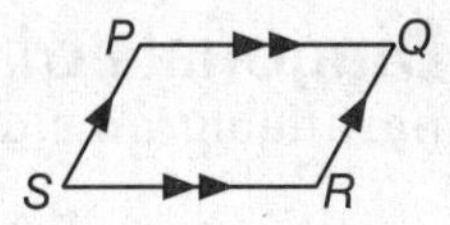

Sides and Angles of Parallelograms A quadrilateral with both pairs of opposite sides parallel is a **parallelogram**. Here are four important properties of parallelograms.

	If *PQRS* is a parallelogram, then
If a quadrilateral is a parallelogram, then its opposite sides are congruent.	$\overline{PQ} \cong \overline{SR}$ and $\overline{PS} \cong \overline{QR}$
If a quadrilateral is a parallelogram, then its opposite angles are congruent.	$\angle P \cong \angle R$ and $\angle S \cong \angle Q$
If a quadrilateral is a parallelogram, then its consecutive angles are supplementary.	$\angle P$ and $\angle S$ are supplementary; $\angle S$ and $\angle R$ are supplementary; $\angle R$ and $\angle Q$ are supplementary; $\angle Q$ and $\angle P$ are supplementary.
If a parallelogram has one right angle, then it has four right angles.	If $m\angle P = 90$, then $m\angle Q = 90$, $m\angle R = 90$, and $m\angle S = 90$.

Example **If *ABCD* is a parallelogram, find the value of each variable.**

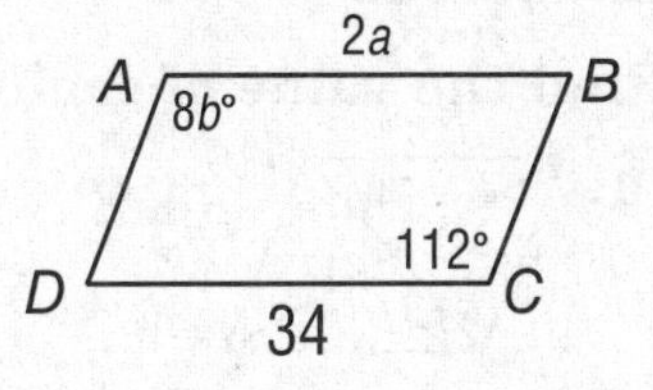

$\overline{AB}$ and $\overline{CD}$ are opposite sides, so $\overline{AB} \cong \overline{CD}$.

$2a = 34$

$a = 17$

$\angle A$ and $\angle C$ are opposite angles, so $\angle A \cong \angle C$.

$8b = 112$

$b = 14$

Exercises

Find the value of each variable.

1.

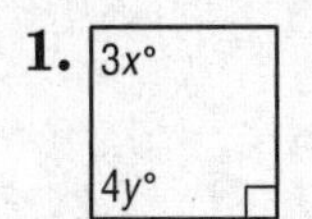

2.

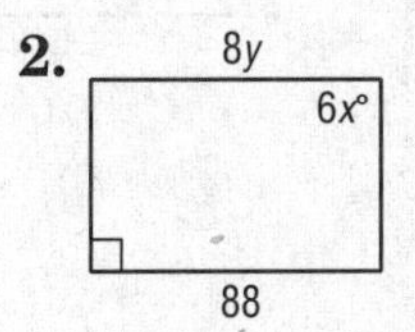

3.

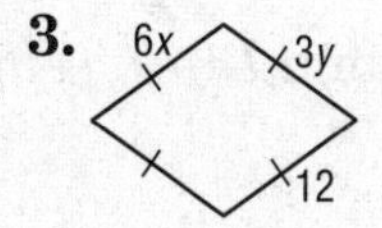

4. 3y° 6x° 12x°

5.

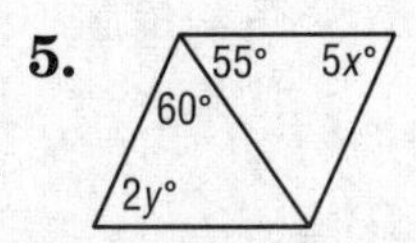

6. 2y 30x 150 72x

6-2 Study Guide and Intervention *(continued)*

Parallelograms

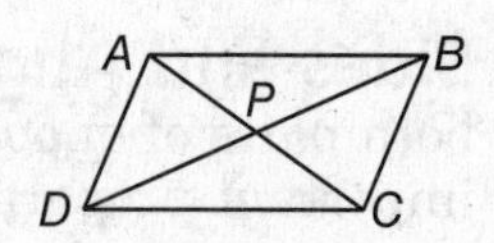

Diagonals of Parallelograms Two important properties of parallelograms deal with their diagonals.

	If *ABCD* is a parallelogram, then
If a quadrilateral is a parallelogram, then its diagonals bisect each other.	$AP = PC$ and $DP = PB$
If a quadrilateral is a parallelogram, then each diagonal separates the parallelogram into two congruent triangles.	$\triangle ACD \cong \triangle CAB$ and $\triangle ADB \cong \triangle CBD$

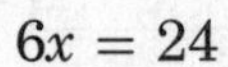

Find the value of x and y in parallelogram *ABCD*.

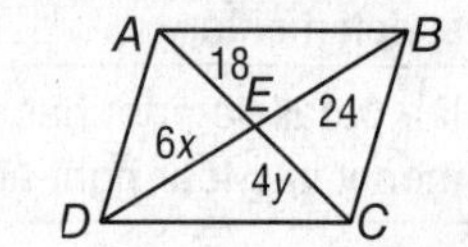

The diagonals bisect each other, so $AE = CE$ and $DE = BE$.

$6x = 24$ $\quad$ $4y = 18$

$x = 4$ $\quad$ $y = 4.5$

Exercises

Find the value of each variable.

1. 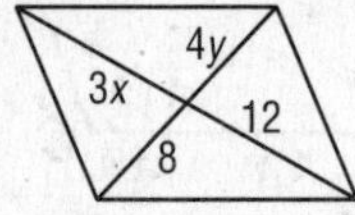

2. 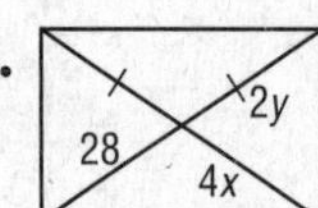

3. 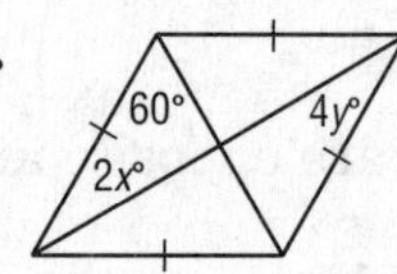

4. 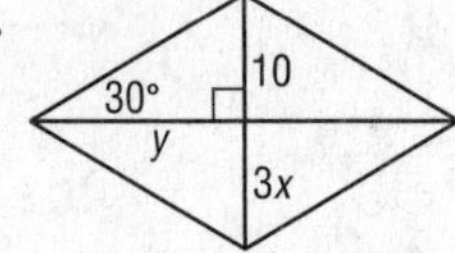

5. 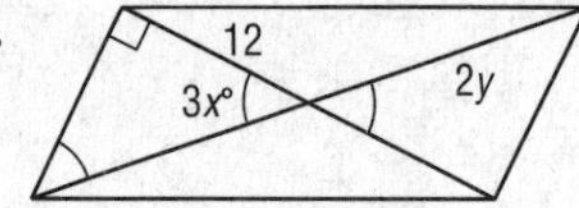

6. 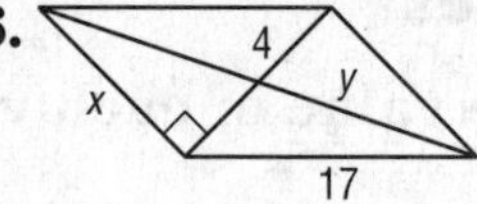

COORDINATE GEOMETRY Find the coordinates of the intersection of the diagonals of ▱*ABCD* with the given vertices.

7. $A(3, 6)$, $B(5, 8)$, $C(3, -2)$, and $D(1, -4)$

8. $A(-4, 3)$, $B(2, 3)$, $C(-1, -2)$, and $D(-7, -2)$

9. PROOF Write a paragraph proof of the following.

Given: ▱*ABCD*

Prove: $\triangle AED \cong \triangle BEC$

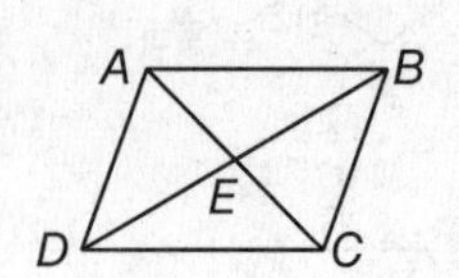

NAME ______________________ DATE __________ PERIOD __________

6-3 Study Guide and Intervention

Tests for Parallelograms

Conditions for Parallelograms There are many ways to establish that a quadrilateral is a parallelogram.

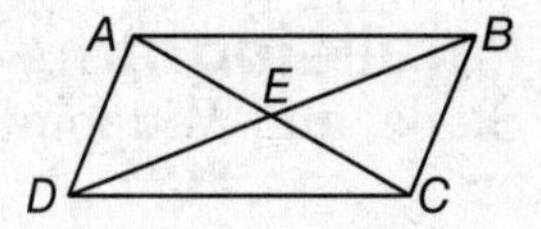

If:	If:
both pairs of opposite sides are parallel,	$\overline{AB} \parallel \overline{DC}$ and $\overline{AD} \parallel \overline{BC}$,
both pairs of opposite sides are congruent,	$\overline{AB} \cong \overline{DC}$ and $\overline{AD} \cong \overline{BC}$,
both pairs of opposite angles are congruent,	$\angle ABC \cong \angle ADC$ and $\angle DAB \cong \angle BCD$,
the diagonals bisect each other,	$\overline{AE} \cong \overline{CE}$ and $\overline{DE} \cong \overline{BE}$,
one pair of opposite sides is congruent and parallel,	$\overline{AB} \parallel \overline{CD}$ and $\overline{AB} \cong \overline{CD}$, or $\overline{AD} \parallel \overline{BC}$ and $\overline{AD} \cong \overline{BC}$,
then: the figure is a parallelogram.	**then:** ABCD is a parallelogram.

Example **Find x and y so that $FGHJ$ is a parallelogram.**

F, G, H, J; $6x + 3$; $4x - 2y$; 2; 15

$FGHJ$ is a parallelogram if the lengths of the opposite sides are equal.

$$6x + 3 = 15 \qquad 4x - 2y = 2$$
$$6x = 12 \qquad 4(2) - 2y = 2$$
$$x = 2 \qquad 8 - 2y = 2$$
$$-2y = -6$$
$$y = 3$$

Exercises

Find x and y so that the quadrilateral is a parallelogram.

1. 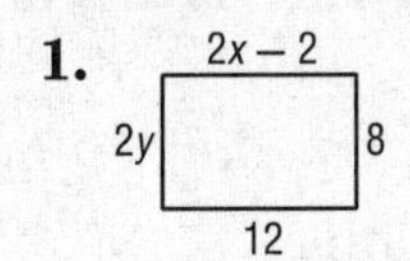

2. 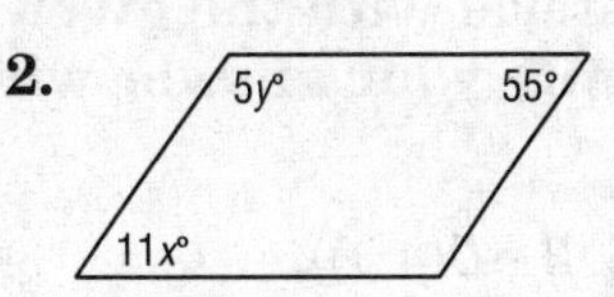

3. 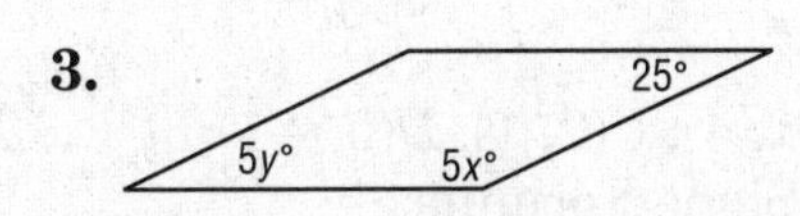

4. 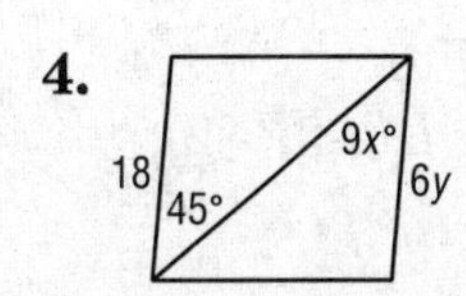

5. 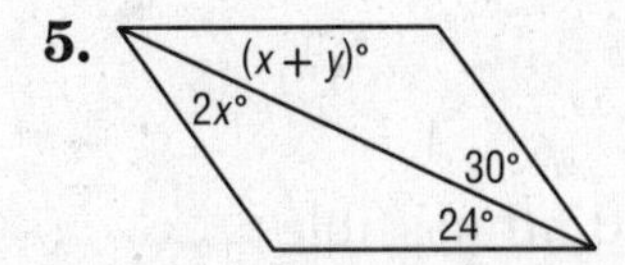

6. 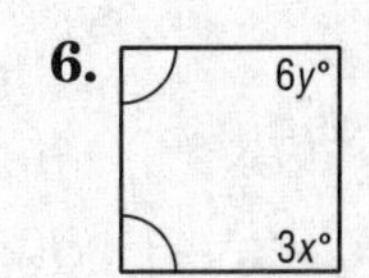

NAME ______________________ DATE ____________ PERIOD ________

6-3 Study Guide and Intervention *(continued)*

Tests for Parallelograms

Parallelograms on the Coordinate Plane On the coordinate plane, the Distance, Slope, and Midpoint Formulas can be used to test if a quadrilateral is a parallelogram.

Example **Determine whether *ABCD* is a parallelogram.**

The vertices are $A(-2, 3)$, $B(3, 2)$, $C(2, -1)$, and $D(-3, 0)$.

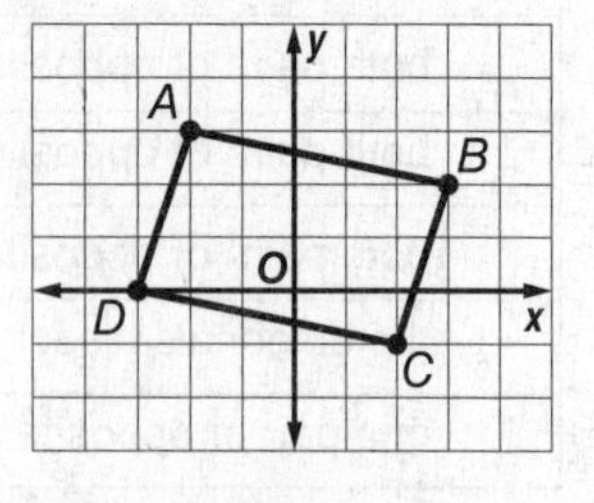

Method 1: Use the Slope Formula, $m = \frac{y_2 - y_1}{x_2 - x_1}$.

slope of $\overline{AD} = \frac{3 - 0}{-2 - (-3)} = \frac{3}{1} = 3$ slope of $\overline{BC} = \frac{2 - (-1)}{3 - 2} = \frac{3}{1} = 3$

slope of $\overline{AB} = \frac{2 - 3}{3 - (-2)} = -\frac{1}{5}$ slope of $\overline{CD} = \frac{-1 - 0}{2 - (-3)} = -\frac{1}{5}$

Since opposite sides have the same slope, $\overline{AB} \parallel \overline{CD}$ and $\overline{AD} \parallel \overline{BC}$. Therefore, *ABCD* is a parallelogram by definition.

Method 2: Use the Distance Formula, $d = \sqrt{(x_2 - x_1)^2 + (y_2 - y_1)^2}$.

$AB = \sqrt{(-2 - 3)^2 + (3 - 2)^2} = \sqrt{25 + 1}$ or $\sqrt{26}$

$CD = \sqrt{(2 - (-3))^2 + (-1 - 0)^2} = \sqrt{25 + 1}$ or $\sqrt{26}$

$AD = \sqrt{(-2 - (-3))^2 + (3 - 0)^2} = \sqrt{1 + 9}$ or $\sqrt{10}$

$BC = \sqrt{(3 - 2)^2 + (2 - (-1))^2} = \sqrt{1 + 9}$ or $\sqrt{10}$

Since both pairs of opposite sides have the same length, $\overline{AB} \cong \overline{CD}$ and $\overline{AD} \cong \overline{BC}$. Therefore, *ABCD* is a parallelogram by Theorem 6.9.

Exercises

Graph each quadrilateral with the given vertices. Determine whether the figure is a parallelogram. Justify your answer with the method indicated.

1. $A(0, 0)$, $B(1, 3)$, $C(5, 3)$, $D(4, 0)$; Slope Formula

2. $D(-1, 1)$, $E(2, 4)$, $F(6, 4)$, $G(3, 1)$; Slope Formula

3. $R(-1, 0)$, $S(3, 0)$, $T(2, -3)$, $U(-3, -2)$; Distance Formula

4. $A(-3, 2)$, $B(-1, 4)$, $C(2, 1)$, $D(0, -1)$; Distance and Slope Formulas

5. $S(-2, 4)$, $T(-1, -1)$, $U(3, -4)$, $V(2, 1)$; Distance and Slope Formulas

6. $F(3, 3)$, $G(1, 2)$, $H(-3, 1)$, $I(-1, 4)$; Midpoint Formula

7. A parallelogram has vertices $R(-2, -1)$, $S(2, 1)$, and $T(0, -3)$. Find all possible coordinates for the fourth vertex.

NAME ______________________ DATE ____________ PERIOD ____________

6-4 Study Guide and Intervention

Rectangles

Properties of Rectangles A **rectangle** is a quadrilateral with four right angles. Here are the properties of rectangles.

A rectangle has all the properties of a parallelogram.
- Opposite sides are parallel.
- Opposite angles are congruent.
- Opposite sides are congruent.
- Consecutive angles are supplementary.
- The diagonals bisect each other.

Also:
- All four angles are right angles. $\angle UTS$, $\angle TSR$, $\angle SRU$, and $\angle RUT$ are right angles.
- The diagonals are congruent. $\overline{TR} \cong \overline{US}$

Example 1 **Quadrilateral *RUTS* above is a rectangle. If $US = 6x + 3$ and $RT = 7x - 2$, find x.**

The diagonals of a rectangle are congruent, so $US = RT$.

$6x + 3 = 7x - 2$

$3 = x - 2$

$5 = x$

Example 2 **Quadrilateral *RUTS* above is a rectangle. If $m\angle STR = 8x + 3$ and $m\angle UTR = 16x - 9$, find $m\angle STR$.**

$\angle UTS$ is a right angle, so $m\angle STR + m\angle UTR = 90$.

$8x + 3 + 16x - 9 = 90$

$24x - 6 = 90$

$24x = 96$

$x = 4$

$m\angle STR = 8x + 3 = 8(4) + 3$ or 35

Exercises

Quadrilateral *ABCD* is a rectangle.

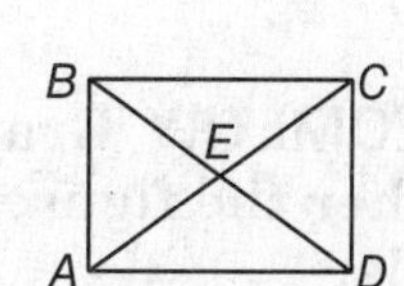

1. If $AE = 36$ and $CE = 2x - 4$, find x.

2. If $BE = 6y + 2$ and $CE = 4y + 6$, find y.

3. If $BC = 24$ and $AD = 5y - 1$, find y.

4. If $m\angle BEA = 62$, find $m\angle BAC$.

5. If $m\angle AED = 12x$ and $m\angle BEC = 10x + 20$, find $m\angle AED$.

6. If $BD = 8y - 4$ and $AC = 7y + 3$, find BD.

7. If $m\angle DBC = 10x$ and $m\angle ACB = 4x^2 - 6$, find $m\angle ACB$.

8. If $AB = 6y$ and $BC = 8y$, find BD in terms of y.

NAME ______________________ DATE __________ PERIOD __________

6-4 Study Guide and Intervention *(continued)*

Rectangles

Prove that Parallelograms Are Rectangles The diagonals of a rectangle are congruent, and the converse is also true.

If the diagonals of a parallelogram are congruent, then the parallelogram is a rectangle.

In the coordinate plane you can use the Distance Formula, the Slope Formula, and properties of diagonals to show that a figure is a rectangle.

Example **Quadrilateral *ABCD* has vertices *A*(−3, 0), *B*(−2, 3), *C*(4, 1), and *D*(3, −2). Determine whether *ABCD* is a rectangle.**

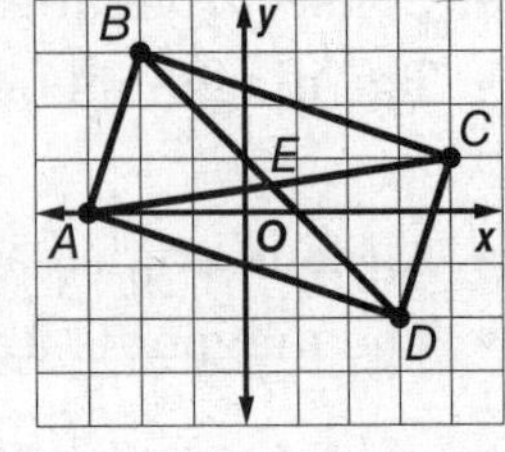

Method 1: Use the Slope Formula.

slope of $\overline{AB} = \dfrac{3-0}{-2-(-3)} = \dfrac{3}{1}$ or 3 slope of $\overline{AD} = \dfrac{-2-0}{3-(-3)} = \dfrac{-2}{6}$ or $-\dfrac{1}{3}$

slope of $\overline{CD} = \dfrac{-2-1}{3-4} = \dfrac{-3}{-1}$ or 3 slope of $\overline{BC} = \dfrac{1-3}{4-(-2)} = \dfrac{-2}{6}$ or $-\dfrac{1}{3}$

Opposite sides are parallel, so the figure is a parallelogram. Consecutive sides are perpendicular, so *ABCD* is a rectangle.

Method 2: Use the Distance Formula.

$AB = \sqrt{(-3-(-2))^2 + (0-3)^2}$ or $\sqrt{10}$ $BC = \sqrt{(-2-4)^2 + (3-1)^2}$ or $\sqrt{40}$

$CD = \sqrt{(4-3)^2 + (1-(-2))^2}$ or $\sqrt{10}$ $AD = \sqrt{(-3-3)^2 + (0-(-2))^2}$ or $\sqrt{40}$

Opposite sides are congruent, thus *ABCD* is a parallelogram.

$AC = \sqrt{(-3-4)^2 + (0-1)^2}$ or $\sqrt{50}$ $BD = \sqrt{(-2-3)^2 + (3-(-2))^2}$ or $\sqrt{50}$

ABCD is a parallelogram with congruent diagonals, so *ABCD* is a rectangle.

Exercises

COORDINATE GEOMETRY Graph each quadrilateral with the given vertices. Determine whether the figure is a rectangle. Justify your answer using the indicated formula.

1. *A*(−3, 1), *B*(−3, 3), *C*(3, 3), *D*(3, 1); Distance Formula

2. *A*(−3, 0), *B*(−2, 3), *C*(4, 5), *D*(3, 2); Slope Formula

3. *A*(−3, 0), *B*(−2, 2), *C*(3, 0), *D*(2 −2); Distance Formula

4. *A*(−1, 0), *B*(0, 2), *C*(4, 0), *D*(3, −2); Distance Formula

6-5 Study Guide and Intervention

Rhombi and Squares

Properties of Rhombi and Squares A **rhombus** is a quadrilateral with four congruent sides. Opposite sides are congruent, so a rhombus is also a parallelogram and has all of the properties of a parallelogram. Rhombi also have the following properties.

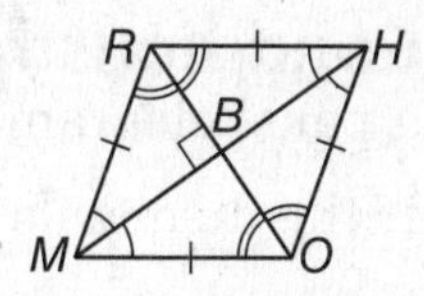

The diagonals are perpendicular.	$\overline{MH} \perp \overline{RO}$
Each diagonal bisects a pair of opposite angles.	$\overline{MH}$ bisects $\angle RMO$ and $\angle RHO$. $\overline{RO}$ bisects $\angle MRH$ and $\angle MOH$.

A **square** is a parallelogram with four congruent sides and four congruent angles. A square is both a rectangle and a rhombus; therefore, all properties of parallelograms, rectangles, and rhombi apply to squares.

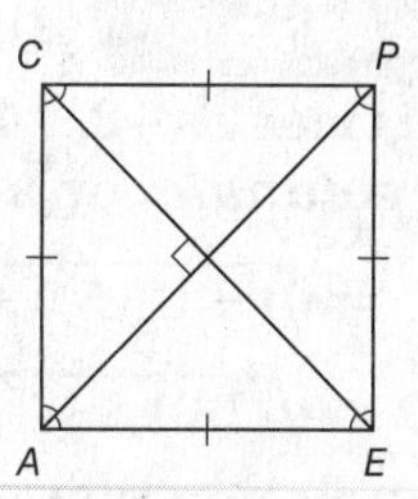

Example **In rhombus *ABCD*, $m\angle BAC = 32$. Find the measure of each numbered angle.**

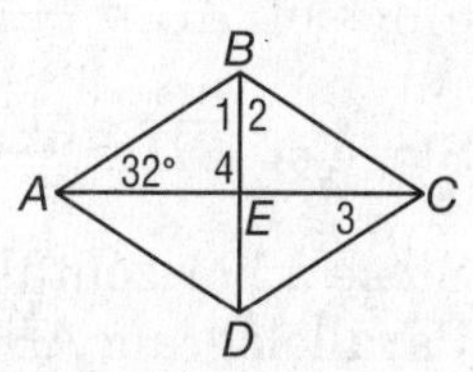

ABCD is a rhombus, so the diagonals are perpendicular and $\triangle ABE$ is a right triangle. Thus $m\angle 4 = 90$ and $m\angle 1 = 90 - 32$ or 58. The diagonals in a rhombus bisect the vertex angles, so $m\angle 1 = m\angle 2$. Thus, $m\angle 2 = 58$.

A rhombus is a parallelogram, so the opposite sides are parallel. $\angle BAC$ and $\angle 3$ are alternate interior angles for parallel lines, so $m\angle 3 = 32$.

Exercises

Quadrilateral *ABCD* is a rhombus. Find each value or measure.

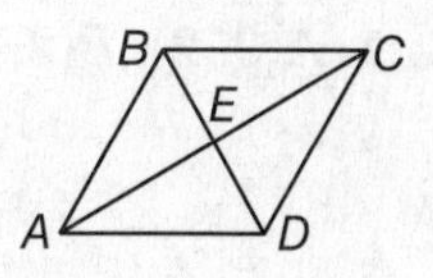

1. If $m\angle ABD = 60$, find $m\angle BDC$.

2. If $AE = 8$, find AC.

3. If $AB = 26$ and $BD = 20$, find AE.

4. Find $m\angle CEB$.

5. If $m\angle CBD = 58$, find $m\angle ACB$.

6. If $AE = 3x - 1$ and $AC = 16$, find x.

7. If $m\angle CDB = 6y$ and $m\angle ACB = 2y + 10$, find y.

8. If $AD = 2x + 4$ and $CD = 4x - 4$, find x.

6-5 Study Guide and Intervention *(continued)*

Rhombi and Squares

Conditions for Rhombi and Squares The theorems below can help you prove that a parallelogram is a rectangle, rhombus, or square.

If the diagonals of a parallelogram are perpendicular, then the parallelogram is a rhombus.

If one diagonal of a parallelogram bisects a pair of opposite angles, then the parallelogram is a rhombus.

If one pair of consecutive sides of a parallelogram are congruent, the parallelogram is a rhombus.

If a quadrilateral is both a rectangle and a rhombus, then it is a square.

Example **Determine whether parallelogram *ABCD* with vertices *A*(–3, –3), *B*(1, 1), *C*(5, –3), *D*(1, –7) is a *rhombus*, a *rectangle*, or a *square*.**

$AC = \sqrt{(-3-5)^2 + ((-3-(-3))^2} = \sqrt{64} = 8$

$BD = \sqrt{(1-1)^2 + (-7-1)^2} = \sqrt{64} = 8$

The diagonals are the same length; the figure is a rectangle.

Slope of $\overline{AC} = \frac{-3-(-3)}{-3-5} = \frac{0}{-8} = 8$ The line is horizontal.

Slope of $\overline{BD} = \frac{1-(-7)}{1-1} = \frac{8}{0} = \textit{undefined}$ The line is vertical.

Since a horizontal and vertical line are perpendicular, the diagonals are perpendicular. Parallelogram *ABCD* is a square which is also a rhombus and a rectangle.

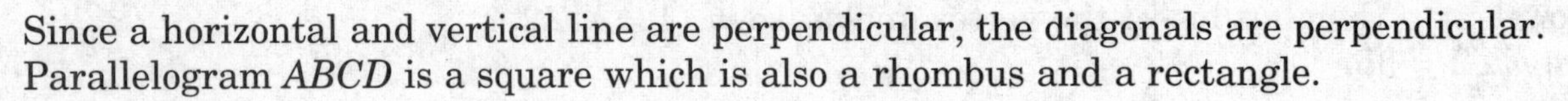

Exercises

Given each set of vertices, determine whether ▱*ABCD* is a *rhombus*, *rectangle*, or *square*. List all that apply. Explain.

1. *A*(0, 2), *B*(2, 4), *C*(4, 2), *D*(2, 0)

2. *A*(–2, 1), *B*(–1, 3), *C*(3, 1), *D*(2, –1)

3. *A*(–2, –1), *B*(0, 2), *C*(2, –1), *D*(0, –4)

4. *A*(–3, 0), *B*(–1, 3), *C*(5, –1), *D*(3, –4)

5. PROOF Write a two-column proof.
Given: Parallelogram *RSTU*. $\overline{RS} \cong \overline{ST}$
Prove: *RSTU* is a rhombus.

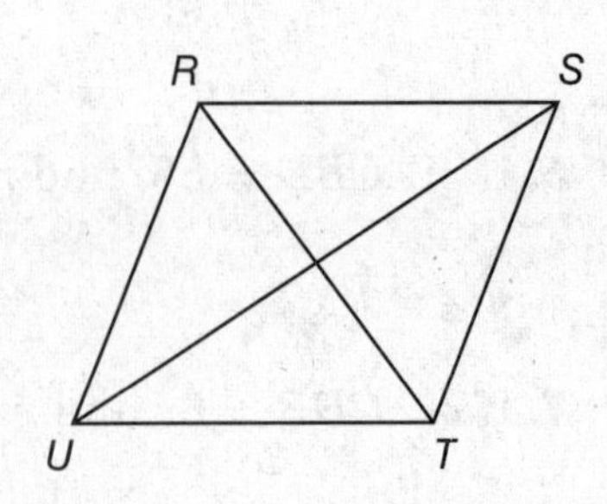

NAME ______________________ DATE ____________ PERIOD ____________

6-6 Study Guide and Intervention

Trapezoids and Kites

Properties of Trapezoids A **trapezoid** is a quadrilateral with exactly one pair of parallel sides. The **midsegment** or **median** of a trapezoid is the segment that connects the midpoints of the legs of the trapezoid. Its measure is equal to one-half the sum of the lengths of the bases. If the legs are congruent, the trapezoid is an **isosceles trapezoid**. In an isosceles trapezoid both pairs of **base angles** are congruent and the diagonals are congruent.

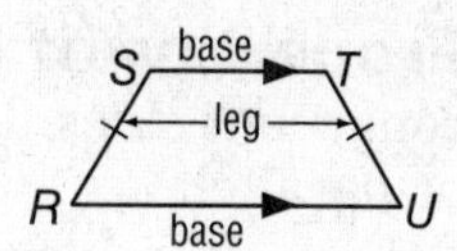

STUR is an isosceles trapezoid.
$\overline{SR} \cong \overline{TU}$; $\angle R \cong \angle U$, $\angle S \cong \angle T$

Example **The vertices of *ABCD* are *A*(−3, −1), *B*(−1, 3), *C*(2, 3), and *D*(4, −1). Show that *ABCD* is a trapezoid and determine whether it is an isosceles trapezoid.**

slope of $\overline{AB} = \dfrac{3-(-1)}{-1-(-3)} = \dfrac{4}{2} = 2$

slope of $\overline{AD} = \dfrac{-1-(-1)}{4-(-3)} = \dfrac{0}{7} = 0$

slope of $\overline{BC} = \dfrac{3-3}{2-(-1)} = \dfrac{0}{3} = 0$

slope of $\overline{CD} = \dfrac{-1-3}{4-2} = \dfrac{-4}{2} = -2$

$AB = \sqrt{(-3-(-1))^2 + (-1-3)^2}$

$= \sqrt{4+16} = \sqrt{20} = 2\sqrt{5}$

$CD = \sqrt{(2-4)^2 + (3-(-1))^2}$

$= \sqrt{4+16} = \sqrt{20} = 2\sqrt{5}$

Exactly two sides are parallel, $\overline{AD}$ and $\overline{BC}$, so *ABCD* is a trapezoid. *AB* = *CD*, so *ABCD* is an isosceles trapezoid.

Exercises

Find each measure.

1. $m\angle D$

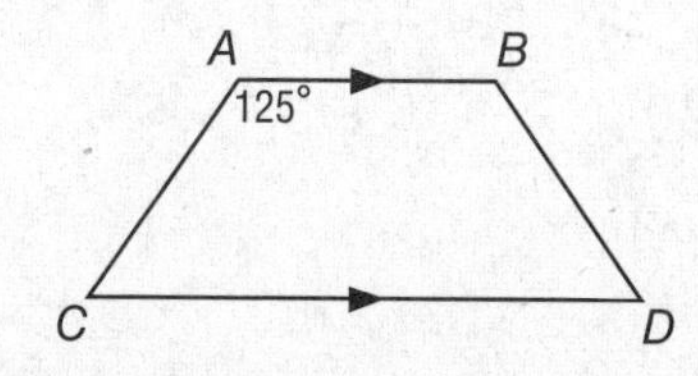

2. $m\angle L$

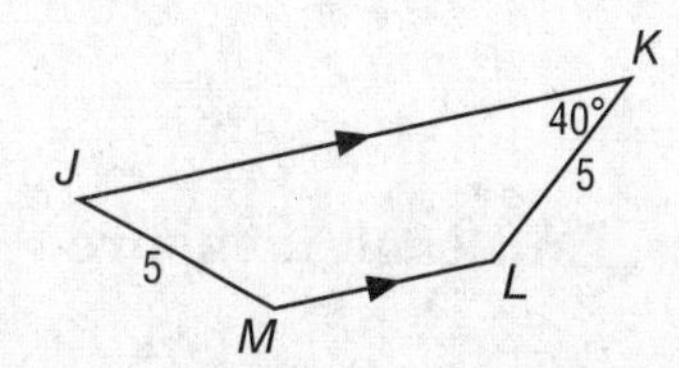

COORDINATE GEOMETRY For each quadrilateral with the given vertices, verify that the quadrilateral is a trapezoid and determine whether the figure is an isosceles trapezoid.

3. *A*(−1, 1), *B*(3, 2), *C*(1,−2), *D*(−2, −1)

4. *J*(1, 3), *K*(3, 1), *L*(3, −2), *M*(−2, 3)

For trapezoid *HJKL*, *M* and *N* are the midpoints of the legs.

5. If *HJ* = 32 and *LK* = 60, find *MN*.

6. If *HJ* = 18 and *MN* = 28, find *LK*.

6-6 Study Guide and Intervention *(continued)*

Trapezoids and Kites

Properties of Kites A **kite** is a quadrilateral with exactly two pairs of consecutive congruent sides. Unlike a parallelogram, the opposite sides of a kite are not congruent or parallel.

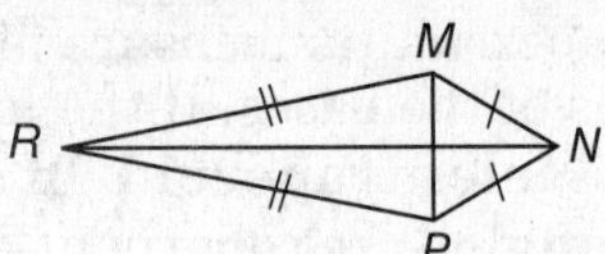

The diagonals of a kite are perpendicular.
For kite *RMNP*, $\overline{MP} \perp \overline{RN}$.

In a kite, exactly one pair of opposite angles is congruent.
For kite *RMNP*, $\angle M \cong \angle P$.

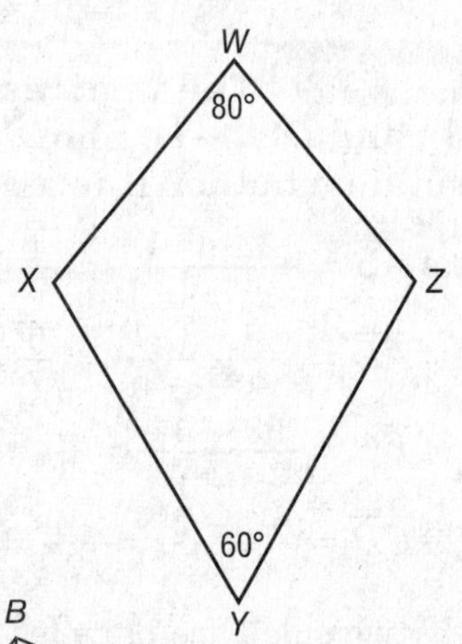

Example 1 **If *WXYZ* is a kite, find $m\angle Z$.**

The measures of $\angle Y$ and $\angle W$ are not congruent, so $\angle X \cong \angle Z$.

$m\angle X + m\angle Y + m\angle Z + m\angle W = 360$

$m\angle X + 60 + m\angle Z + 80 = 360$

$m\angle X + m\angle Z = 220$

$m\angle X = 110, m\angle Z = 110$

Example 2 **If ABCD is a kite, find *BC*.**

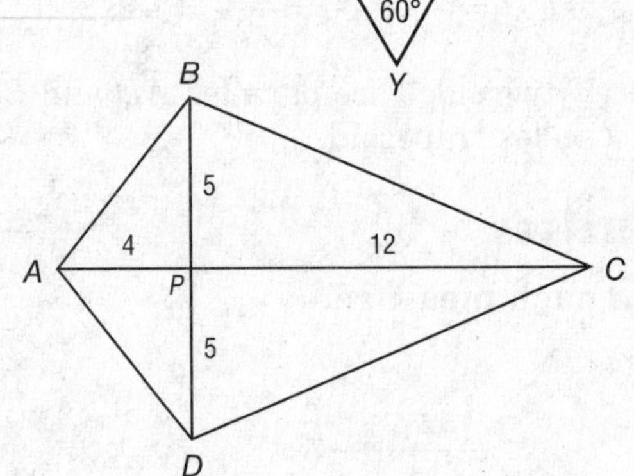

The diagonals of a kite are perpendicular. Use the Pythagorean Theorem to find the missing length.

$BP^2 + PC^2 = BC^2$

$5^2 + 12^2 = BC^2$

$169 = BC^2$

$13 = BC$

Exercises

If *GHJK* is a kite, find each measure.

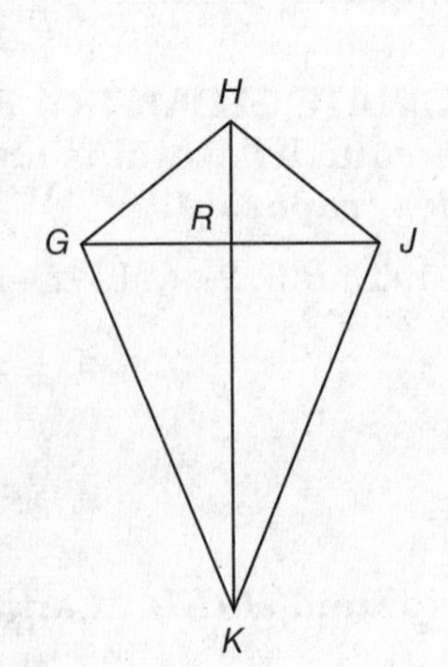

1. Find $m\angle JRK$.

2. If $RJ = 3$ and $RK = 10$, find JK.

3. If $m\angle GHJ = 90$ and $m\angle GKJ = 110$, find $m\angle HGK$.

4. If $HJ = 7$, find HG.

5. If $HG = 7$ and $GR = 5$, find HR.

6. If $m\angle GHJ = 52$ and $m\angle GKJ = 95$, find $m\angle HGK$.

NAME ______________________ DATE __________ PERIOD __________

7-1 Study Guide and Intervention

Ratios and Proportions

Write and Use Ratios A **ratio** is a comparison of two quantities by divisions. The ratio a to b, where b is not zero, can be written as $\frac{a}{b}$ or $a:b$.

Example 1 **In 2007 the Boston RedSox baseball team won 96 games out of 162 games played. Write a ratio for the number of games won to the total number of games played.**

To find the ratio, divide the number of games won by the total number of games played. The result is $\frac{96}{162}$, which is about 0.59. The Boston RedSox won about 59% of their games in 2007.

Example 2 **The ratio of the measures of the angles in $\triangle JHK$ is 2:3:4. Find the measures of the angles.**

The extended ratio 2:3:4 can be rewritten $2x{:}3x{:}4x$.

Sketch and label the angle measures of the triangle.
Then write and solve an equation to find the value of x.

K, 3x°, H, 4x°, J, 2x°

$2x + 3x + 4x = 180$ Triangle Sum Theorem

$9x = 180$ Combine like terms.

$x = 20$ Divide each side by 9.

The measures of the angles are 2(20) or 40°, 3(20) or 60°, and 4(20) or 80°.

Exercises

1. In the 2007 Major League Baseball season, Alex Rodriguez hit 54 home runs and was at bat 583 times. What is the ratio of home runs to the number of times he was at bat?

2. There are 182 girls in the sophomore class of 305 students. What is the ratio of girls to total students?

3. The length of a rectangle is 8 inches and its width is 5 inches. What is the ratio of length to width?

4. The ratio of the sides of a triangle is 8:15:17. Its perimeter is 480 inches. Find the length of each side of the triangle.

5. The ratio of the measures of the three angles of a triangle is 7:9:20. Find the measure of each angle of the triangle.

7-1 Study Guide and Intervention *(continued)*

Ratios and Proportions

Use Properties of Proportions A statement that two ratios are equal is called a **proportion**. In the proportion $\frac{a}{b} = \frac{c}{d}$, where b and d are not zero, the values a and d are the **extremes** and the values b and c are the **means**. In a proportion, the product of the means is equal to the product of the extremes, so $ad = bc$. This is the Cross Product Property.

$$\frac{a}{b} = \frac{c}{d}$$

$$a \cdot d = b \cdot c$$

↑ extremes ↑ means

Example 1 **Solve $\frac{9}{16} = \frac{27}{x}$.**

$\frac{9}{16} = \frac{27}{x}$

$9 \cdot x = 16 \cdot 27$ Cross Products Property

$9x = 432$ Multiply.

$x = 48$ Divide each side by 9.

Example 2 **POLITICS Mayor Hernandez conducted a random survey of 200 voters and found that 135 approve of the job she is doing. If there are 48,000 voters in Mayor Hernandez's town, predict the total number of voters who approve of the job she is doing.**

Write and solve a proportion that compares the number of registered voters and the number of registered voters who approve of the job the mayor is doing.

$\frac{135}{200} = \frac{x}{48,000}$ ← voters who approve / ← all voters

$135 \cdot 48,000 = 200 \cdot x$ Cross Products Property

$6,480,000 = 200x$ Simplify.

$32,400 = x$ Divide each side by 200.

Based on the survey, about 32,400 registered voters approve of the job the mayor is doing.

Exercises

Solve each proportion.

1. $\frac{1}{2} = \frac{28}{x}$

2. $\frac{3}{8} = \frac{y}{24}$

3. $\frac{x + 22}{x + 2} = \frac{30}{10}$

4. $\frac{3}{18.2} = \frac{9}{y}$

5. $\frac{2x + 3}{8} = \frac{5}{4}$

6. $\frac{x + 1}{x - 1} = \frac{3}{4}$

7. If 3 DVDs cost $44.85, find the cost of one DVD.

8. BOTANY Bryon is measuring plants in a field for a science project. Of the first 25 plants he measures, 15 of them are smaller than a foot in height. If there are 4000 plants in the field, predict the total number of plants smaller than a foot in height.

NAME ______________________ DATE ____________ PERIOD ____________

7-2 Study Guide and Intervention

Similar Polygons

Identify Similar Polygons Similar polygons have the same shape but not necessarily the same size.

Example 1 **If $\triangle ABC \sim \triangle XYZ$, list all pairs of congruent angles and write a proportion that relates the corresponding sides.**

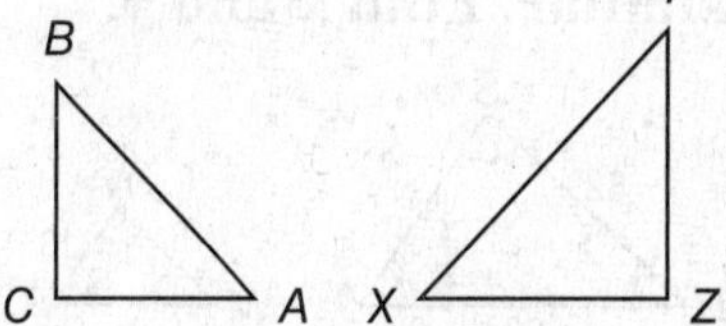

Use the similarity statement.

Congruent angles: $\angle A \cong \angle X$, $\angle B \cong \angle Y$, $\angle C \cong \angle Z$

Proportion: $\frac{AB}{XY} = \frac{BC}{YZ} = \frac{CA}{ZX}$

Example 2 **Determine whether the pair of figures is similar. If so, write the similarity statement and scale factor. Explain your reasoning.**

Step 1 Compare corresponding angles.

$\angle W \cong \angle P$, $\angle X \cong \angle Q$, $\angle Y \cong \angle R$, $\angle Z \cong \angle S$

Corresponding angles are congruent.

Step 2 Compare corresponding sides.

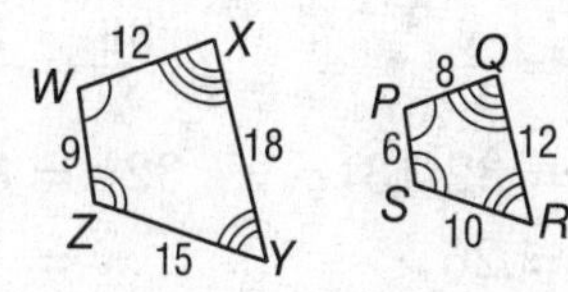

$\frac{WX}{PQ} = \frac{12}{8} = \frac{3}{2}$, $\frac{XY}{QR} = \frac{18}{12} = \frac{3}{2}$, $\frac{YZ}{RS} = \frac{15}{10} = \frac{3}{2}$, and

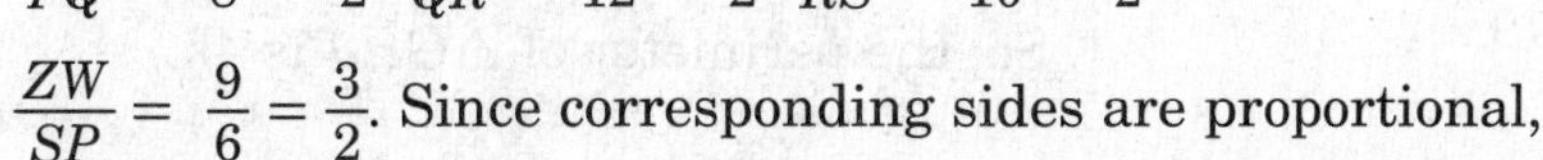

$\frac{ZW}{SP} = \frac{9}{6} = \frac{3}{2}$. Since corresponding sides are proportional,

$WXYZ \sim PQRS$. The polygons are similar with a scale factor of $\frac{3}{2}$.

Exercises

List all pairs of congruent angles, and write a proportion that relates the corresponding sides for each pair of similar polygons.

1. $\triangle DEF \sim \triangle GHJ$

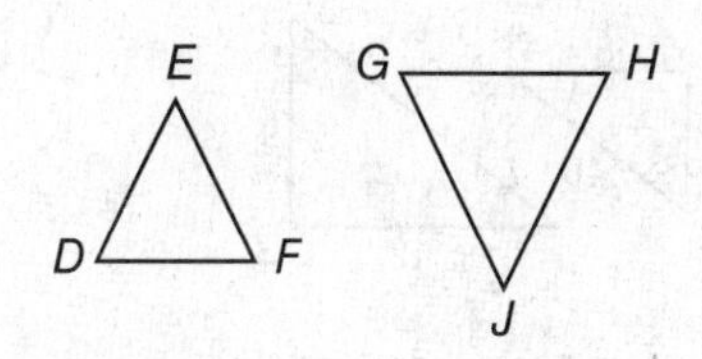

2. $PQRS \sim TUWX$

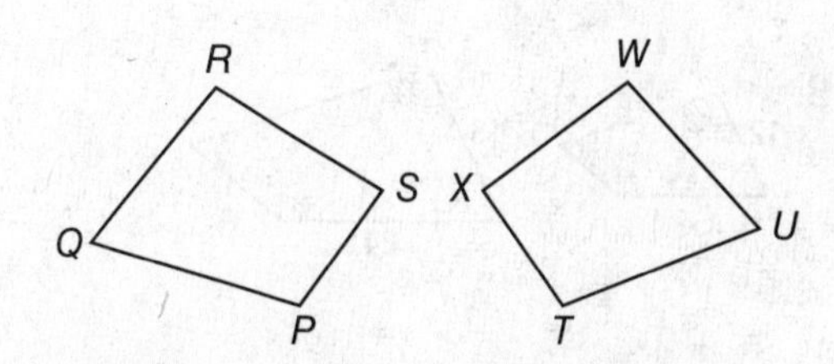

Determine whether each pair of figures is similar. If so, write the similarity statement and scale factor. If not, explain your reasoning.

3.

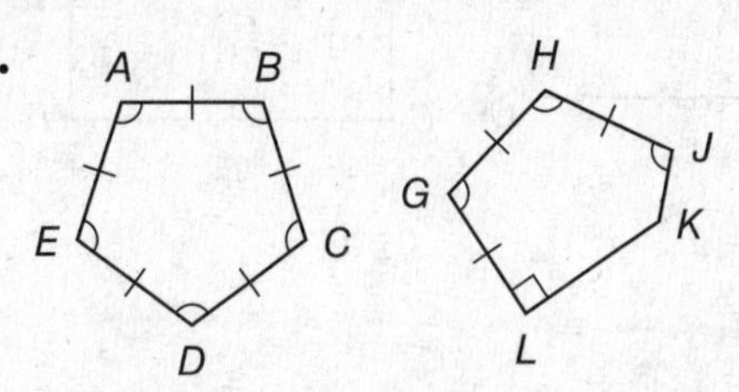

4.

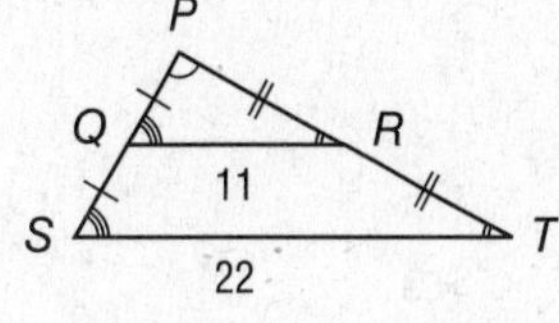

NAME ______________________________ DATE __________ PERIOD __________

7-2 Study Guide and Intervention *(continued)*

Similar Polygons

Use Similar Figures You can use scale factors and proportions to find missing side lengths in similar polygons.

Example 1 **The two polygons are similar. Find x and y.**

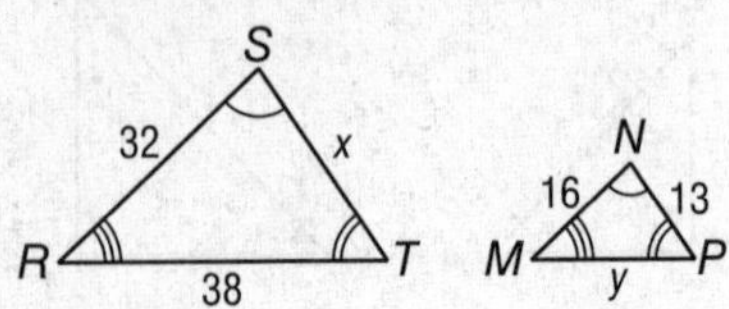

Use the congruent angles to write the corresponding vertices in order.

$\triangle RST \sim \triangle MNP$

Write proportions to find x and y.

$\frac{32}{16} = \frac{x}{13}$ $\qquad$ $\frac{38}{y} = \frac{32}{16}$

$16x = 32(13)$ $\qquad$ $32y = 38(16)$

$x = 26$ $\qquad$ $y = 19$

Example 2 **If $\triangle DEF \sim \triangle GHJ$, find the scale factor of $\triangle DEF$ to $\triangle GHJ$ and the perimeter of each triangle.**

The scale factor is $\frac{EF}{HJ} = \frac{8}{12} = \frac{2}{3}$.

The perimeter of $\triangle DEF$ is 10 + 8 + 12 or 30.

$\frac{2}{3} = \frac{\text{Perimeter of } \triangle DEF}{\text{Perimeter of } \triangle GHJ}$	Theorem 7.1
$\frac{2}{3} = \frac{30}{x}$	Substitution
$(3)(30) = 2x$	Cross Products Property
$45 = x$	Solve.

So, the perimeter of $\triangle GHJ$ is 45.

Exercises

Each pair of polygons is similar. Find the value of x.

1.

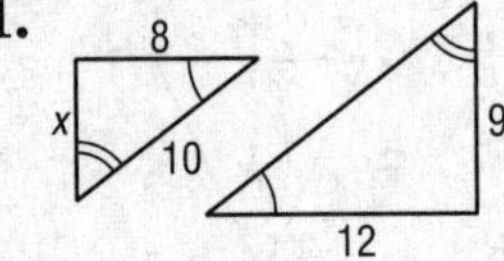

2.

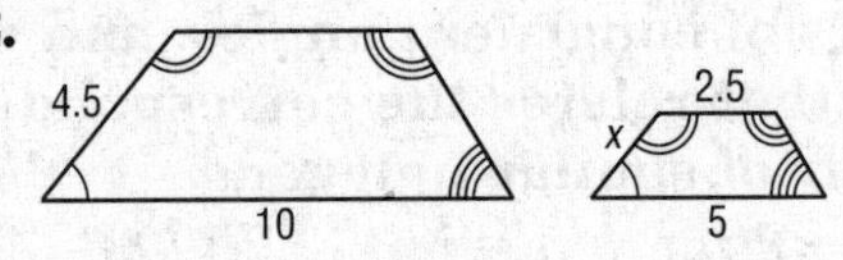

3.

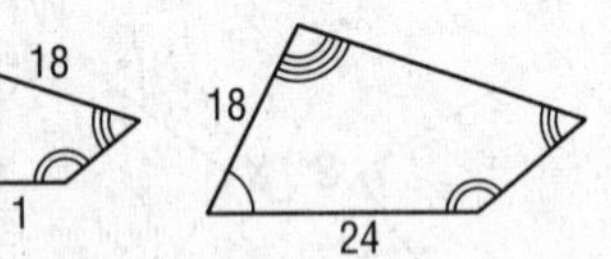

4.

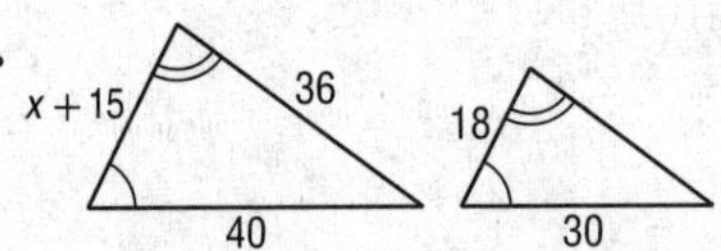

5. If $ABCD \sim PQRS$, find the scale factor of $ABCD$ to $PQRS$ and the perimeter of each polygon.

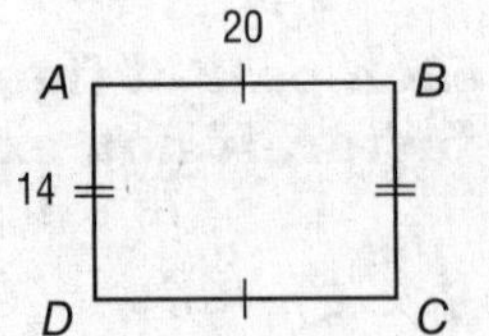

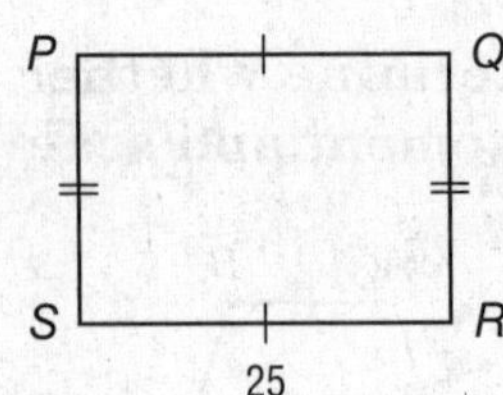

7-3 Study Guide and Intervention

Similar Triangles

Identify Similar Triangles Here are three ways to show that two triangles are similar.

AA Similarity	Two angles of one triangle are congruent to two angles of another triangle.
SSS Similarity	The measures of the corresponding side lengths of two triangles are proportional.
SAS Similarity	The measures of two side lengths of one triangle are proportional to the measures of two corresponding side lengths of another triangle, and the included angles are congruent.

Example 1 **Determine whether the triangles are similar.**

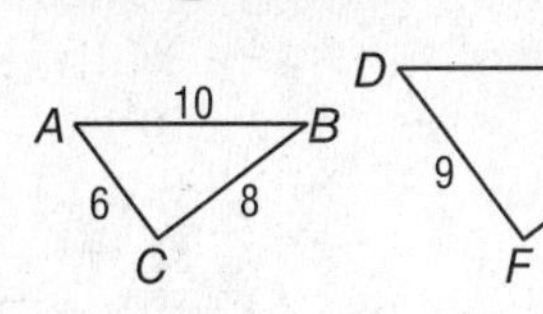

$\frac{AC}{DF} = \frac{6}{9} = \frac{2}{3}$

$\frac{BC}{EF} = \frac{8}{12} = \frac{2}{3}$

$\frac{AB}{DE} = \frac{10}{15} = \frac{2}{3}$

$\triangle ABC \sim \triangle DEF$ by SSS Similarity.

Example 2 **Determine whether the triangles are similar.**

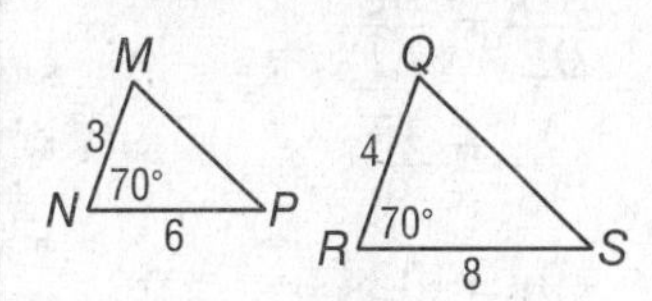

$\frac{3}{4} = \frac{6}{8}$, so $\frac{MN}{QR} = \frac{NP}{RS}$.

$m\angle N = m\angle R$, so $\angle N \cong \angle R$.

$\triangle NMP \sim \triangle RQS$ by SAS Similarity.

Exercises

Determine whether the triangles are similar. If so, write a similarity statement. Explain your reasoning.

1.

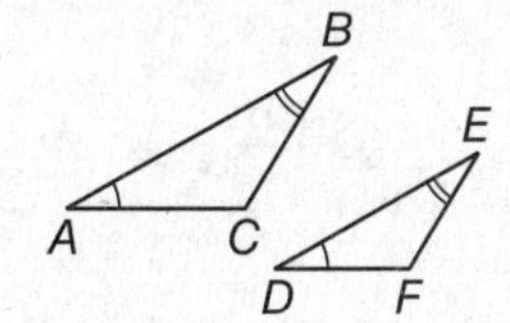

2.

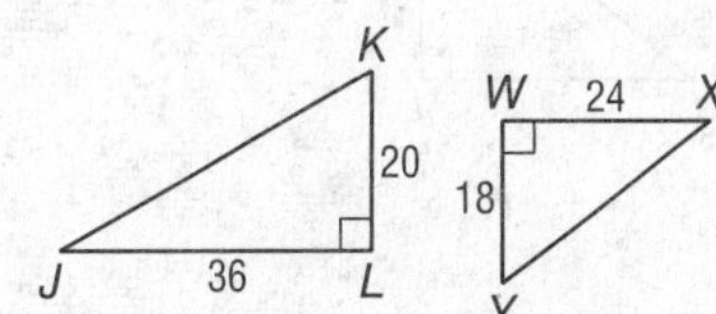

3.

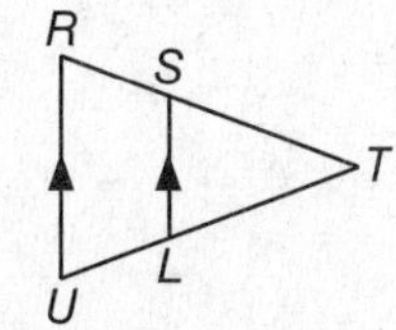

4.

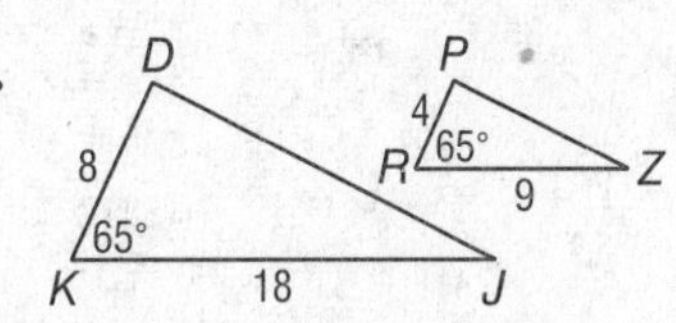

5.

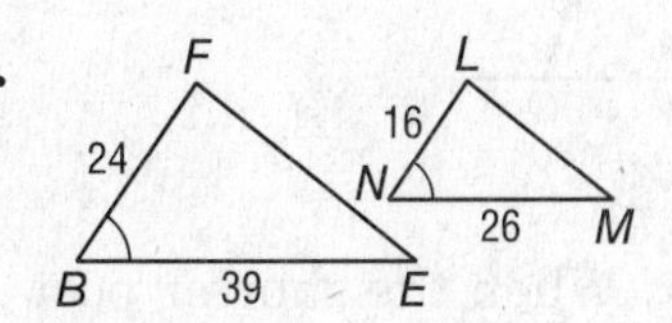

6.

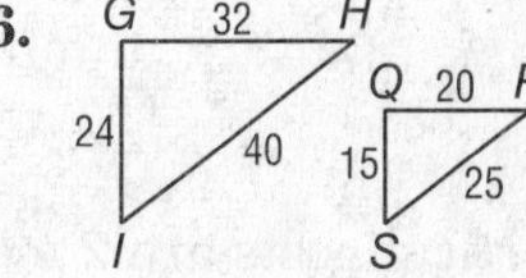

NAME ______________________ DATE ____________ PERIOD ____________

7-3 Study Guide and Intervention *(continued)*

Similar Triangles

Use Similar Triangles Similar triangles can be used to find measurements.

Example 1 **$\triangle ABC \sim \triangle DEF$. Find the values of x and y.**

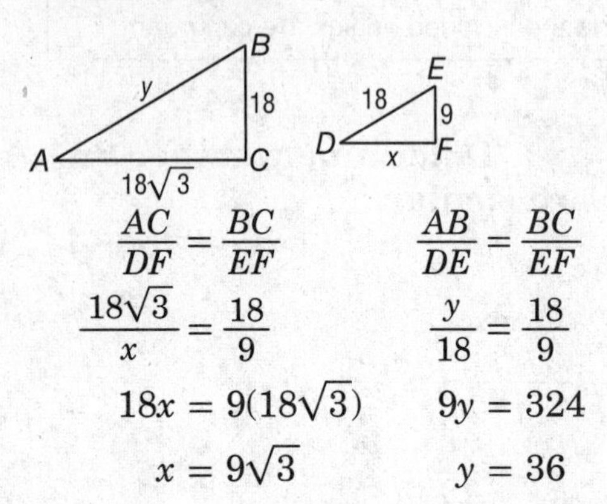

$$\frac{AC}{DF} = \frac{BC}{EF} \qquad \frac{AB}{DE} = \frac{BC}{EF}$$

$$\frac{18\sqrt{3}}{x} = \frac{18}{9} \qquad \frac{y}{18} = \frac{18}{9}$$

$$18x = 9(18\sqrt{3}) \qquad 9y = 324$$

$$x = 9\sqrt{3} \qquad y = 36$$

Example 2 **A person 6 feet tall casts a 1.5-foot-long shadow at the same time that a flagpole casts a 7-foot-long shadow. How tall is the flagpole?**

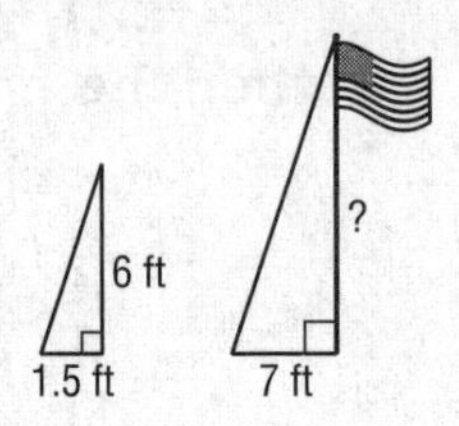

The Sun's rays form similar triangles. Using x for the height of the pole, $\frac{6}{x} = \frac{1.5}{7}$, so $1.5x = 42$ and $x = 28$.
The flagpole is 28 feet tall.

Exercises

ALGEBRA Identify the similar triangles. Then find each measure.

1. JL

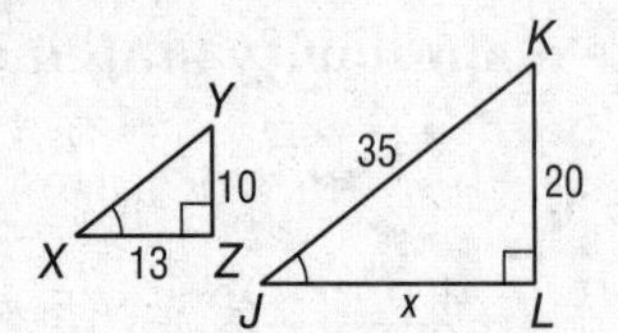

2. IU

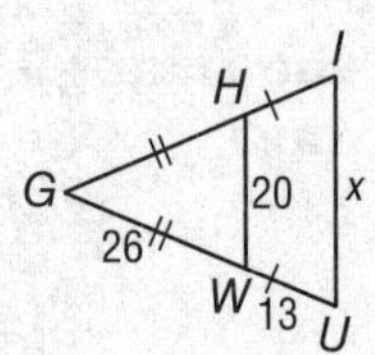

3. QR

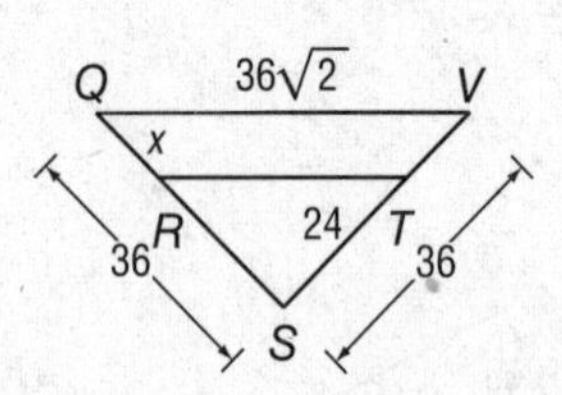

4. BC

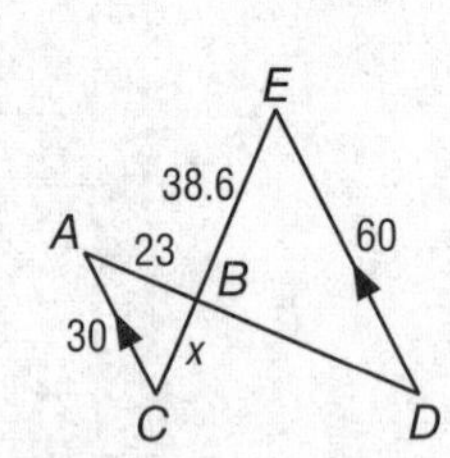

5. LM

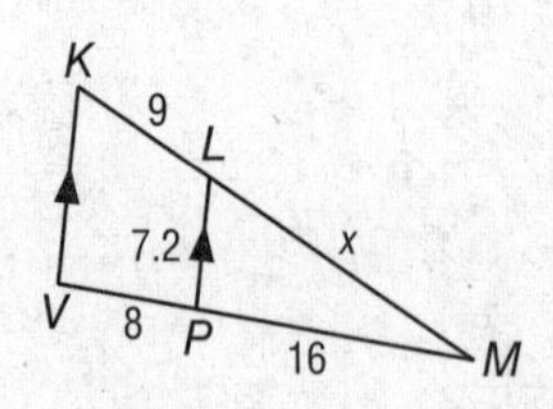

6. QP

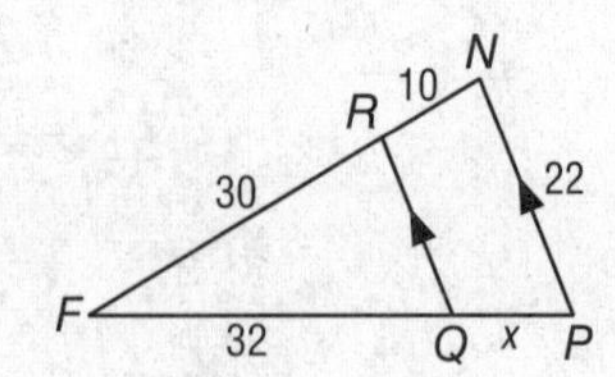

7. The heights of two vertical posts are 2 meters and 0.45 meter. When the shorter post casts a shadow that is 0.85 meter long, what is the length of the longer post's shadow to the nearest hundredth?

NAME ______________________________ DATE ____________ PERIOD ____________

7-4 Study Guide and Intervention

Parallel Lines and Proportional Parts

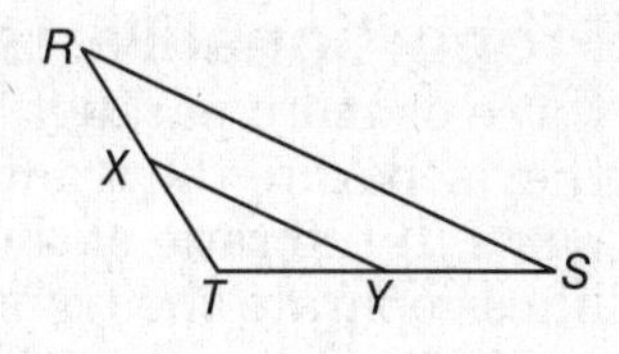

Proportional Parts within Triangles In any triangle, a line parallel to one side of a triangle separates the other two sides proportionally. This is the Triangle Proportionality Theorem. The converse is also true.

If $\overleftrightarrow{XY} \parallel \overleftrightarrow{RS}$, then $\frac{RX}{XT} = \frac{SY}{YT}$. If $\frac{RX}{XT} = \frac{SY}{YT}$, then $\overleftrightarrow{XY} \parallel \overleftrightarrow{RS}$.

If X and Y are the midpoints of $\overline{RT}$ and $\overline{ST}$, then $\overline{XY}$ is a **midsegment** of the triangle. The Triangle Midsegment Theorem states that a midsegment is parallel to the third side and is half its length.

If $\overline{XY}$ is a midsegment, then $\overleftrightarrow{XY} \parallel \overleftrightarrow{RS}$ and $XY = \frac{1}{2}RS$.

Example 1 **In $\triangle ABC$, $\overline{EF} \parallel \overline{CB}$. Find x.**

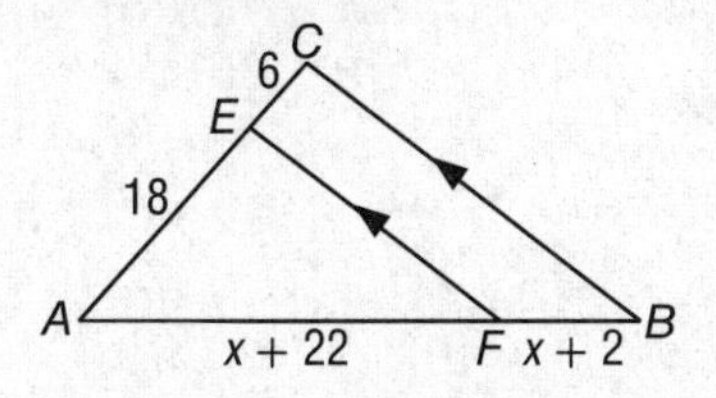

Since $\overline{EF} \parallel \overline{CB}$, $\frac{AF}{FB} = \frac{AE}{EC}$.

$$\frac{x + 22}{x + 2} = \frac{18}{6}$$
$$6x + 132 = 18x + 36$$
$$96 = 12x$$
$$8 = x$$

Example 2 **In $\triangle GHJ$, $HK = 5$, $KG = 10$, and JL is one-half the length of $\overline{LG}$. Is $\overline{HK} \parallel \overline{KL}$?**

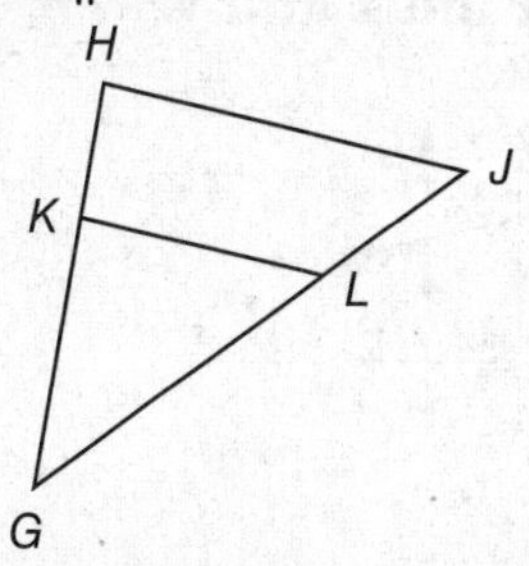

Using the converse of the Triangle Proportionality Theorem, show that $\frac{HK}{KG} = \frac{JL}{LG}$.

Let $JL = x$ and $LG = 2x$.

$\frac{HK}{KG} = \frac{5}{10} = \frac{1}{2}$ $\qquad$ $\frac{JL}{LG} = \frac{x}{2x} = \frac{1}{2}$

Since $\frac{1}{2} = \frac{1}{2}$, the sides are proportional and $\overline{HJ} \parallel \overline{KL}$.

Exercises

ALGEBRA Find the value of x.

1.

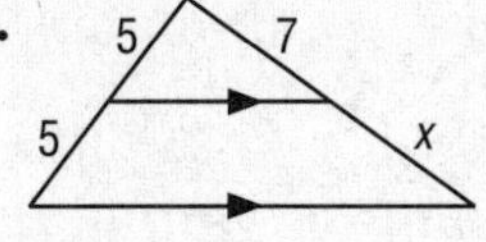

2.

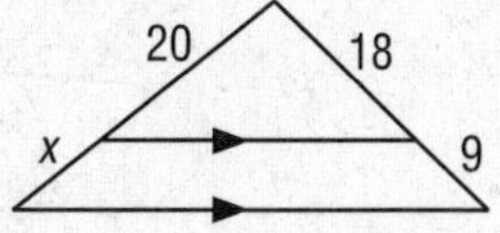

3.

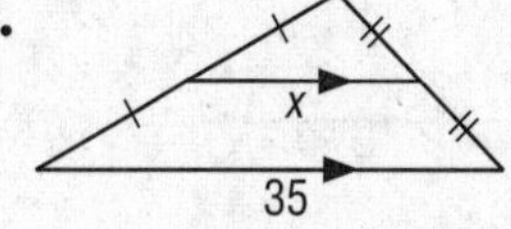

4.

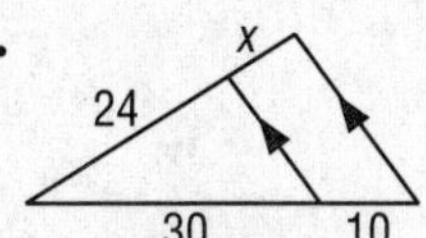

5.

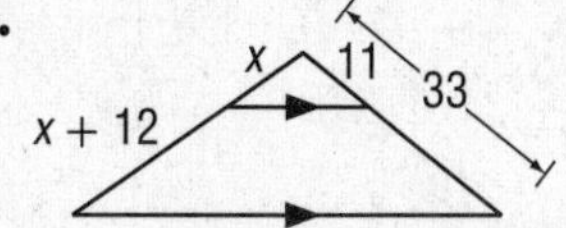

6.

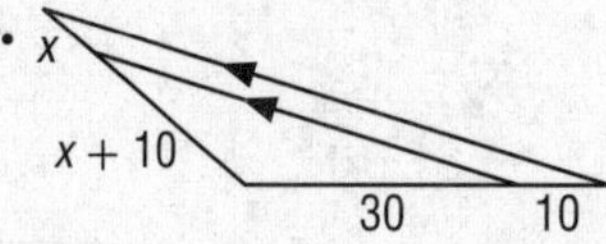

7-4 Study Guide and Intervention *(continued)*

Parallel Lines and Proportional Parts

Proportional Parts with Parallel Lines When three or more parallel lines cut two transversals, they separate the transversals into proportional parts. If the ratio of the parts is 1, then the parallel lines separate the transversals into congruent parts.

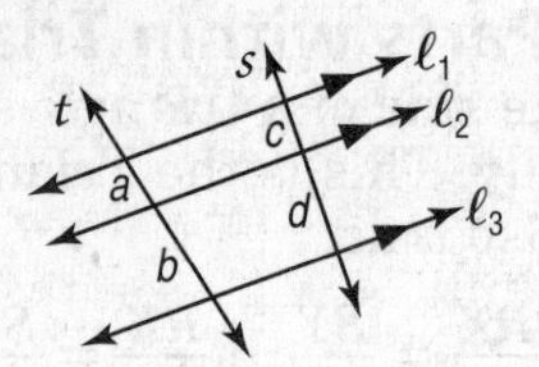

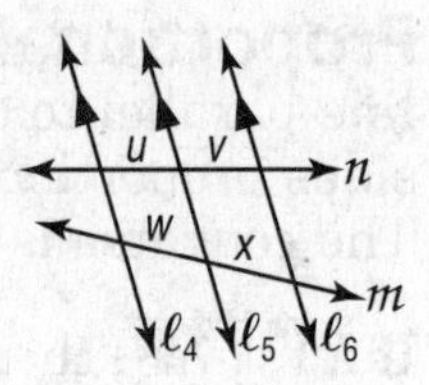

If $\ell_1 \parallel \ell_2 \parallel \ell_3$, then $\frac{a}{b} = \frac{c}{d}$.

If $\ell_4 \parallel \ell_5 \parallel \ell_6$ and $\frac{u}{v} = 1$, then $\frac{w}{x} = 1$.

Example **Refer to lines ℓ_1, ℓ_2, and ℓ_3 above. If $a = 3$, $b = 8$, and $c = 5$, find d.**

$\ell_1 \parallel \ell_2 \parallel \ell_3$ so $\frac{3}{8} = \frac{5}{d}$. Then $3d = 40$ and $d = 13\frac{1}{3}$.

Exercises

ALGEBRA Find x and y.

1.

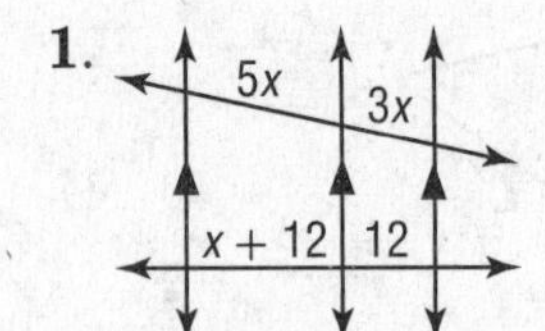

2.

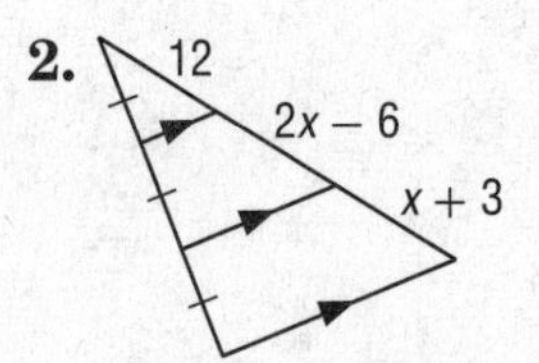

3.

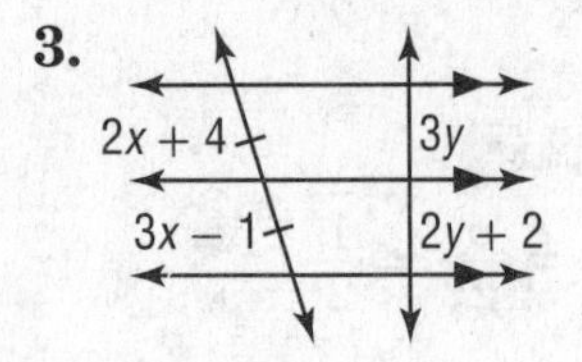

4.

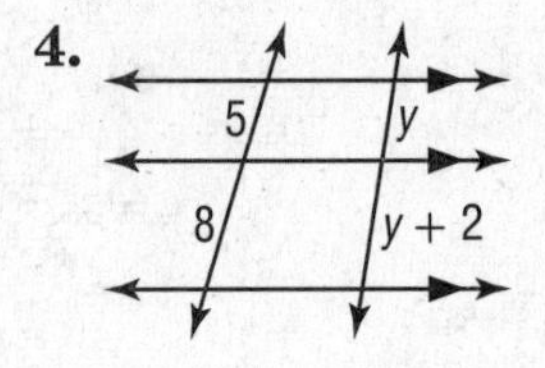

5.

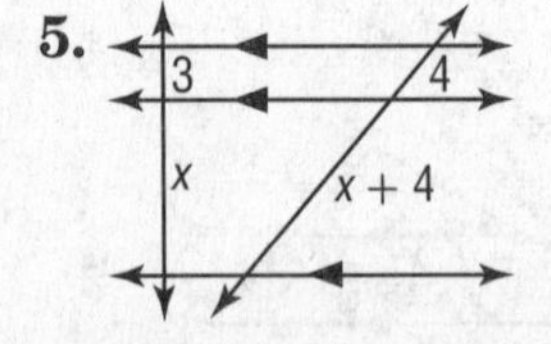

6.

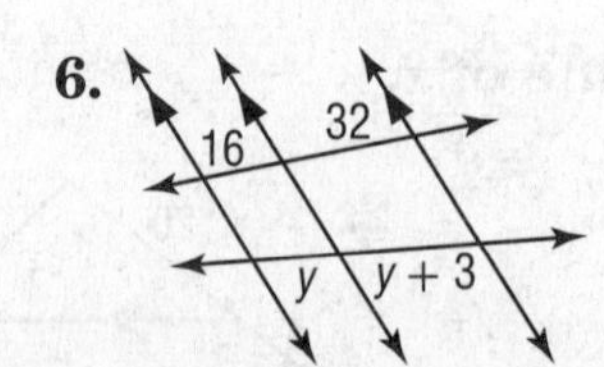

NAME ______________________ DATE ____________ PERIOD ____________

7-5 Study Guide and Intervention

Parts of Similar Triangles

Special Segments of Similar Triangles When two triangles are similar, corresponding altitudes, angle bisectors, and medians are proportional to the corresponding sides.

Example **In the figure, $\triangle ABC \sim \triangle XYZ$, with angle bisectors as shown. Find x.**

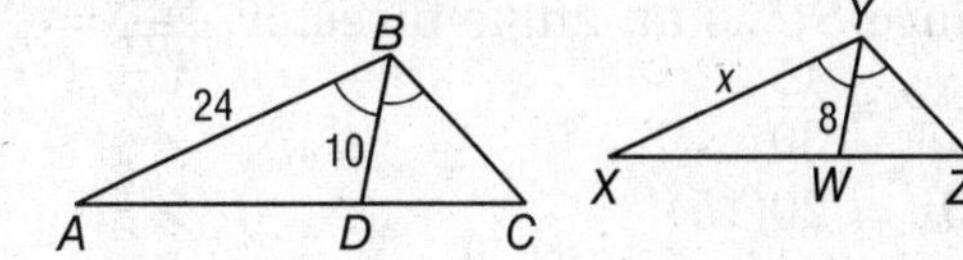

Since $\triangle ABC \sim \triangle XYZ$, the measures of the angle bisectors are proportional to the measures of a pair of corresponding sides.

$$\frac{AB}{XY} = \frac{BD}{YW}$$

$$\frac{24}{x} = \frac{10}{8}$$

$$10x = 24(8)$$

$$10x = 192$$

$$x = 19.2$$

Exercises

Find x.

1.

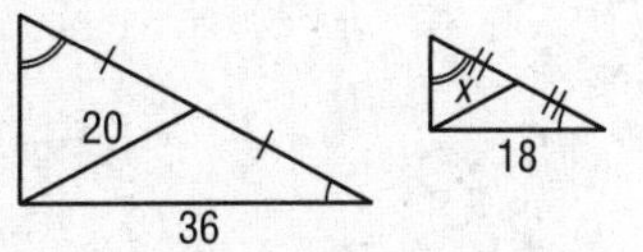

2.

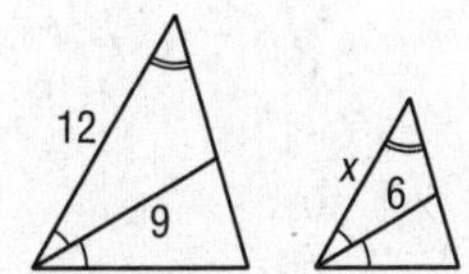

3.

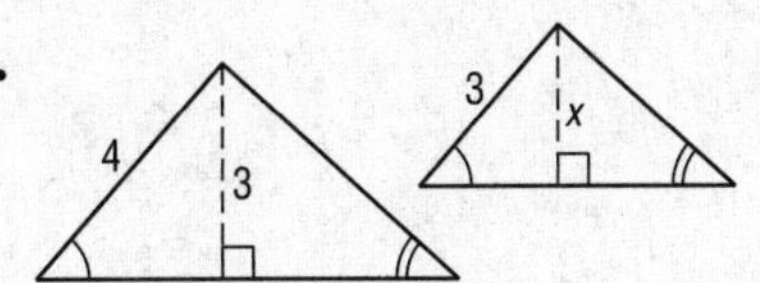

4.

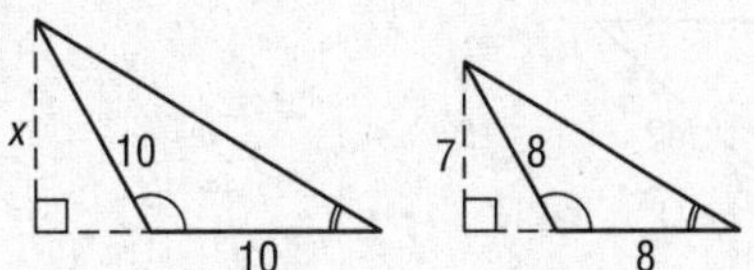

5.

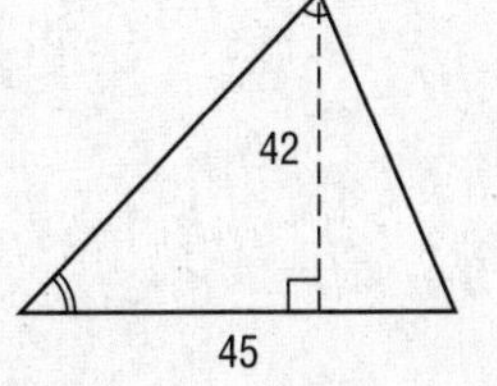

6.

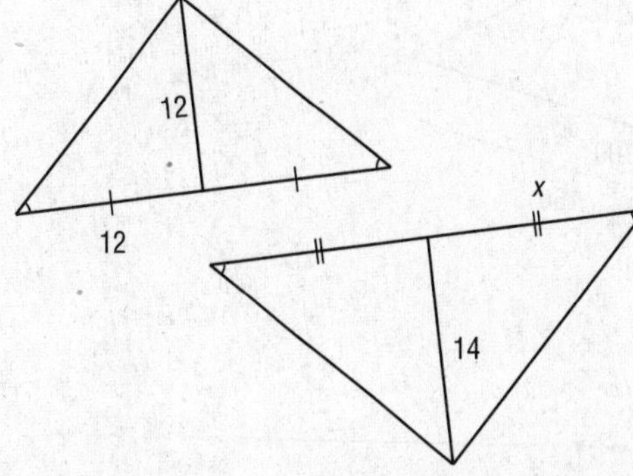

NAME ______________________________ DATE ____________ PERIOD ____________

7-5 Study Guide and Intervention *(continued)*

Parts of Similar Triangles

Triangle Angle Bisector Theorem An angle bisector in a triangle separates the opposite side into two segments that are proportional to the lengths of the other two sides.

Example **Find x.**

Since $\overline{SU}$ is an angle bisector, $\frac{RU}{TU} = \frac{RS}{TS}$.

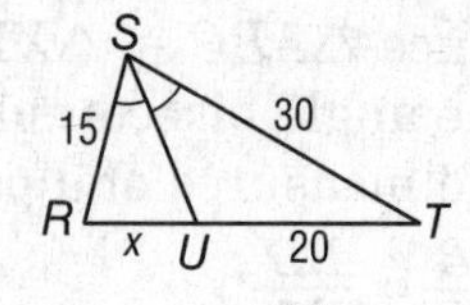

$\frac{x}{20} = \frac{15}{30}$

$30x = 20(15)$

$30x = 300$

$x = 10$

Exercises

Find the value of each variable.

1.

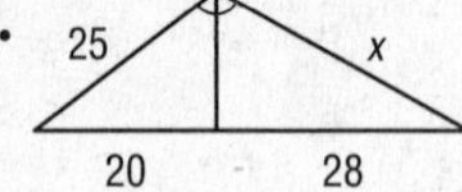

2.

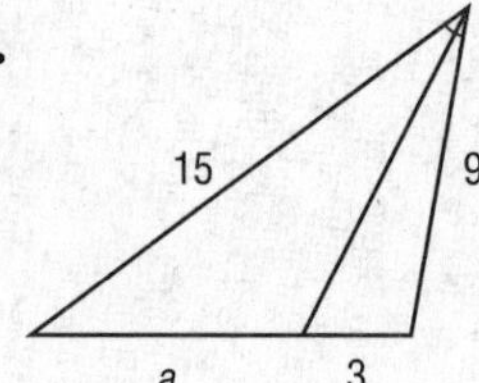

3.

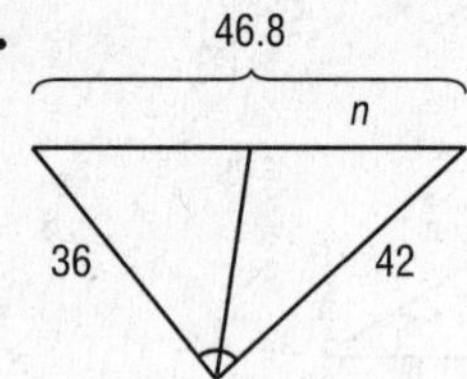

4.

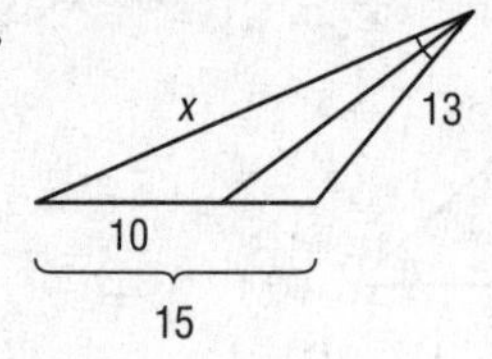

5.

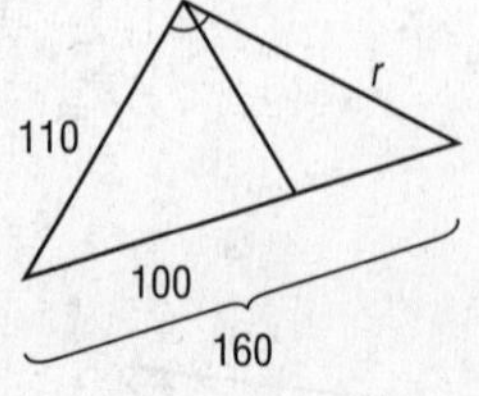

6.

11
17
x
x + 7

7-6 Study Guide and Intervention

Similarity Transformations

Identify Similarity Transformations A **dilation** is a transformation that enlarges or reduces the original figure proportionally. The **scale factor of a dilation,** *k,* is the ratio of a length on the image to a corresponding length on the preimage. A dilation with $k > 1$ is an **enlargement**. A dilation with $0 < k < 1$ is a **reduction**.

Example **Determine whether the dilation from *A* to *B* is an *enlargement* or a *reduction*. Then find the scale factor of the dilation.**

a.

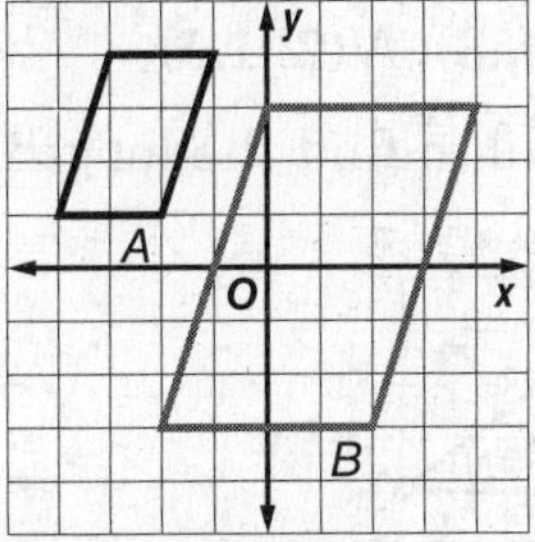

B is larger than *A*, so the dilation is an enlargement.

The distance between the vertices at (−3, 4) and (−1, 4) for *A* is 2. The distance between the vertices at (0, 3) and (4, 3) for *B* is 4.

The scale factor is $\frac{4}{2}$ or 2.

b.

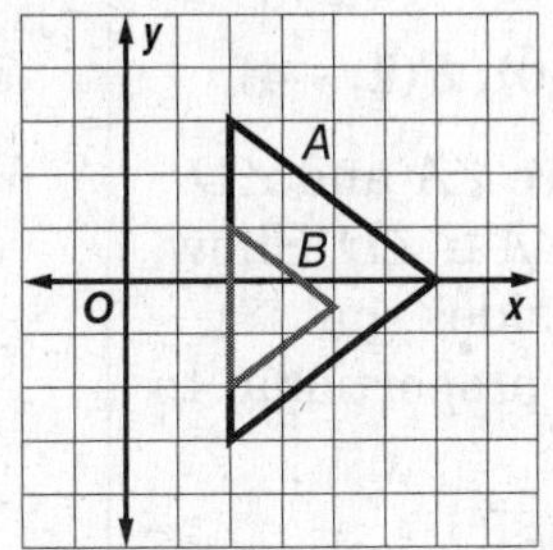

B is smaller than *A*, so the dilation is a reduction.

The distance between the vertices at (2, 3) and (2, −3) for *A* is 6. The distance between the vertices at (2, 1) and (2, −2) for *B* is 3.

The scale factor is $\frac{3}{6}$ or $\frac{1}{2}$.

Exercises

Determine whether the dilation from *A* to *B* is an *enlargement* or a *reduction*. Then find the scale factor of the dilation.

1.

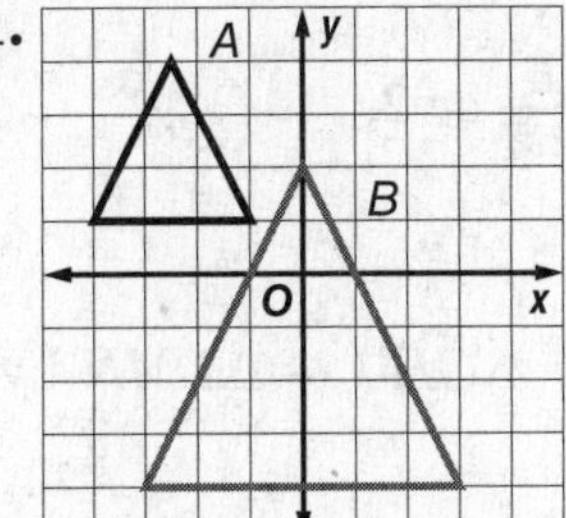

2.

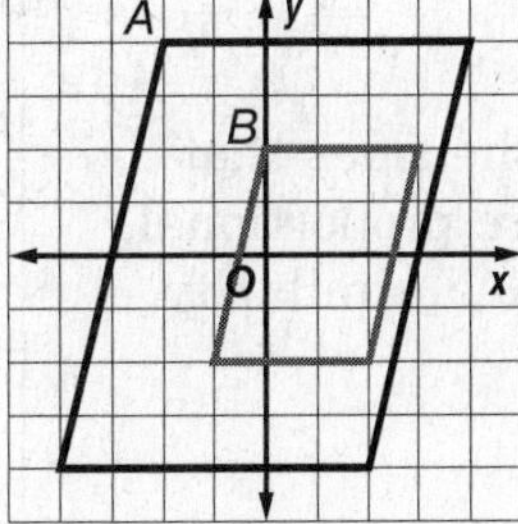

3.

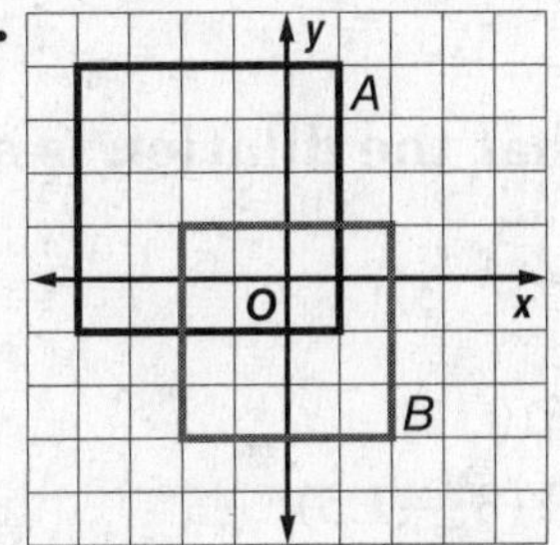

4.

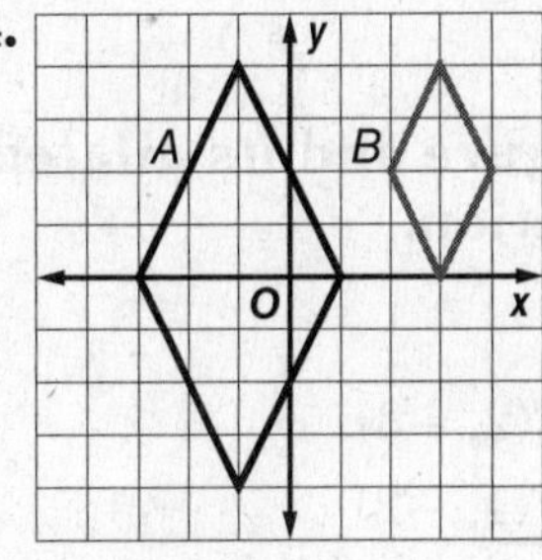

7-6 Study Guide and Intervention *(continued)*

Similarity Transformations

Verify Similarity You can verify that a dilation produces a similar figure by comparing corresponding sides and angles. For triangles, you can also use SAS Similarity.

Example **Graph the original figure and its dilated image. Then verify that the dilation is a similarity transformation.**

a. original: $A(-3, 4)$, $B(2, 4)$, $C(-3, -4)$

image: $D(1, 0)$, $E(3.5, 0)$, $F(1, -4)$

Graph each figure. Since $\angle A$ and $\angle D$ are both right angles, $\angle A \cong \angle D$. Show that the lengths of the sides that include $\angle A$ and $\angle D$ are proportional to prove similarity by SAS.

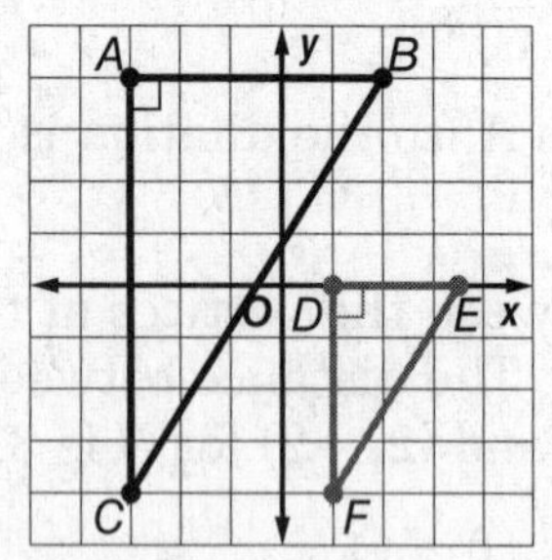

Use the coordinate grid to find the lengths of vertical segments AC and DF and horizontal segments AB and DE.

$\frac{AC}{DF} = \frac{8}{4} = 2$ and $\frac{AB}{DE} = \frac{5}{2.5} = 2$,

so $\frac{AC}{DF} = \frac{AB}{DE}$.

Since the lengths of the sides that include $\angle A$ and $\angle D$ are proportional, $\triangle ABC \sim \triangle DEF$ by SAS similarity.

b. original: $G(-4, 1)$, $H(0, 4)$, $J(4, 1)$

image: $L(-2, 1.5)$, $M(0, 3)$, $N(2, 1.5)$

Use the distance formula to find the length of each side.

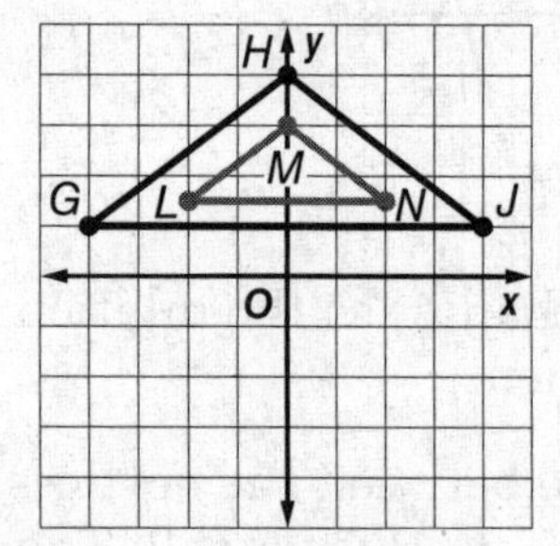

$GH = \sqrt{4^2 + 3^2} = \sqrt{25}$ or 5
$HJ = \sqrt{4^2 + 3^2} = \sqrt{25}$ or 5
$GJ = \sqrt{8^2 + 0^2} = \sqrt{64}$ or 8
$LM = \sqrt{2^2 + 1.5^2} = \sqrt{6.25}$ or 2.5
$MN = \sqrt{2^2 + 1.5^2} = \sqrt{6.25}$ or 2.5
$LN = \sqrt{4^2 + 0^2} = \sqrt{16}$ or 4

Find and compare the ratios of corresponding sides.

$\frac{GH}{LM} = \frac{5}{2.5}$ or 2, $\frac{HJ}{MN} = \frac{5}{2.5}$ or 2,

$\frac{GJ}{LN} = \frac{8}{4}$ or 2,

Since $\frac{GH}{LM} = \frac{HJ}{MN} = \frac{GJ}{LN}$, $\triangle GHJ \sim \triangle LMN$ by SSS similarity.

Exercises

Graph the original figure and its dilated image. Then verify that the dilation is a similarity transformation.

1. $A(-4, -3)$, $B(2, 5)$, $C(2, -3)$;
$D(-2, -2)$, $E(1, 3)$, $F(1, -2)$

2. $P(-4, 1)$, $Q(-2, 4)$, $R(0, 1)$;
$W(1, -1.5)$, $X(2, 0)$, $Y(3, -1.5)$

NAME ______________________ DATE ____________ PERIOD ________

7-7 Study Guide and Intervention

Scale Drawings and Models

Scale Models A **scale model** or a **scale drawing** is an object or drawing with lengths proportional to the object it represents. The **scale** of a model or drawing is the ratio of the length of the model or drawing to the actual length of the object being modeled or drawn.

Example **MAPS The scale on the map shown is 0.75 inches : 6 miles. Find the actual distance from Pineham to Menlo Fields.**

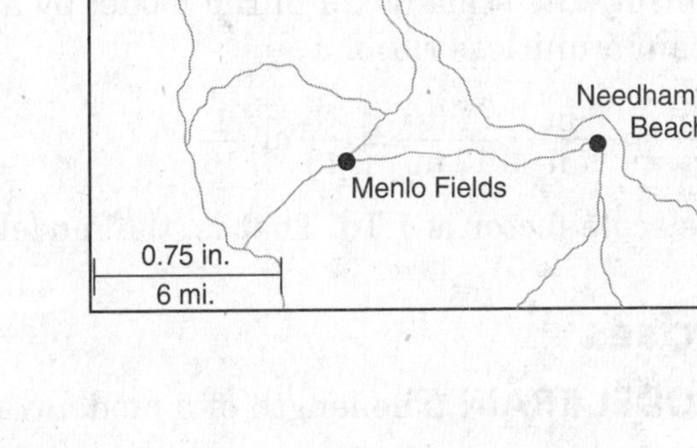

Use a ruler. The distance between Pineham and Menlo Fields is about $1\frac{1}{16}$ or 1.0625 inches.

Method 1: Write and solve a proportion.

Let x represent the distance between cities.

$\frac{0.75 \text{ in.}}{6 \text{ mi}} = \frac{1.0625 \text{ in.}}{x \text{ mi}}$ ← map ← actual

$0.75 \cdot x = 6 \cdot 1.0625$ Cross Products Property

$x = 8.5$ Simplify.

Method 2: Write and solve an equation.

Let a = actual distance and m = map distance in inches. Write the scale as $\frac{6 \text{ mi}}{0.75 \text{ in.}}$, which is $6 \div 0.75$ or 8 miles per inch.

$a = 8 \cdot m$ Write an equation.

$= 8 \cdot 1.0625$ $m = 1.0625$ in.

$= 8.5$ Solve.

The distance between Pineham and Menlo Fields is 8.5 miles.

Exercises

Use the map above and a customary ruler to find the actual distance between each pair of cities. Measure to the nearest sixteenth of an inch.

1. Eastwich and Needham Beach

2. North Park and Menlo Fields

3. North Park and Eastwich

4. Denville and Pineham

5. Pineham and Eastwich

7-7 Study Guide and Intervention *(continued)*

Scale Drawings and Models

Use Scale Factors The **scale factor** of a drawing or scale model is the scale written as a unitless ratio in simplest form. Scale factors are always written so that the model length in the ratio comes first.

Example **SCALE MODEL A doll house that is 15 inches tall is a scale model of a real house with a height of 20 feet.**

a. What is the scale of the model?

To find the scale, write the ratio of a model length to an actual length.

$$\frac{\text{model length}}{\text{actual length}} = \frac{15 \text{ in.}}{20 \text{ ft}} \text{ or } \frac{3 \text{ in.}}{4 \text{ ft}}$$

The scale of the model is 3 in.:4 ft

b. How many times as tall as the actual house is the model?

Multiply the scale factor of the model by a conversion factor that relates inches to feet to obtain a unitless ratio.

$$\frac{3 \text{ in.}}{4 \text{ ft}} = \frac{3 \text{ in.}}{4 \text{ ft}} \cdot \frac{1 \text{ ft}}{12 \text{ in.}} = \frac{3}{48} \text{ or } \frac{1}{16}$$

The scale factor is 1:16. That is, the model is $\frac{1}{16}$ as tall as the actual house.

Exercises

1. **MODEL TRAIN** The length of a model train is 18 inches. It is a scale model of a train that is 48 feet long. Find the scale factor.

2. **ART** An artist in Portland, Oregon, makes bronze sculptures of dogs. The ratio of the height of a sculpture to the actual height of the dog is 2:3. If the height of the sculpture is 14 inches, find the height of the dog.

3. **BRIDGES** The span of the Benjamin Franklin suspension bridge in Philadelphia, Pennsylvania, is 1750 feet. A model of the bridge has a span of 42 inches. What is the scale factor of the model to the span of the actual Benjamin Franklin Bridge?

NAME ______________________ DATE __________ PERIOD __________

8-1 Study Guide and Intervention

Geometric Mean

Geometric Mean The **geometric mean** between two numbers is the positive square root of their product. For two positive numbers a and b, the geometric mean of a and b is the positive number x in the proportion $\frac{a}{x} = \frac{x}{b}$. Cross multiplying gives $x^2 = ab$, so $x = \sqrt{ab}$.

Example **Find the geometric mean between each pair of numbers.**

a. 12 and 3

$x = \sqrt{ab}$ Definition of geometric mean

$= \sqrt{12 \cdot 3}$ $a = 12$ and $b = 3$

$= \sqrt{(2 \cdot 2 \cdot 3) \cdot 3}$ Factor.

$= 6$ Simplify.

The geometric mean between 12 and 3 is 6.

b. 8 and 4

$x = \sqrt{ab}$ Definition of geometric mean

$= \sqrt{8 \cdot 4}$ $a = 8$ and $b = 4$

$= \sqrt{(2 \cdot 4) \cdot 4}$ Factor.

$= \sqrt{16 \cdot 2}$ Associative Property

$= 4\sqrt{2}$ Simplify.

The geometric mean between 8 and 4 is $4\sqrt{2}$ or about 5.7.

Exercises

Find the geometric mean between each pair of numbers.

1. 4 and 4

2. 4 and 6

3. 6 and 9

4. $\frac{1}{2}$ and 2

5. 12 and 20

6. 4 and 25

7. 16 and 30

8. 10 and 100

9. $\frac{1}{2}$ and $\frac{1}{4}$

10. 17 and 3

11. 4 and 16

12. 3 and 24

8-1 Study Guide and Intervention *(continued)*

Geometric Mean

Geometric Means in Right Triangles

In the diagram, $\triangle ABC \sim \triangle ADB \sim \triangle BDC$. An altitude to the hypotenuse of a right triangle forms two right triangles. The two triangles are similar and each is similar to the original triangle.

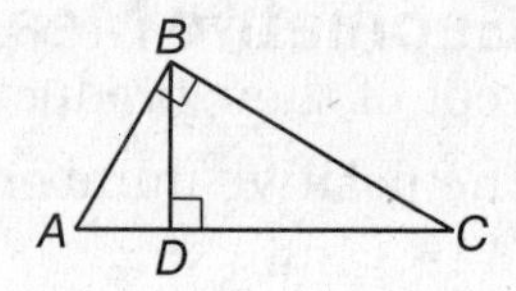

Example 1 **Use right $\triangle ABC$ with $\overline{BD} \perp \overline{AC}$. Describe two geometric means.**

a. $\triangle ADB \sim \triangle BDC$ so $\frac{AD}{BD} = \frac{BD}{CD}$.

In $\triangle ABC$, the altitude is the geometric mean between the two segments of the hypotenuse.

b. $\triangle ABC \sim \triangle ADB$ and $\triangle ABC \sim \triangle BDC$, so $\frac{AC}{AB} = \frac{AB}{AD}$ and $\frac{AC}{BC} = \frac{BC}{DC}$.

In $\triangle ABC$, each leg is the geometric mean between the hypotenuse and the segment of the hypotenuse adjacent to that leg.

Example 2 **Find x, y, and z.**

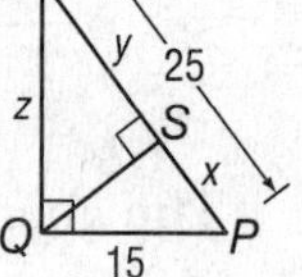

$15 = \sqrt{RP \cdot SP}$ Geometric Mean (Leg) Theorem

$15 = \sqrt{25x}$ $RP = 25$ and $SP = x$

$225 = 25x$ Square each side.

$9 = x$ Divide each side by 25.

Then

$y = RP - SP$

$= 25 - 9$

$= 16$

$z = \sqrt{RS \cdot RP}$ Geometric Mean (Leg) Theorem

$= \sqrt{16 \cdot 25}$ $RS = 16$ and $RP = 25$

$= \sqrt{400}$ Multiply.

$= 20$ Simplify.

Exercises

Find x, y, and z to the nearest tenth.

1.

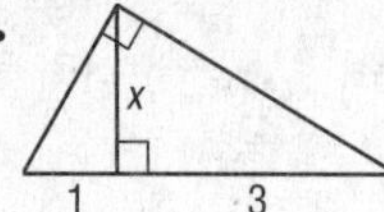

2.

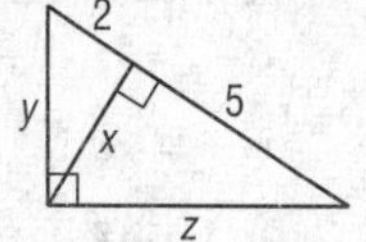

3.

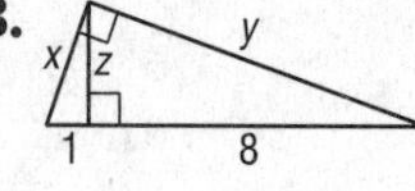

4.

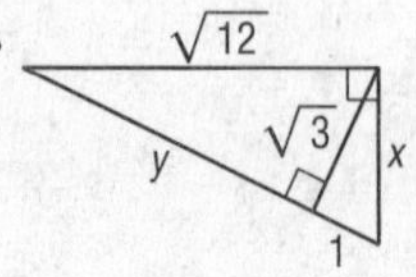

5.

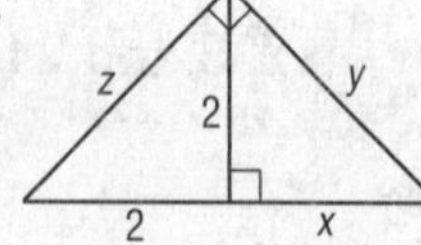

6.

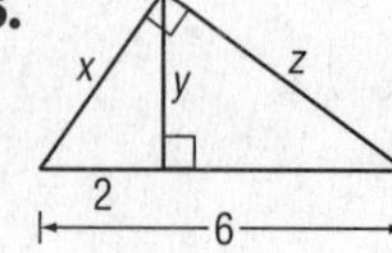

NAME _______________________________ DATE ____________ PERIOD ____________

8-2 Study Guide and Intervention

The Pythagorean Theorem and Its Converse

The Pythagorean Theorem In a right triangle, the sum of the squares of the lengths of the legs equals the square of the length of the hypotenuse. If the three whole numbers a, b, and c satisfy the equation $a^2 + b^2 = c^2$, then the numbers a, b, and c form a **Pythagorean triple.**

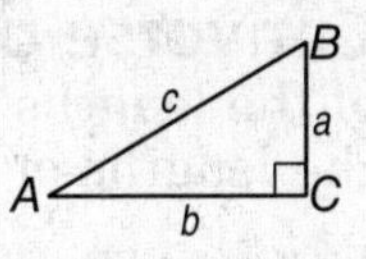

$\triangle ABC$ is a right triangle.

so $a^2 + b^2 = c^2$.

Example

a. Find a.

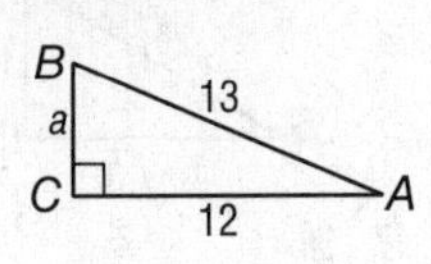

$a^2 + b^2 = c^2$	Pythagorean Theorem
$a^2 + 12^2 = 13^2$	$b = 12$, $c = 13$
$a^2 + 144 = 169$	Simplify.
$a^2 = 25$	Subtract.
$a = 5$	Take the positive square root of each side.

b. Find c.

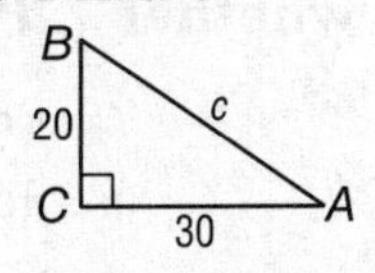

$a^2 + b^2 = c^2$	Pythagorean Theorem
$20^2 + 30^2 = c^2$	$a = 20$, $b = 30$
$400 + 900 = c^2$	Simplify.
$1300 = c^2$	Add.
$\sqrt{1300} = c$	Take the positive square root of each side.
$36.1 \approx c$	Use a calculator.

Exercises

Find x.

1.

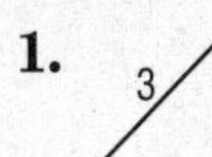

2.

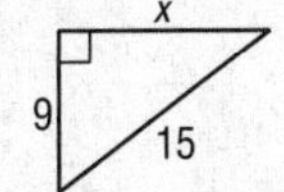

3.

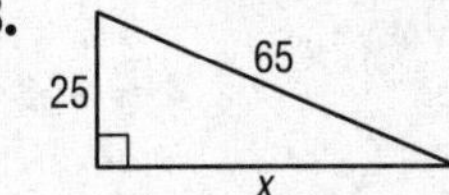

4.

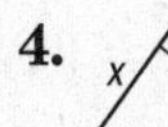

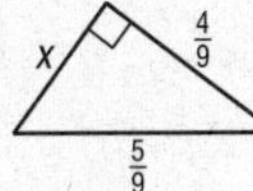

5.

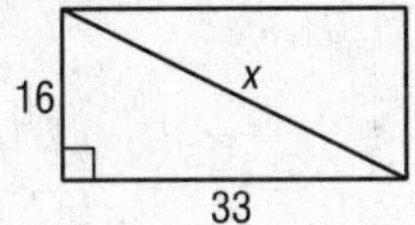

6.

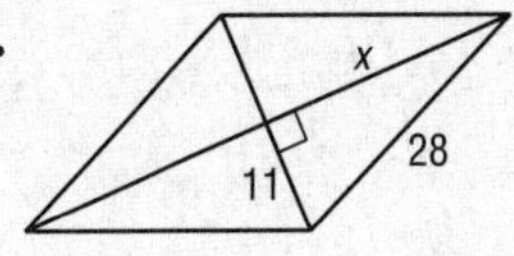

Use a Pythagorean Triple to find x.

7.

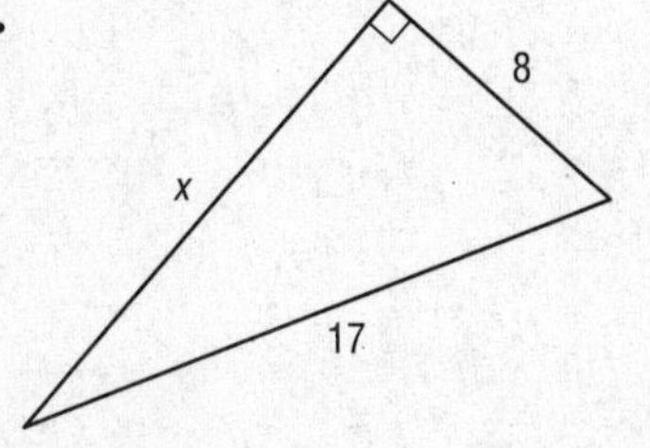

8.

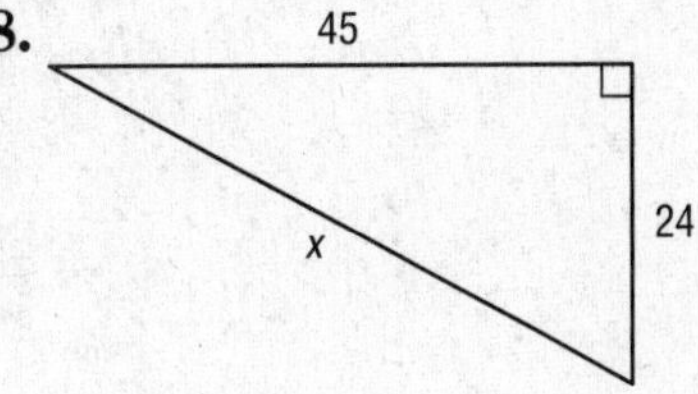

9.

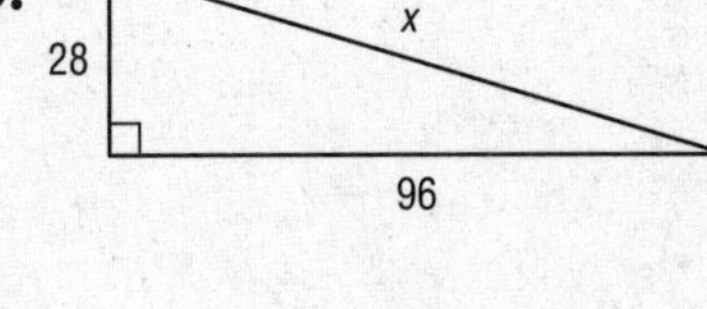

8-2 Study Guide and Intervention *(continued)*

The Pythagorean Theorem and Its Converse

Converse of the Pythagorean Theorem If the sum of the squares of the lengths of the two shorter sides of a triangle equals the square of the lengths of the longest side, then the triangle is a right triangle.

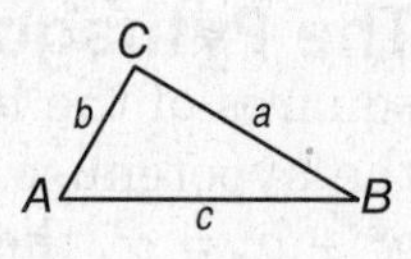

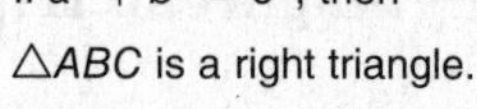
If $a^2 + b^2 = c^2$, then $\triangle ABC$ is a right triangle.

You can also use the lengths of sides to classify a triangle.

if $a^2 + b^2 = c^2$ then $\triangle ABC$ is a right triangle.
if $a^2 + b^2 > c^2$ then $\triangle ABC$ is acute.
if $a^2 + b^2 < c^2$ then $\triangle ABC$ is obtuse.

Example **Determine whether $\triangle PQR$ is a right triangle.**

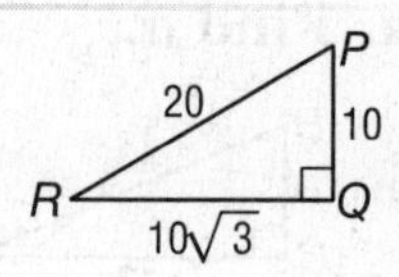

$a^2 + b^2 \stackrel{?}{=} c^2$	Compare c^2 and $a^2 + b^2$
$10^2 + (10\sqrt{3})^2 \stackrel{?}{=} 20^2$	$a = 10$, $b = 10\sqrt{3}$, $c = 20$
$100 + 300 \stackrel{?}{=} 400$	Simplify.
$400 = 400$✓	Add.

Since $c^2 =$ and $a^2 + b^2$, the triangle is a right triangle.

Exercises

Determine whether each set of measures can be the measures of the sides of a triangle. If so, classify the triangle as *acute*, *obtuse*, or *right*. Justify your answer.

1. 30, 40, 50

2. 20, 30, 40

3. 18, 24, 30

4. 6, 8, 9

5. 6, 12, 18

6. 10, 15, 20

7. $\sqrt{5}, \sqrt{12}, \sqrt{13}$

8. $2, \sqrt{8}, \sqrt{12}$

9. 9, 40, 41

8-3 Study Guide and Intervention

Special Right Triangles

Properties of 45°-45°-90° Triangles The sides of a 45°-45°-90° right triangle have a special relationship.

Example 1 **If the leg of a 45°-45°-90° right triangle is x units, show that the hypotenuse is $x\sqrt{2}$ units.**

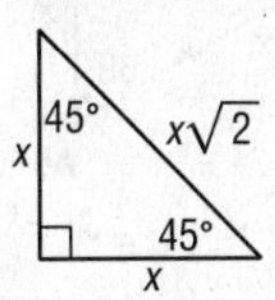

Using the Pythagorean Theorem with $a = b = x$, then

$$c^2 = a^2 + b^2$$
$$c^2 = x^2 + x^2$$
$$c^2 = 2x^2$$
$$c = \sqrt{2x^2}$$
$$c = x\sqrt{2}$$

Example 2 **In a 45°-45°-90° right triangle the hypotenuse is $\sqrt{2}$ times the leg. If the hypotenuse is 6 units, find the length of each leg.**

The hypotenuse is $\sqrt{2}$ times the leg, so divide the length of the hypotenuse by $\sqrt{2}$.

$$a = \frac{6}{\sqrt{2}}$$
$$= \frac{6}{\sqrt{2}} \cdot \frac{\sqrt{2}}{\sqrt{2}}$$
$$= \frac{6\sqrt{2}}{2}$$
$$= 3\sqrt{2} \text{ units}$$

Exercises

Find x.

1.

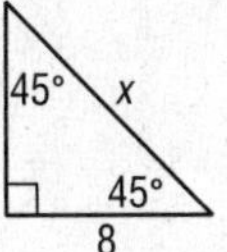

2.

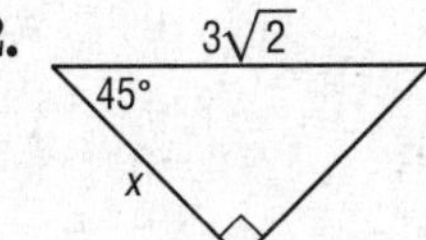

3.

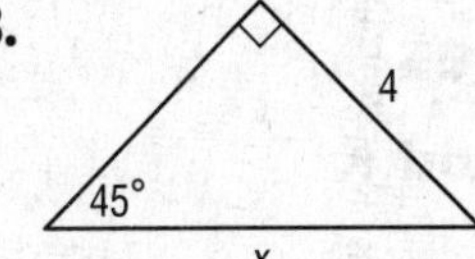

4.

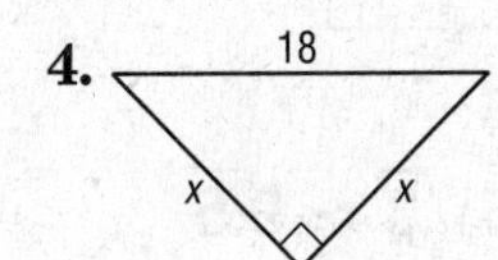

5.

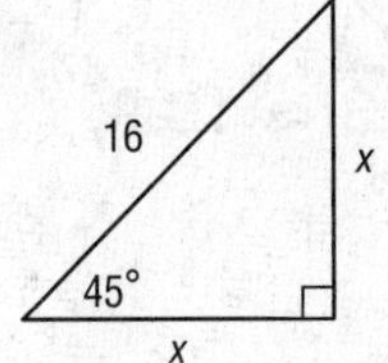

6.

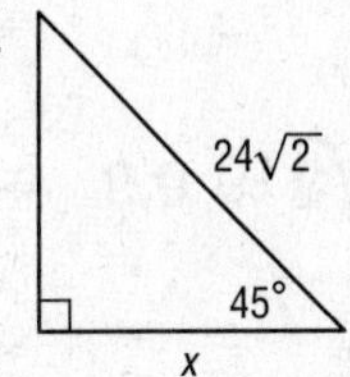

7. If a 45°-45°-90° triangle has a hypotenuse length of 12, find the leg length.

8. Determine the length of the leg of 45°-45°-90° triangle with a hypotenuse length of 25 inches.

9. Find the length of the hypotenuse of a 45°-45°-90° triangle with a leg length of 14 centimeters.

NAME ______________________ DATE __________ PERIOD __________

8-3 Study Guide and Intervention *(continued)*

Special Right Triangles

Properties of 30°-60°-90° Triangles The sides of a 30°-60°-90° right triangle also have a special relationship.

Example 1 **In a 30°-60°-90° right triangle, show that the hypotenuse is twice the shorter leg and the longer leg is $\sqrt{3}$ times the shorter leg.**

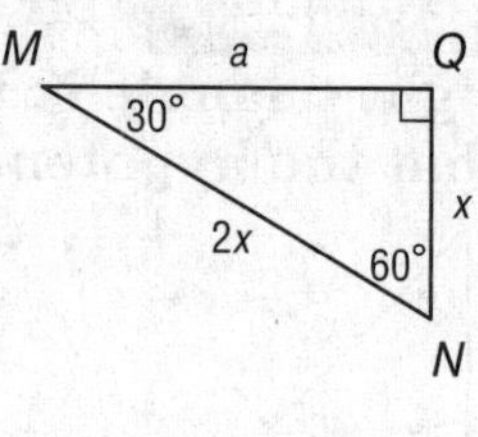

$\triangle MNQ$ is a 30°-60°-90° right triangle, and the length of the hypotenuse $\overline{MN}$ is two times the length of the shorter side $\overline{NQ}$.
Use the Pythagorean Theorem.

$a^2 = (2x)^2 - x^2$

$a^2 = 4x^2 - x^2$

$a^2 = 3x^2$

$a = \sqrt{3x^2}$

$a = x\sqrt{3}$

Example 2 **In a 30°-60°-90° right triangle, the hypotenuse is 5 centimeters. Find the lengths of the other two sides of the triangle.**

If the hypotenuse of a 30°-60°-90° right triangle is 5 centimeters, then the length of the shorter leg is one-half of 5, or 2.5 centimeters. The length of the longer leg is $\sqrt{3}$ times the length of the shorter leg, or $(2.5)(\sqrt{3})$ centimeters.

Exercises

Find *x* and *y*.

1.

$x = 1$;
$y = 0.5\sqrt{3} \approx 0.9$

2. 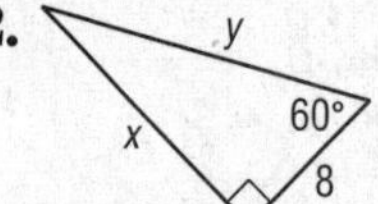

$x = 8\sqrt{3} \approx 13.9$;
$y = 16$

3. 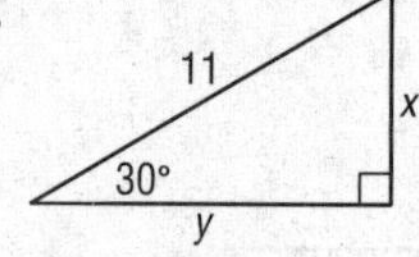

$x = 5.5$;
$y = 5.5\sqrt{3} \approx 9.5$

4.

$x = 9$;
$y = 18$

5. 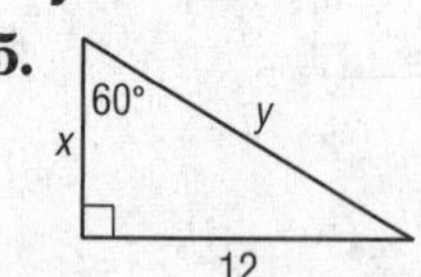

$x = 4\sqrt{3} \approx 6.9$;
$y = 8\sqrt{3} \approx 13.9$

6. 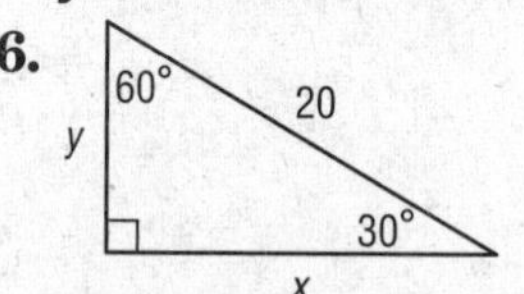

$x = 10\sqrt{3} \approx 17.3$;
$y = 10$

7. An equilateral triangle has an altitude length of 36 feet. Determine the length of a side of the triangle.

$24\sqrt{3}$ feet $\approx$ 41.6 ft

8. Find the length of the side of an equilateral triangle that has an altitude length of 45 centimeters.

$30\sqrt{3}$ cm $\approx$ 52 cm

8-4 Study Guide and Intervention

Trigonometry

Trigonometric Ratios The ratio of the lengths of two sides of a right triangle is called a **trigonometric ratio.** The three most common ratios are **sine, cosine,** and **tangent,** which are abbreviated *sin*, *cos*, and *tan*, respectively.

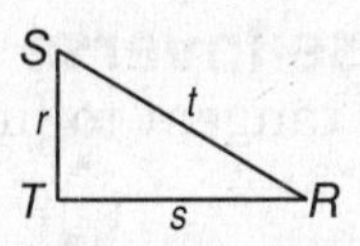

$\sin R = \frac{\text{leg opposite } \angle R}{\text{hypotenuse}} = \frac{r}{t}$ $\qquad \cos R = \frac{\text{leg adjacent to } \angle R}{\text{hypotenuse}} = \frac{s}{t}$ $\qquad \tan R = \frac{\text{leg opposite } \angle R}{\text{leg adjacent to } \angle R} = \frac{r}{s}$

Example **Find sin *A*, cos *A*, and tan *A*. Express each ratio as a fraction and a decimal to the nearest hundredth.**

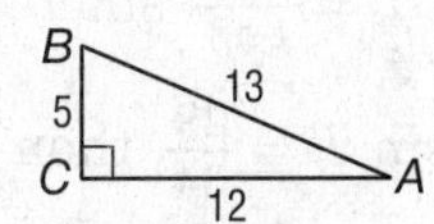

$\sin A = \frac{\text{opposite leg}}{\text{hypotenuse}} = \frac{BC}{BA} = \frac{5}{13} \approx 0.38$

$\cos A = \frac{\text{adjacent leg}}{\text{hypotenuse}} = \frac{AC}{AB} = \frac{12}{13} \approx 0.92$

$\tan A = \frac{\text{opposite leg}}{\text{adjacent leg}} = \frac{BC}{AC} = \frac{5}{12} \approx 0.42$

Exercises

Find sin *J*, cos *J*, tan *J*, sin *L*, cos *L*, and tan *L*. Express each ratio as a fraction and as a decimal to the nearest hundredth if necessary.

1.

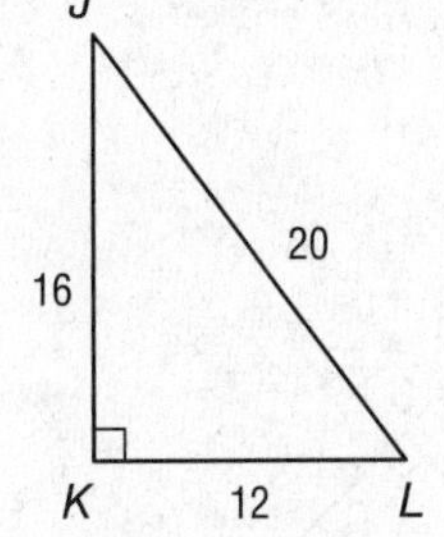

2.

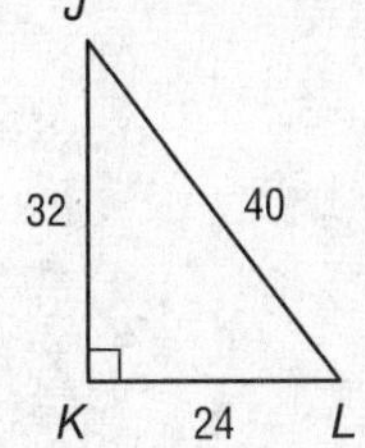

3.

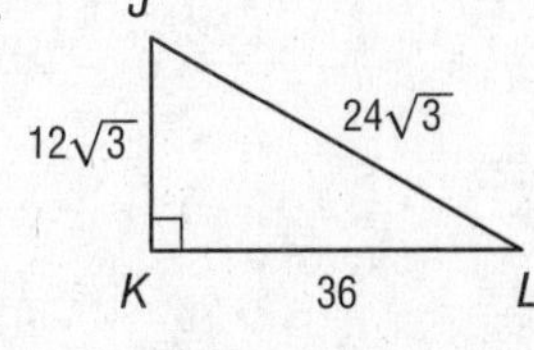

8-4 Study Guide and Intervention *(continued)*

Trigonometry

Use Inverse Trigonometric Ratios You can use a calculator and the sine, cosine, or tangent to find the measure of the angle, called the **inverse** of the trigonometric ratio.

Example **Use a calculator to find the measure of $\angle T$ to the nearest tenth.**

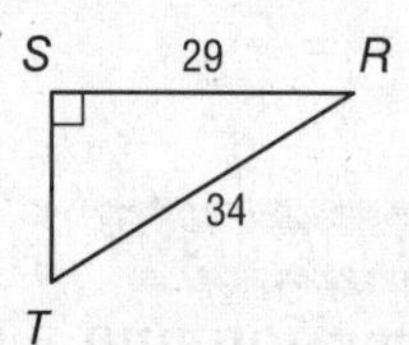

The measures given are those of the leg opposite $\angle T$ and the hypotenuse, so write an equation using the sine ratio.

$\sin T = \frac{opp}{hyp}$ $\sin T = \frac{29}{34}$

If $\sin T = \frac{29}{34}$, then $\sin^{-1} \frac{29}{34} = m\angle T$.

Use a calculator. So, $m\angle T \approx 58.5$.

Exercises

Use a calculator to find the measure of $\angle T$ to the nearest tenth.

1.

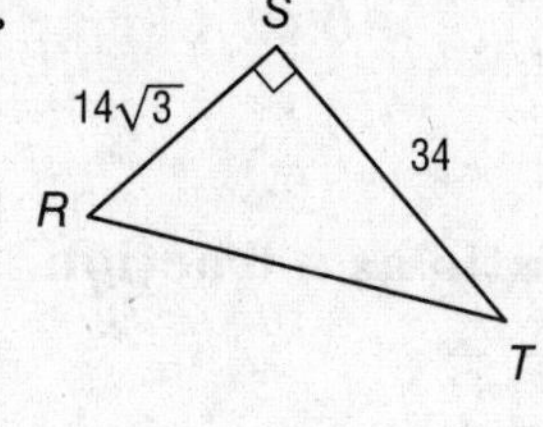

2.

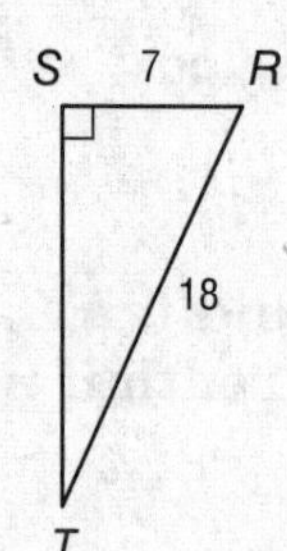

3.

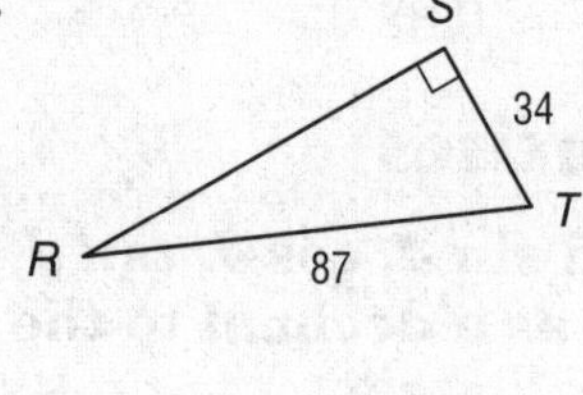

4.

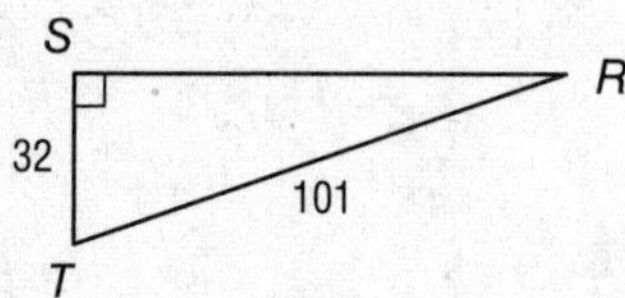

5.

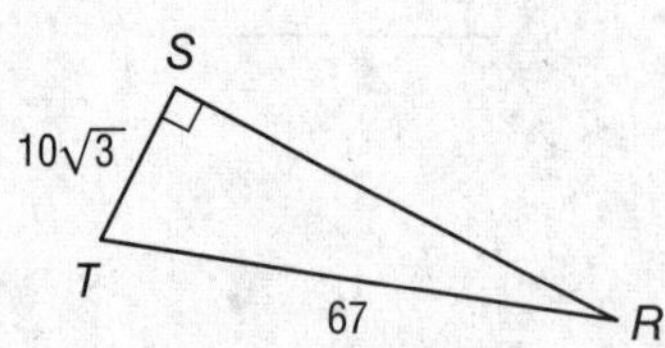

6.

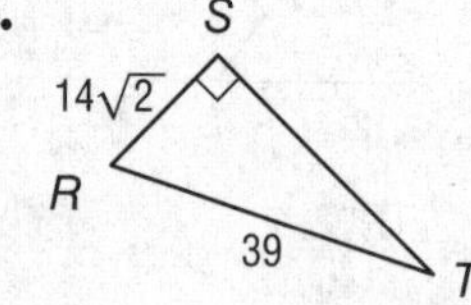

NAME ______________________ DATE __________ PERIOD __________

8-5 Study Guide and Intervention

Angles of Elevation and Depression

Angles of Elevation and Depression Many real-world problems that involve looking up to an object can be described in terms of an **angle of elevation,** which is the angle between an observer's line of sight and a horizontal line.

When an observer is looking down, the **angle of depression** is the angle between the observer's line of sight and a horizontal line.

Example **The angle of elevation from point *A* to the top of a cliff is 34°. If point *A* is 1000 feet from the base of the cliff, how high is the cliff?**

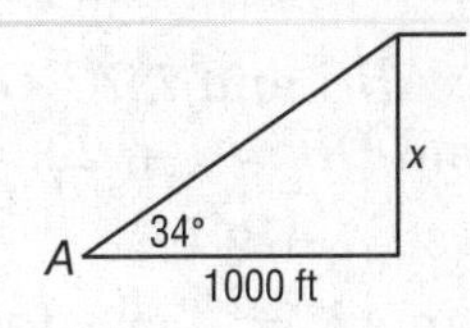

Let x = the height of the cliff.

$\tan 34° = \frac{x}{1000}$ tan = $\frac{\text{opposite}}{\text{adjacent}}$

$1000(\tan 34°) = x$ Multiply each side by 1000.

$674.5 = x$ Use a calculator.

The height of the cliff is about 674.5 feet.

Exercises

1. **HILL TOP** The angle of elevation from point *A* to the top of a hill is 49°. If point *A* is 400 feet from the base of the hill, how high is the hill?

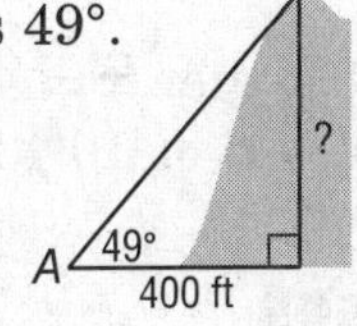

2. **SUN** Find the angle of elevation of the Sun when a 12.5-meter-tall telephone pole casts an 18-meter-long shadow.

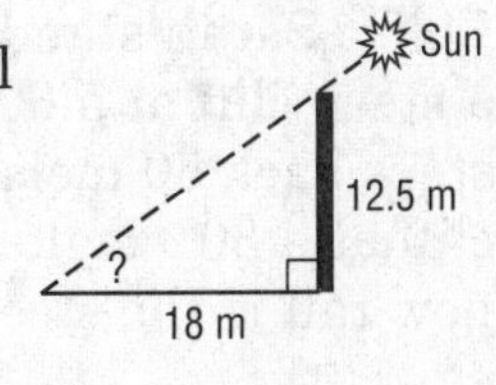

3. **SKIING** A ski run is 1000 yards long with a vertical drop of 208 yards. Find the angle of depression from the top of the ski run to the bottom.

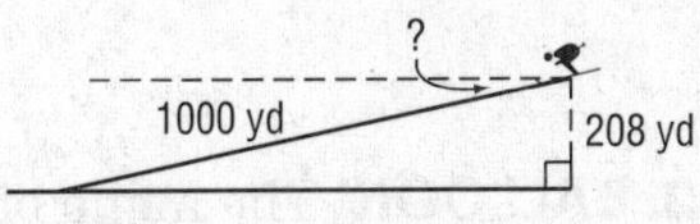

4. **AIR TRAFFIC** From the top of a 120-foot-high tower, an air traffic controller observes an airplane on the runway at an angle of depression of 19°. How far from the base of the tower is the airplane?

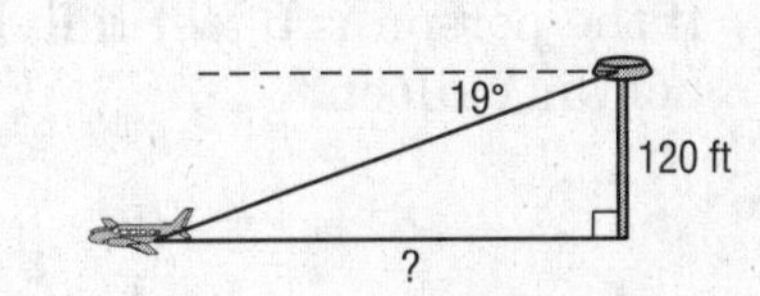

8-5 Study Guide and Intervention *(continued)*

Angles of Elevation and Depression

Two Angles of Elevation or Depression Angles of elevation or depression to two different objects can be used to estimate distance between those objects. The angles from two different positions of observation to the same object can be used to estimate the height of the object.

Example **To estimate the height of a garage, Jason sights the top of the garage at a 42° angle of elevation. He then steps back 20 feet and sites the top at a 10° angle. If Jason is 6 feet tall, how tall is the garage to the nearest foot?**

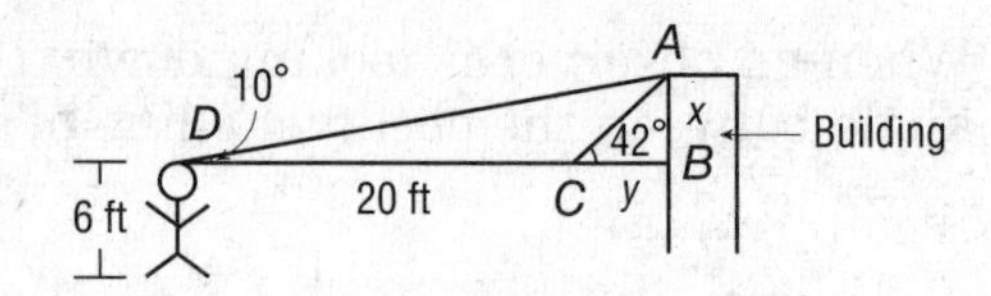

$\triangle ABC$ and $\triangle ABD$ are right triangles. We can determine $AB = x$ and $CB = y$, and $DB = y + 20$.

Use $\triangle ABC$.

$\tan 42° = \frac{x}{y}$ or $y \tan 42° = x$

Use $\triangle ABD$.

$\tan 10° = \frac{x}{y + 20}$ or $(y + 20) \tan 10° = x$

Substitute the value for x from $\triangle ABD$ in the equation for $\triangle ABC$ and solve for y.

$$y \tan 42° = (y + 20) \tan 10°$$
$$y \tan 42° = y \tan 10° + 20 \tan 10°$$
$$y \tan 42° - y \tan 10° = 20 \tan 10°$$
$$y (\tan 42° - \tan 10°) = 20 \tan 10°$$
$$y = \frac{20 \tan 10°}{\tan 42° - \tan 10°} \approx 4.87$$

If $y = 4.87$, then $x = 4.87 \tan 42°$ or about 4.4 feet. Add Jason's height, so the garage is about $4.4 + 6$ or 10.4 feet tall.

Exercises

1. **CLIFF** Sarah stands on the ground and sights the top of a steep cliff at a 60° angle of elevation. She then steps back 50 meters and sights the top of the steep cliff at a 30° angle. If Sarah is 1.8 meters tall, how tall is the steep cliff to the nearest meter?

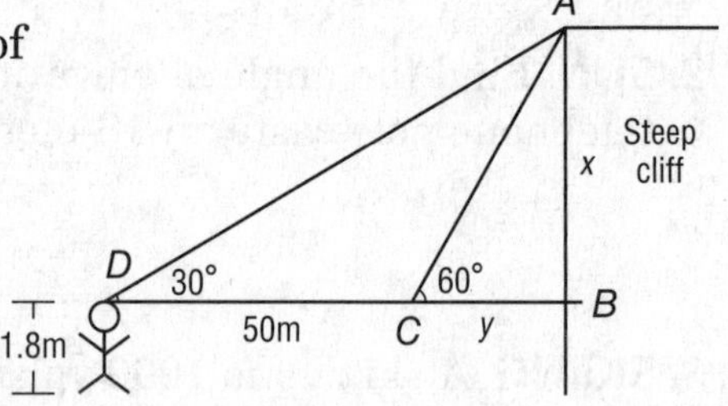

2. **BALLOON** The angle of depression from a hot air balloon in the air to a person on the ground is 36°. If the person steps back 10 feet, the new angle of depression is 25°. If the person is 6 feet tall, how far off the ground is the hot air balloon?

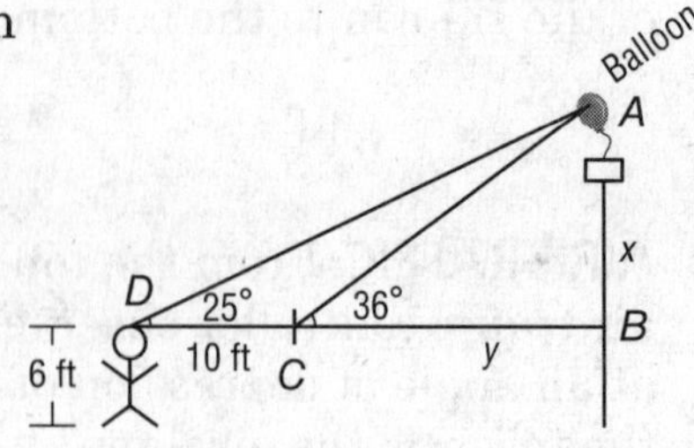

NAME ______________________ DATE __________ PERIOD __________

8-6 Study Guide and Intervention

The Law of Sines and Law of Cosines

The Law of Sines In any triangle, there is a special relationship between the angles of the triangle and the lengths of the sides opposite the angles.

Law of Sines	$\frac{\sin A}{a} = \frac{\sin B}{b} = \frac{\sin C}{c}$

Example 1 **Find b. Round to the nearest tenth.**

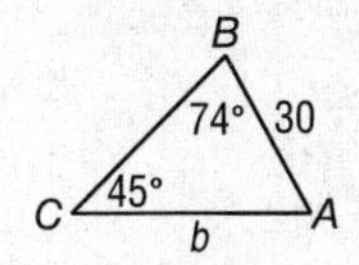

$\frac{\sin C}{c} = \frac{\sin B}{b}$ Law of Sines

$\frac{\sin 45°}{30} = \frac{\sin 74°}{b}$ $m\angle C = 45$, $c = 30$, $m\angle B = 74$

$b \sin 45° = 30 \sin 74°$ Cross Products Property

$b = \frac{30 \sin 74°}{\sin 45°}$ Divide each side by sin 45°.

$b \approx 40.8$ Use a calculator.

Example 2 **Find $m\angle E$. Round to the nearest degree.**

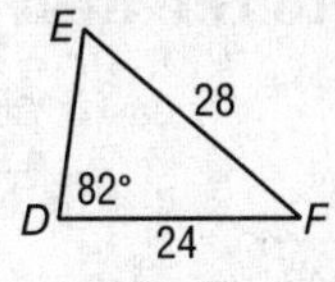

$\frac{\sin D}{d} = \frac{\sin E}{e}$ Law of Sines

$\frac{\sin 82°}{28} = \frac{\sin E}{24}$ $d = 28$, $m\angle D = 82$, $e = 24$

$24 \sin 82° = 28 \sin E$ Cross Products Property

$\sin E = \frac{24 \sin 82°}{28}$ Divide each side by 28.

$E = \sin^{-1} \frac{24 \sin 82°}{28}$ Use the inverse sine.

$E \approx 24°$ Use a calculator.

Exercises

Find x. Round angle measures to the nearest degree and side measures to the nearest tenth.

1.

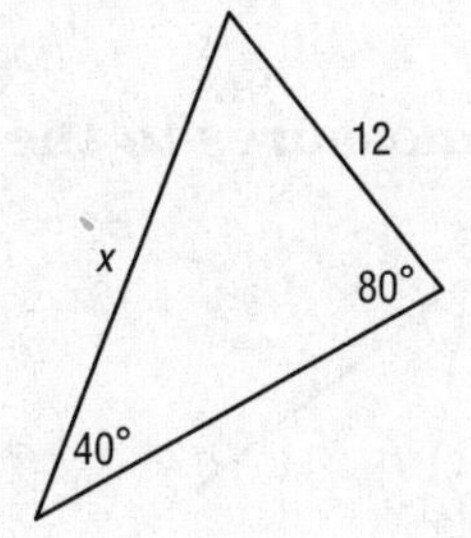

2.

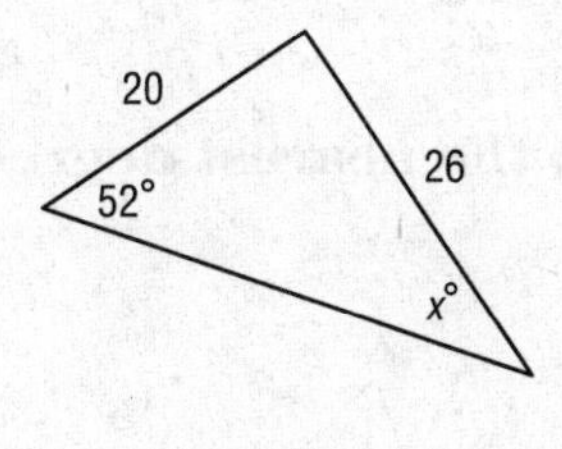

3.

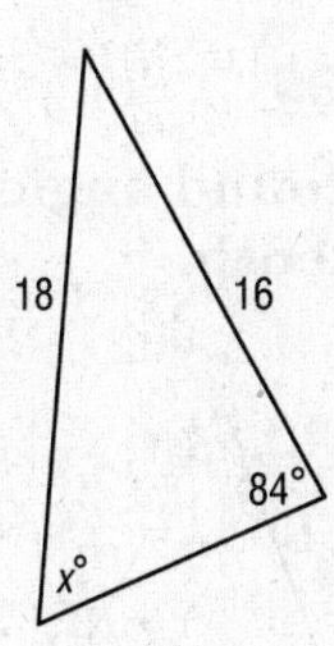

4.

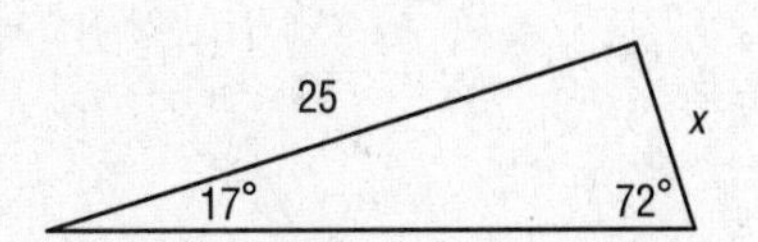

5.

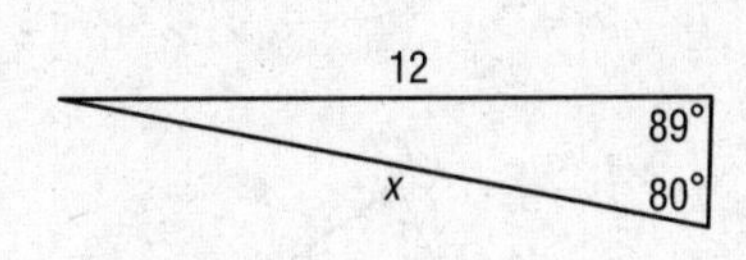

6.

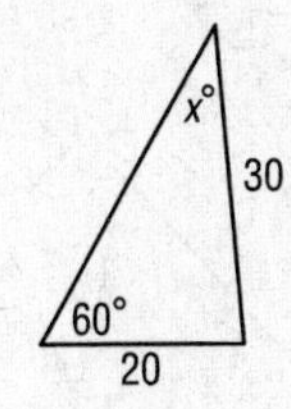

8-6 Study Guide and Intervention *(continued)*

The Law of Sines and Law of Cosines

The Law of Cosines Another relationship between the sides and angles of any triangle is called the **Law of Cosines.** You can use the Law of Cosines if you know three sides of a triangle or if you know two sides and the included angle of a triangle.

Law of Cosines	Let $\triangle ABC$ be any triangle with a, b, and c representing the measures of the sides opposite the angles with measures A, B, and C, respectively. Then the following equations are true. $a^2 = b^2 + c^2 - 2bc \cos A$ $\quad b^2 = a^2 - c^2 - 2ac \cos B$ $\quad c^2 = a^2 + b^2 - 2ab \cos C$

Example 1 **Find c. Round to the nearest tenth.**

$c^2 = a^2 + b^2 - 2ab \cos C$ — Law of Cosines

$c^2 = 12^2 + 10^2 - 2(12)(10)\cos 48°$ — $a = 12$, $b = 10$, $m\angle C = 48$

$c = \sqrt{12^2 + 10^2 - 2(12)(10)\cos 48°}$ — Take the square root of each side.

$c \approx 9.1$ — Use a calculator.

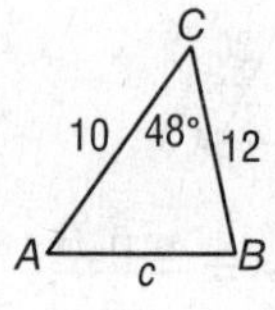

Example 2 **Find $m\angle A$. Round to the nearest degree.**

$a^2 = b^2 + c^2 - 2bc \cos A$ — Law of Cosines

$7^2 = 5^2 + 8^2 - 2(5)(8) \cos A$ — $a = 7$, $b = 5$, $c = 8$

$49 = 25 + 64 - 80 \cos A$ — Multiply.

$-40 = -80 \cos A$ — Subtract 89 from each side.

$\frac{1}{2} = \cos A$ — Divide each side by -80.

$\cos^{-1} \frac{1}{2} = A$ — Use the inverse cosine.

$60° = A$ — Use a calculator.

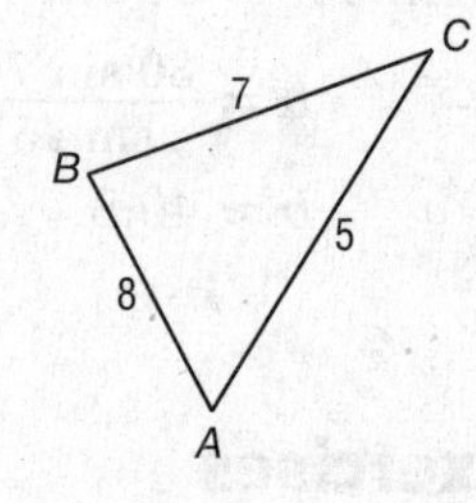

Exercises

Find x. Round angle measures to the nearest degree and side measures to the nearest tenth.

1.

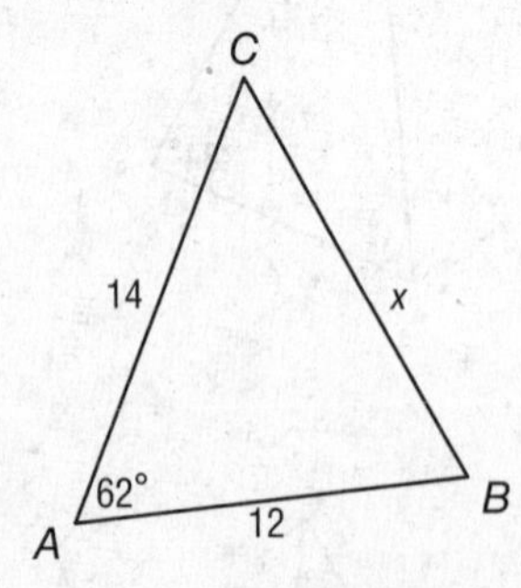

2.

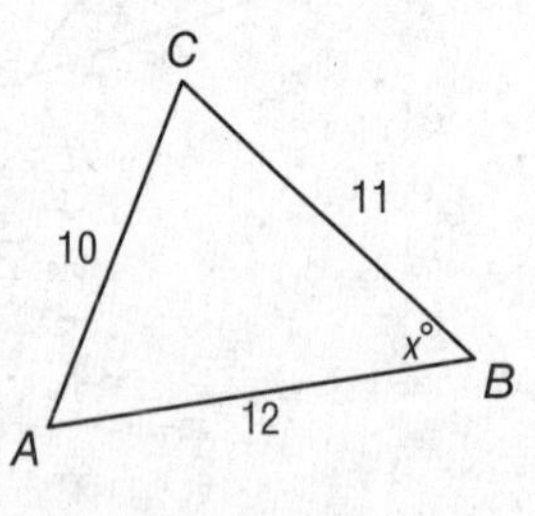

3.

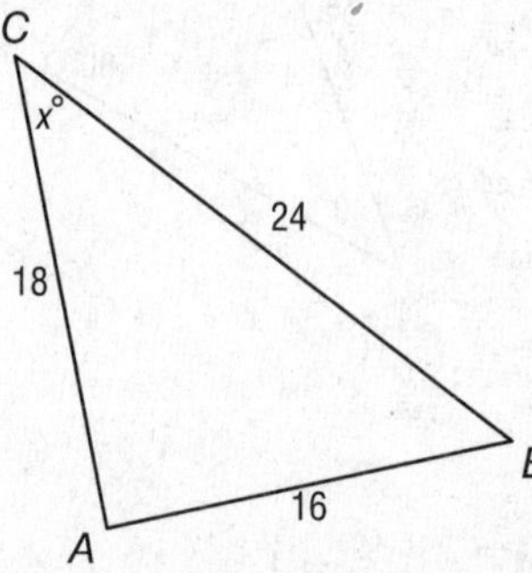

4.

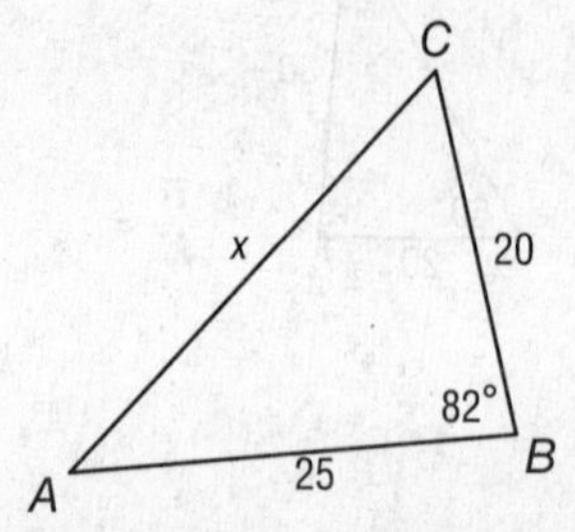

5.

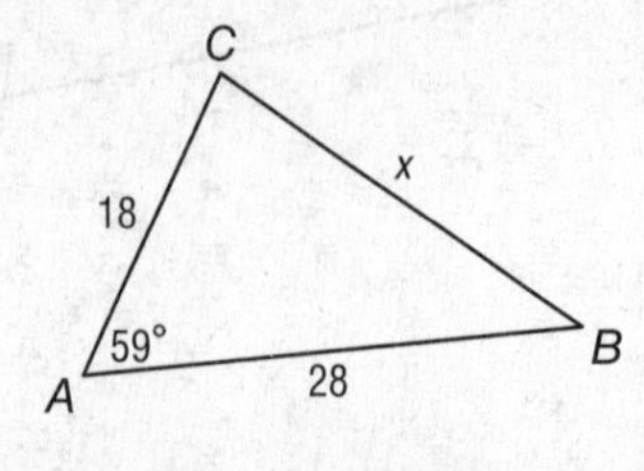

6.

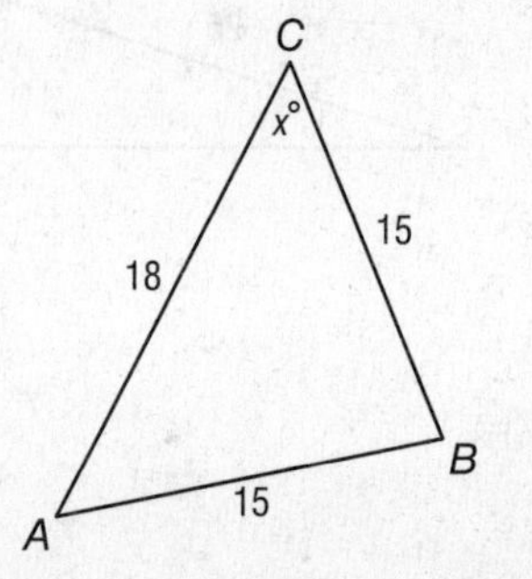

8-7 Study Guide and Intervention

Vectors

Describe Vectors A vector is a directed segment representing a quantity that has both **magnitude,** or length, and **direction.** For example, the speed and direction of an airplane can be represented by a vector. In symbols, a vector is written as $\overrightarrow{AB}$, where A is the initial point and B is the endpoint, or as $\vec{\mathbf{v}}$.

A vector in **standard position** has its initial point at (0, 0) and can be represented by the ordered pair for point B. The vector at the right can be expressed as $\vec{\mathbf{v}} = \langle 5, 3 \rangle$.

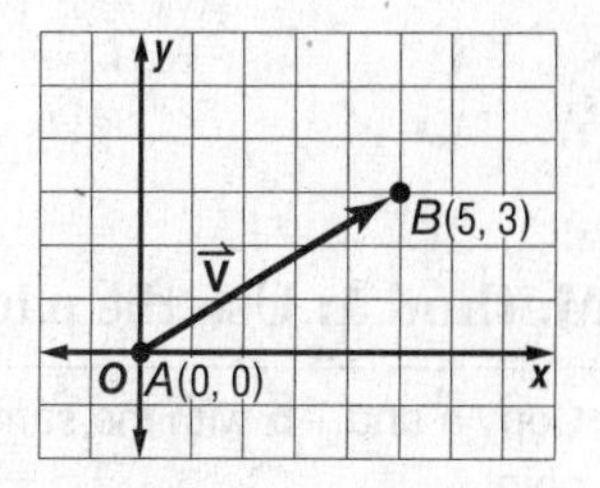

You can use the Distance Formula to find the magnitude $|\overrightarrow{AB}|$ of a vector. You can describe the direction of a vector by measuring the angle that the vector forms with the positive x-axis or with any other horizontal line.

Example **Find the magnitude and direction of $\overrightarrow{AB}$ for $A(5, 2)$ and $B(8, 7)$.**

Find the magnitude.

$$\overrightarrow{AB} = \sqrt{(x_2 - x_1)^2 + (y_2 - y_1)^2}$$
$$= \sqrt{(8 - 5)^2 + (7 - 2)^2}$$
$$= \sqrt{34} \text{ or about } 5.8$$

To find the direction, use the tangent ratio.

$\tan A = \frac{5}{3}$ The tangent ratio is opposite over adjacent.

$m\angle A \approx 59.0$ Use a calculator.

The magnitude of the vector is about 5.8 units and its direction is 59°.

Exercises

Find the magnitude and direction of each vector.

1. $\overrightarrow{AB}$: $A(3, 1)$, $B(-2, 3)$

2. $\overrightarrow{PQ}$: $P(0, 0)$, $Q(-2, 1)$

3. $\overrightarrow{TU}$: $T(0, 1)$, $U(3, 5)$

4. $\overrightarrow{MN}$: $M(-2, 2)$, $N(3, 1)$

5. $\overrightarrow{RS}$: $R(3, 4)$, $S(0, 0)$

6. $\overrightarrow{CD}$: $C(4, 2)$, $D(0, 3)$

NAME ______________ DATE ________ PERIOD ________

8-7 Study Guide and Intervention *(continued)*

Vectors

Vectors Addition The sum of two vectors is called the **resultant.** Subtracting a vector is equivalent to adding its opposite. The resultant of two vectors can be found using the **parallelogram method** or the **triangle method.**

Example **Copy the vectors to find $\vec{a} - \vec{b}$.**

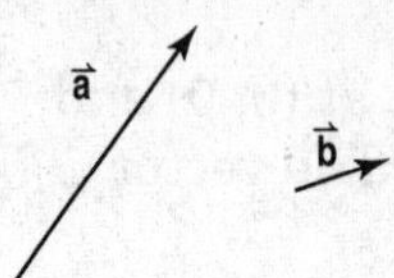

Method 1: Use the parallelogram method.

Copy $\vec{a}$ and $-\vec{b}$ with the same initial point.	Complete the parallelogram.	Draw the diagonal of the parallelogram from the initial point.
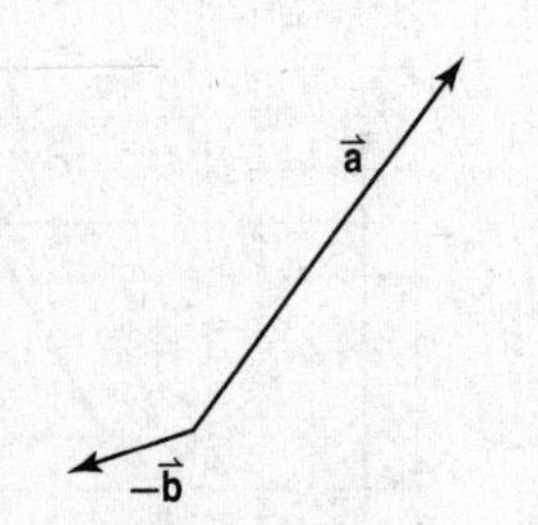	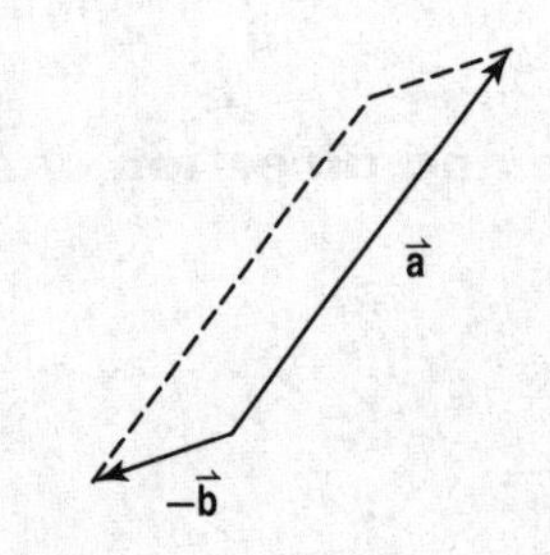	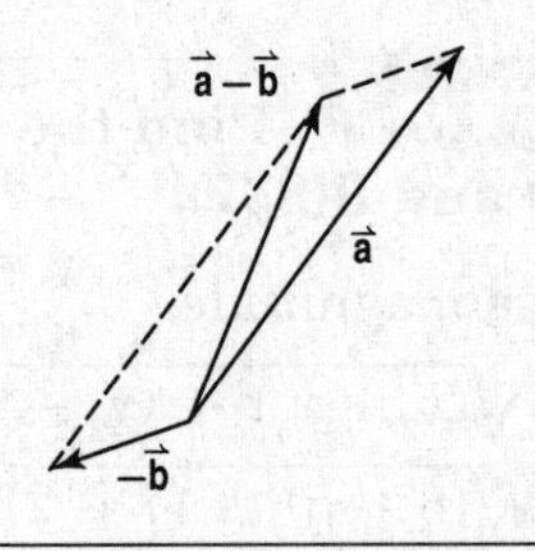

Method 2: Use the triangle method.

Copy $\vec{a}$.	Place the initial point of $-\vec{b}$ at the terminal point of $\vec{a}$.	Draw the vector from the initial point of $\vec{a}$ to the terminal point of $-\vec{b}$.
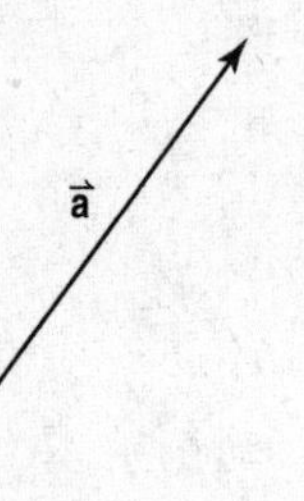	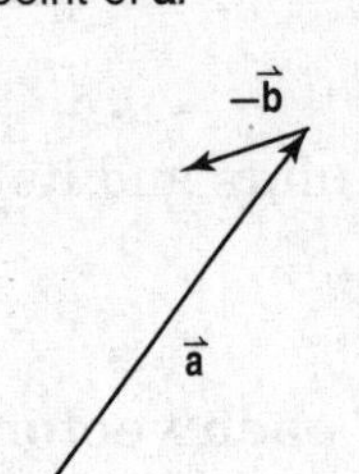	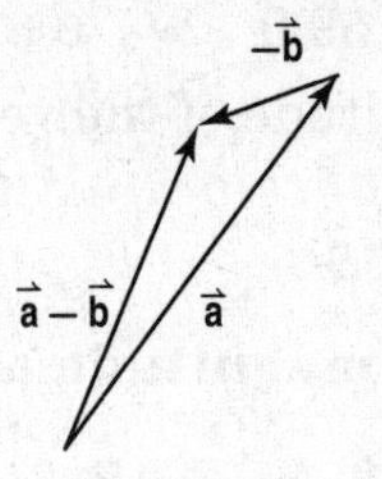

Exercises

Copy the vectors to find each sum or difference.

1. $\vec{c} + \vec{d}$

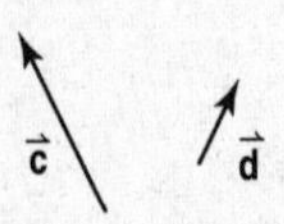

2. $\vec{w} - \vec{z}$

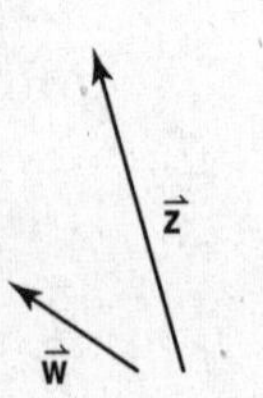

3. $\vec{a} - \vec{b}$

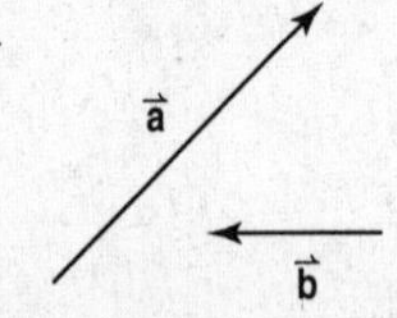

4. $\vec{r} + \vec{t}$

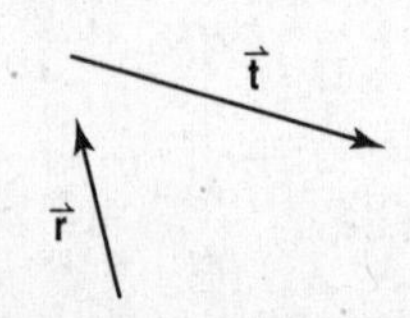

9-1 Study Guide and Intervention

Reflections

Draw Reflections A reflection is a flip of a figure in a line called the **line of reflection.** Each point of the preimage and its corresponding point on the image are the same distance from the line. A reflection maps a point to its image. If the point is on the line of reflection, then the image and preimage are the same point. If the point does not lie on the line of reflection, the line of reflection is the perpendicular bisector of the segment joining the two points.

Example **Construct the image of quadrilateral *ABCD* under a reflection in line *m*.**

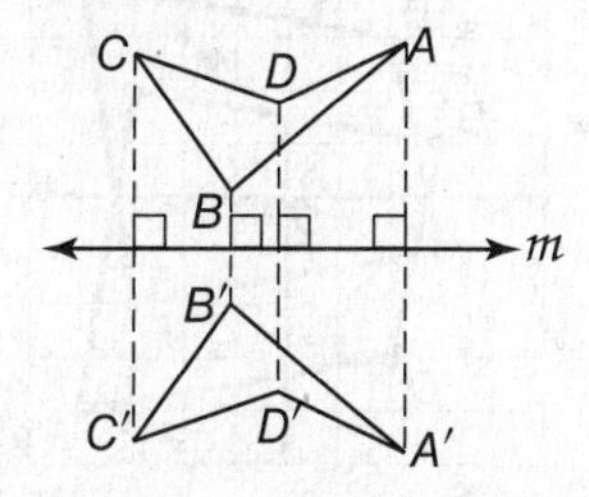

Draw a perpendicular from each vertex of the quadrilateral to m. Find vertices A', B', C', and D' that are the same distance from m on the other side of m. The image is $A'B'C'D'$.

Exercises

Use the figure and given line of reflection. Then draw the reflected image in this line using a ruler.

1.

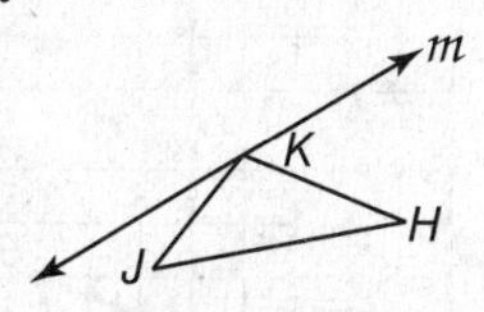

2.

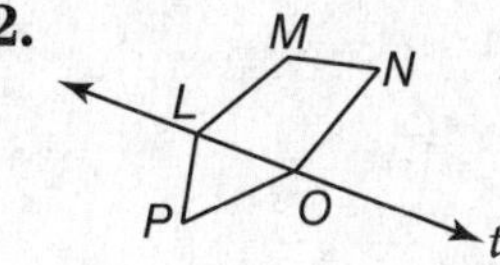

3.

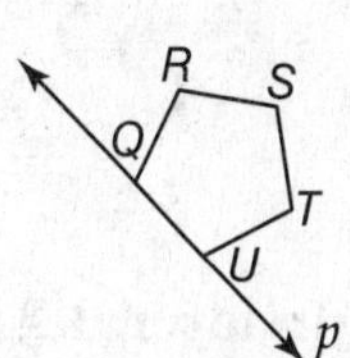

4.

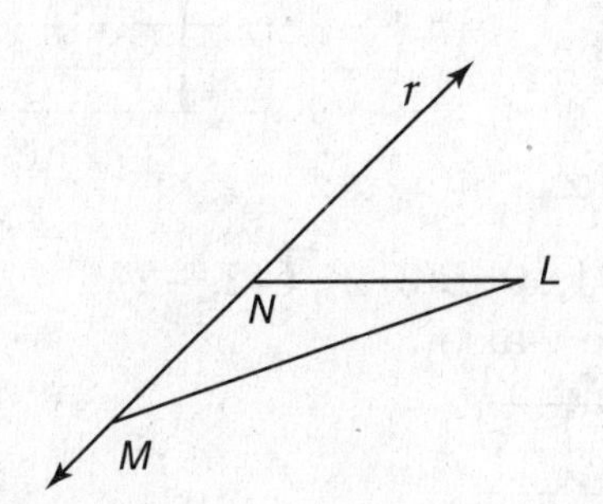

5.

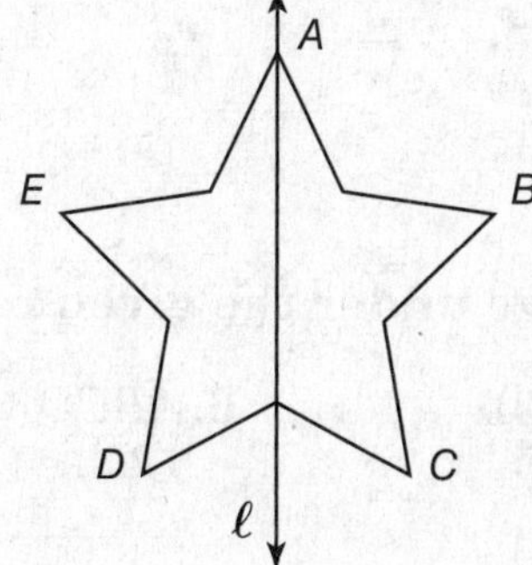

6.

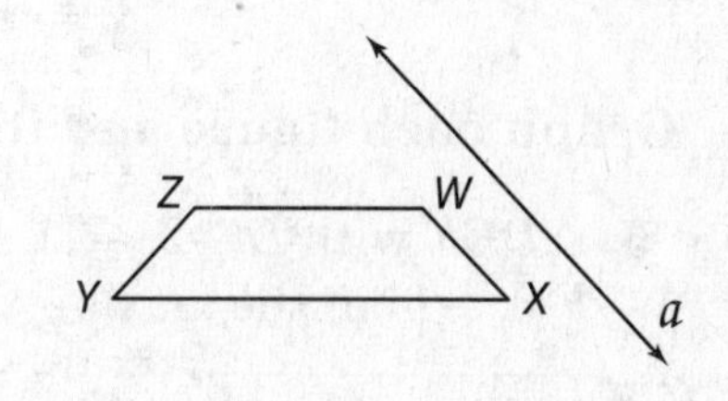

9-1 Study Guide and Intervention *(continued)*

Draw Reflections In The Coordinate Plane Reflections can be performed in the coordinate plane. Each point of the image and its corresponding point on the preimage must be the same distance from the line of reflection.

- To reflect a point in the x-axis, multiply its y-coordinate by -1.
- To reflect a point in the y-axis, multiply its x-coordinate by -1.
- To reflect a point in the line $y = x$, interchange the x-and y-coordinates.

Example **Quadrilateral *DEFG* has vertices D(–2, 3), E(4, 4), F(3, –2), and G(–3, –1). Find the image under reflection in the *x*-axis.**

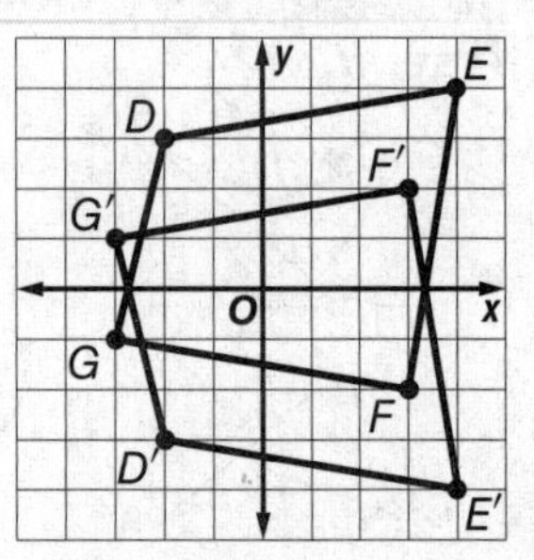

To find an image for a reflection in the x-axis, use the same x-coordinate and multiply the y-coordinate by -1. In symbols, $(a, b) \rightarrow (a, -b)$. The new coordinates are $D'(-2, -3)$, $E'(4, -4)$, $F'(3, 2)$, and $G'(-3, 1)$. The image is $D'E'F'G'$.

Exercises

Graph △*FGH* and its image in the given line.

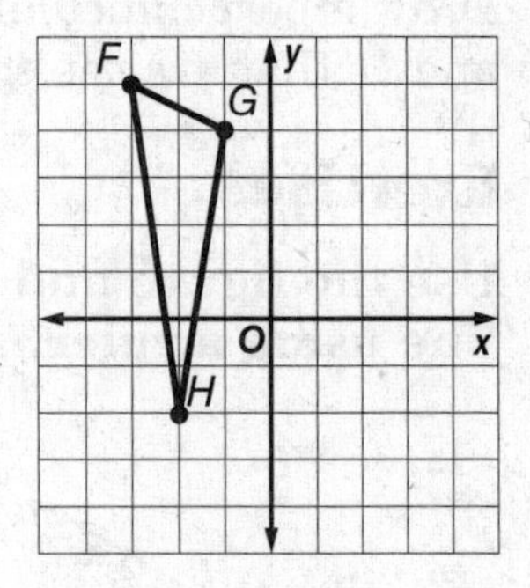

1. $x = -1$

2. $y = 1$

Graph quadrilateral *ABCD* and its image in the given line.

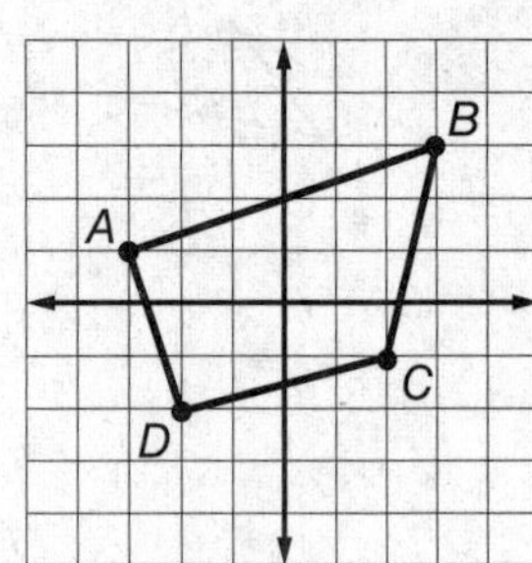

3. $x = 0$

4. $y = 1$

Graph each figure and its image under the given reflection.

5. $\triangle DEF$ with $D(-2, -1)$, $E(-1, 3)$, $F(3, -1)$ in the x-axis

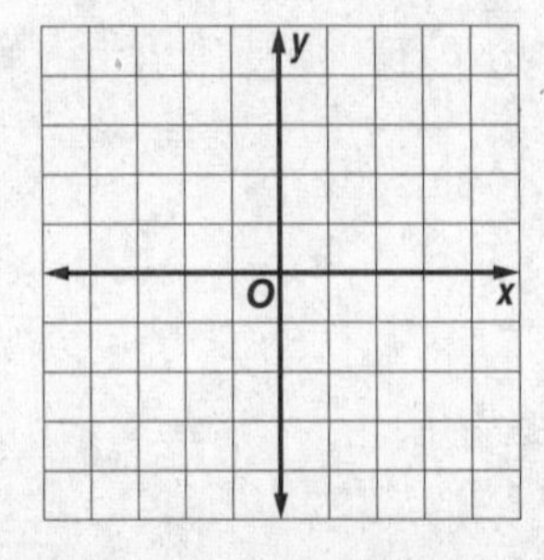

6. $ABCD$ with $A(1, 4)$, $B(3, 2)$, $C(2, -2)$, $D(-3, 1)$ in the y-axis

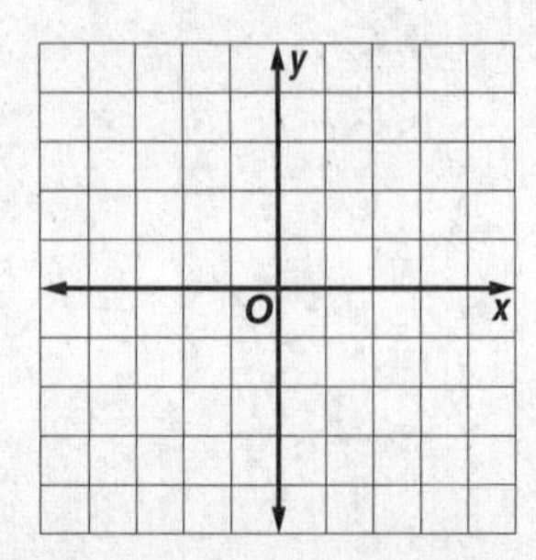

NAME ______________ DATE ________ PERIOD ________

9-2 Study Guide and Intervention

Translations

Draw Translations A **translation** is a transformation that moves all points of a figure the same distance in the same direction. Vectors can be used to describe the distance and direction of the translation.

Example **Draw the translation of the figure along the translation vector.**

Draw a line through each vertex parallel to vector $\vec{u}$. Measure the length of vector $\vec{u}$. Locate the image of each point by marking off this distance along the line through each vertex. Start at the vertex and move in the same direction as the vector.

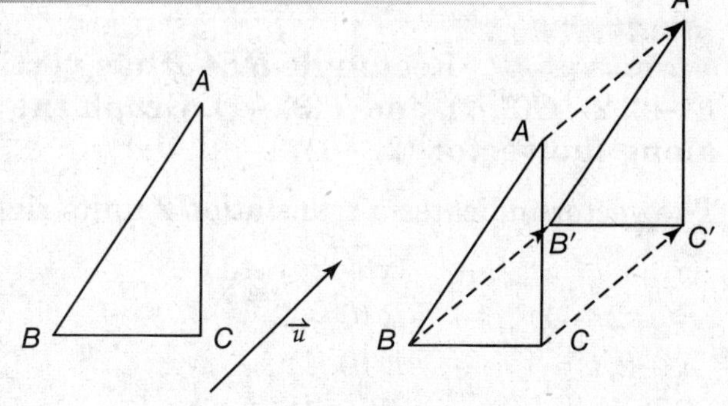

Exercises

Use the figure and the given translation vector. Then draw the translation of the figure along the translation vector.

1.

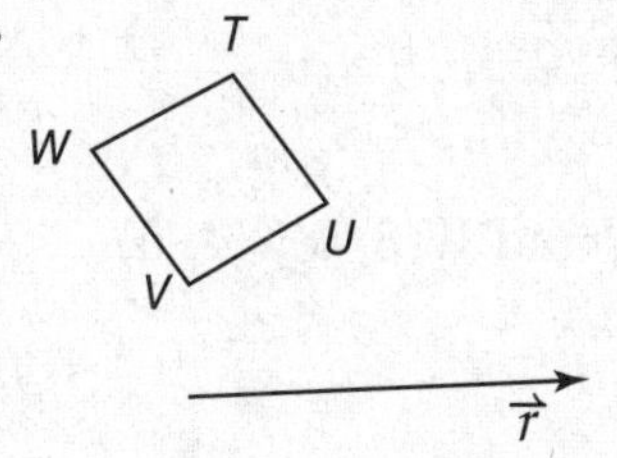

2.

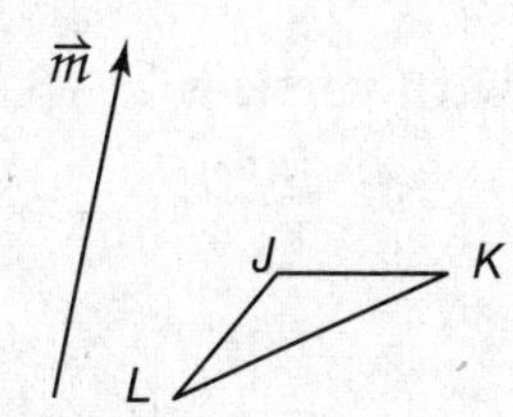

3.

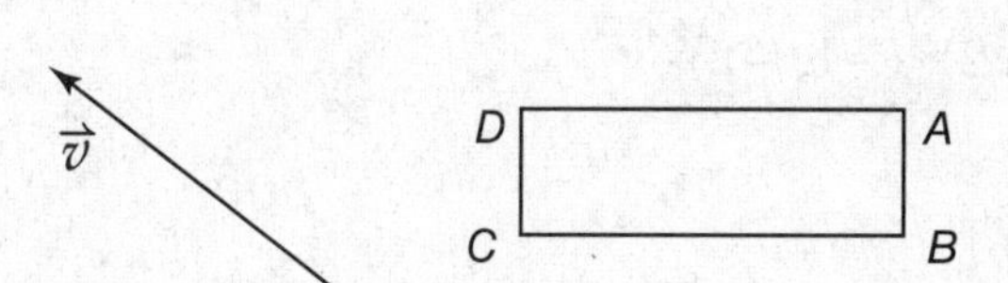

4.

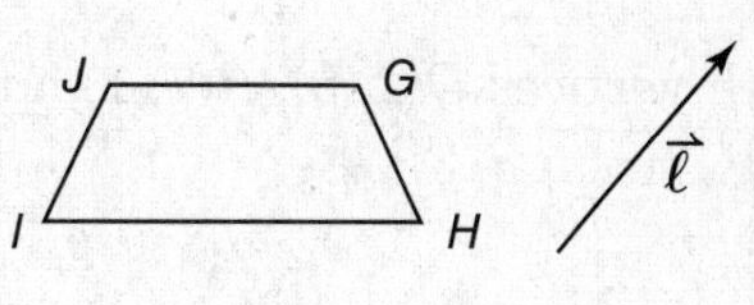

5.

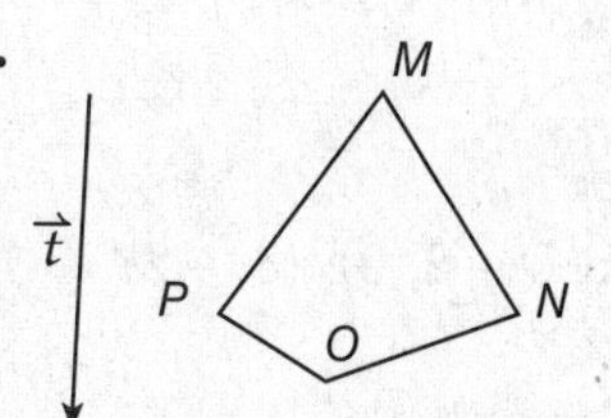

6.

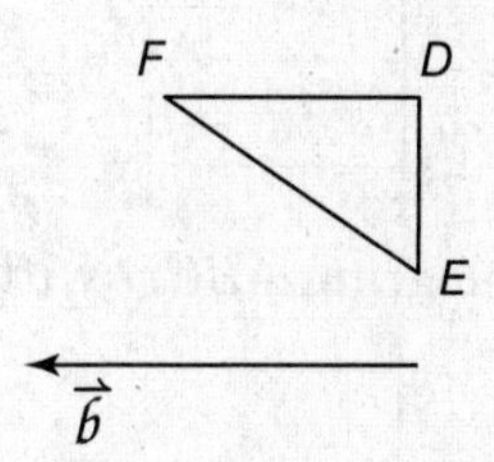

9-2 Study Guide and Intervention *(continued)*

Translations

Translations In The Coordinate Plane A vector can be used to translate a figure on the coordinate plane when written in the form $\langle a, b\rangle$ where a represents the horizontal change and b represents the vertical change from the vector's tip to its tail.

Example **Rectangle $RECT$ has vertices $R(-2, -1)$, $E(-2, 2)$, $C(3, 2)$, and $T(3, -1)$. Graph the figure and its image along the vector $\langle 2, -1\rangle$.**

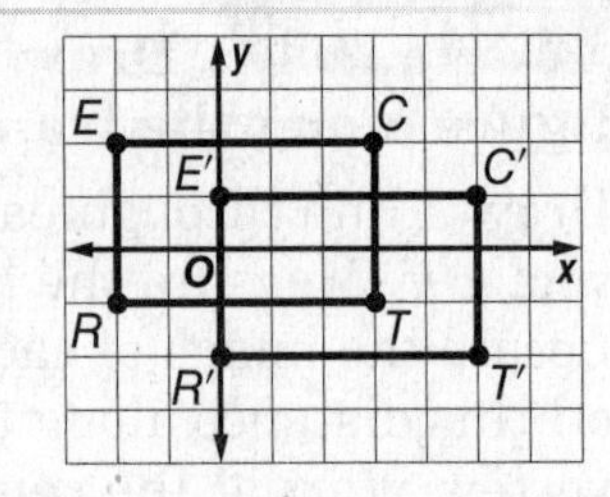

The vector indicates a translation 2 units right and 1 unit down.

$(x, y) \rightarrow (x + 2, y - 1)$
$R(-2, -1) \rightarrow R'(0, -2)$
$E(-2, 2) \rightarrow E'(0, 1)$
$C(3, 2) \rightarrow C'(5, 1)$
$T(3, -1) \rightarrow T'(5, -2)$

Graph $RECT$ and its image $R'E'C'T'$.

Exercises

Graph each figure and its image along the given vector.

1. quadrilateral $TUVW$ with vertices $T(-3, -8)$, $U(-6, 3)$, $V(0, 3)$, and $W(3, 0)$; $\langle 4, 5\rangle$

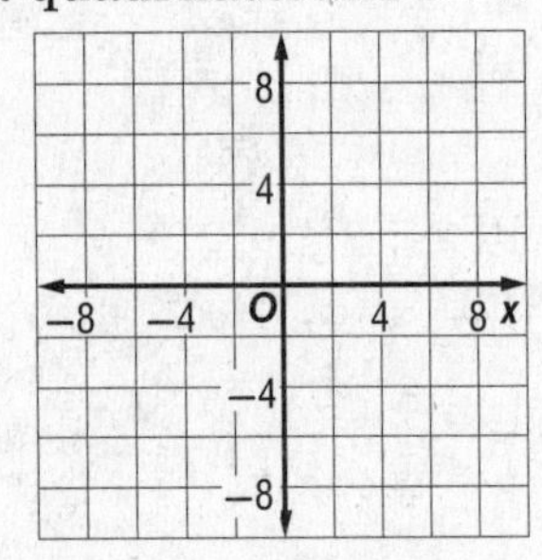

2. $\triangle QRS$ with vertices $Q(2, 5)$, $R(7, 1)$, and $S(-1, 2)$; $\langle -1, -2\rangle$

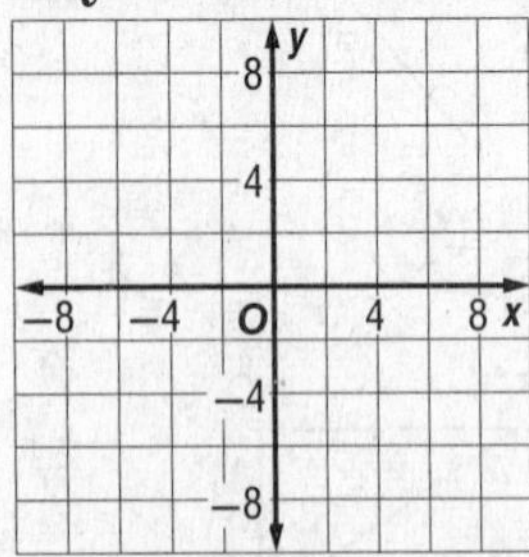

3. parallelogram $ABCD$ with vertices $A(1, 6)$, $B(4, 5)$, $C(1, -1)$, and $D(-2, 0)$; $\langle 3, -2\rangle$

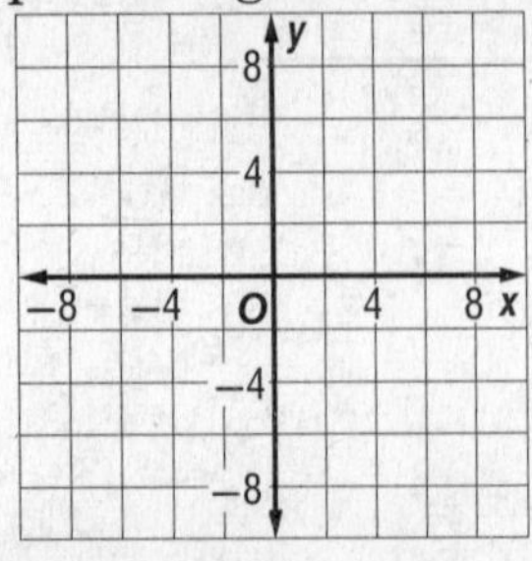

9-3 Study Guide and Intervention

Rotations

Draw Rotations A **rotation** is a transformation that moves every point of the preimage through a specified angle, $x°$, and direction about a fixed point called the **center of rotation**.

- If the point being rotated is the center of rotation, then the image and preimage are the same point.
- If the point being rotated is not the center of rotation, then the image and preimage are the same distance from the center of rotation and the measure of the angle of rotation formed by the preimage, center of rotation, and image points is x.

Example **Use a protractor and ruler to draw a 110° rotation of square *LMNO* about point *P*.**

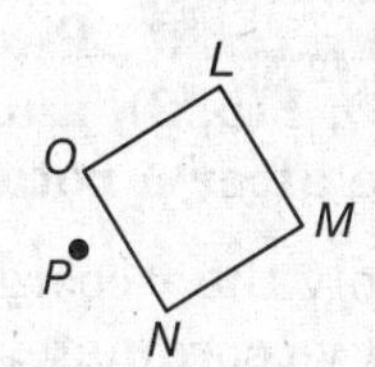

Step 1 Draw a segment from vertex L to point P. 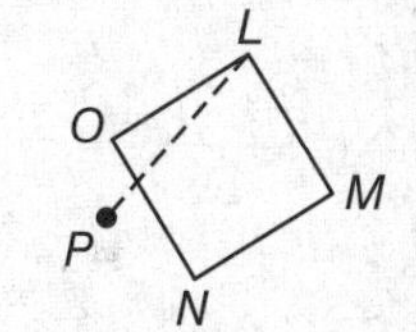	**Step 2** Draw a 110° angle using $\overline{PL}$ as one side. 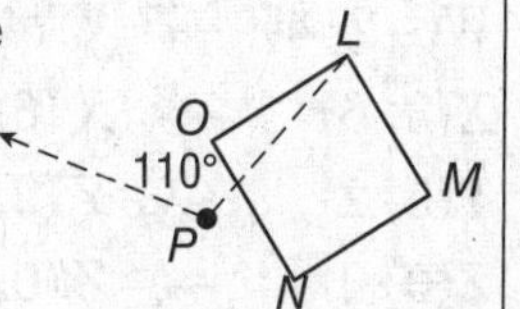
Step 3 Use a ruler to draw L' such that $PL' = PL$. 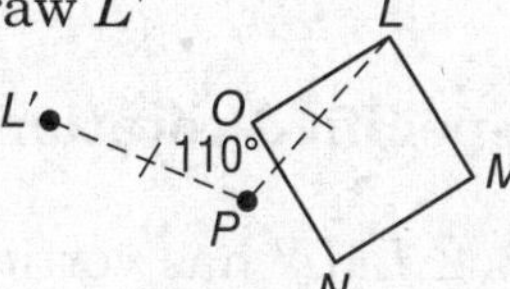	**Step 4** Repeat steps 1–3 for vertices M, N, and O and draw square $L'M'N'O'$. 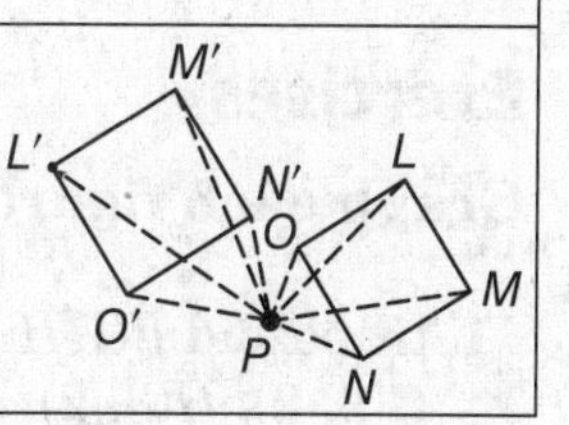

Exercises

Use a protractor and a ruler to draw the specified rotation of each figure about point *K*.

1. 75°

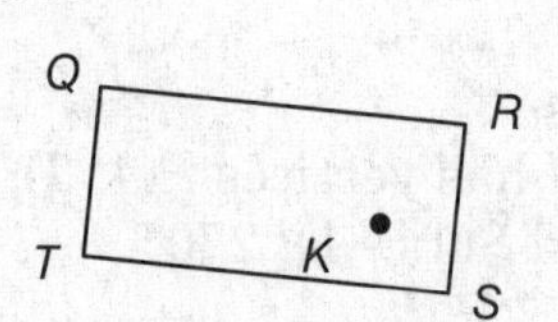

2. 45°

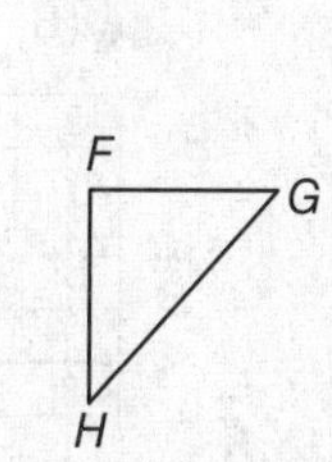

3. 135°

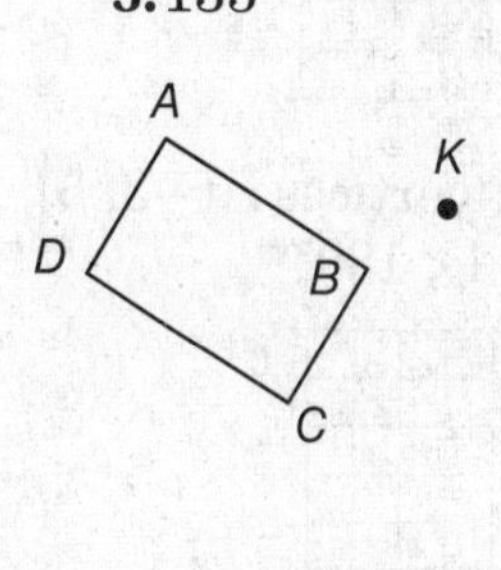

NAME ______________________ DATE ____________ PERIOD ____________

9-3 Study Guide and Intervention *(continued)*

Rotations

Draw Rotations In The Coordinate Plane The following rules can be used to rotate a point 90°, 180°, or 270° counterclockwise about the origin in the coordinate plane.

To rotate	Procedure
90°	Multiply the *y*-coordinate by –1 and then interchange the *x*- and *y*-coordinates.
180°	Multiply the *x*- and *y*-coordinates by –1.
270°	Multiply the *x*-coordinate by –1 and then interchange the *x*-and *y*-coordinates.

Example **Parallelogram *WXYZ* has vertices *W*(–2, 4), *X*(3, 6), *Y*(5, 2), and *Z*(0, 0). Graph parallelogram *WXYZ* and its image after a rotation of 270° about the origin.**

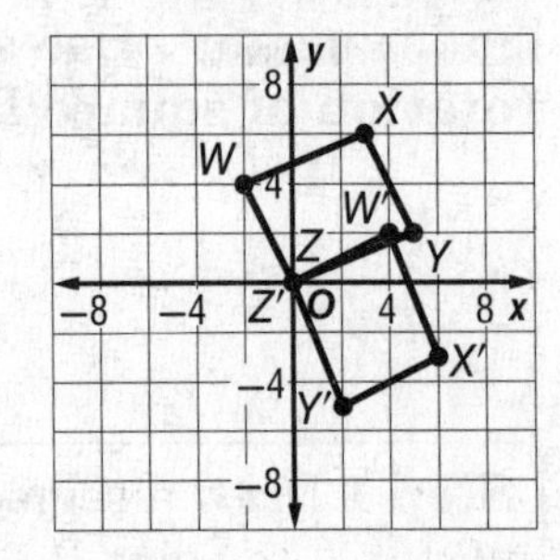

Multiply the x-coordinate by -1 and then interchange the x- and y-coordinates.

$(x, y) \rightarrow (y, -x)$
$W(-2, 4) \rightarrow W'(4, 2)$
$X(3, 6) \rightarrow X'(6, -3)$
$Y(5, 2) \rightarrow Y'(2, -5)$
$Z(0, 0) \rightarrow Z'(0, 0)$

Exercises

Graph each figure and its image after the specified rotation about the origin.

1. trapezoid *FGHI* has vertices *F*(7, 7), *G*(9, 2), *H*(3, 2), and *I*(5, 7); 90°

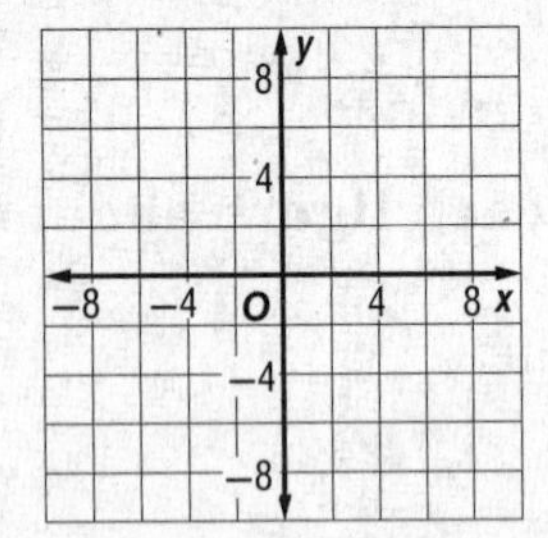

2. △*LMN* has vertices *L*(–1, –1), *M*(0, –4), and *N*(–6, –2); 90°

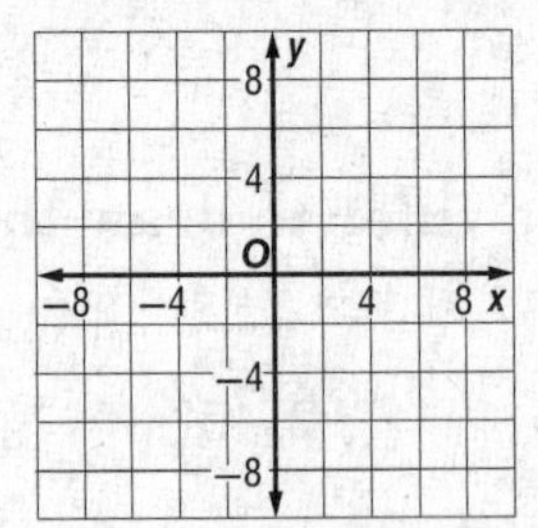

3. △*ABC* has vertices *A*(–3, 5), *B*(0, 2), and *C*(–5, 1); 180°

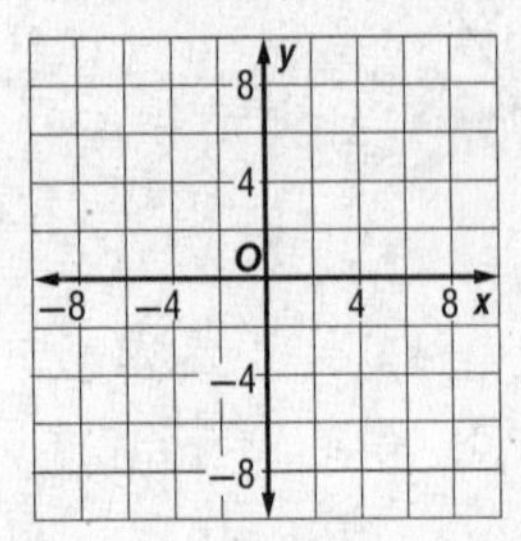

4. parallelogram *PQRS* has vertices *P*(4, 7), *Q*(6, 6), *R*(3, –2), and *S*(1, –1); 270°

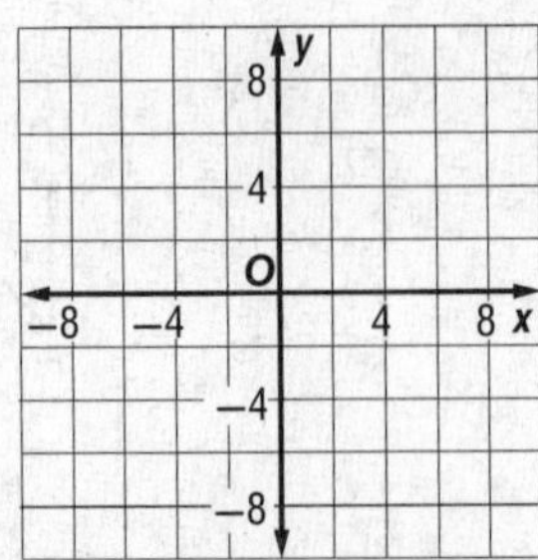

9-4 Study Guide and Intervention

Compositions of Transformations

Glide Reflections When two transformations are applied to a figure, one after another, the total transformation is a **composition of transformations.** A **glide reflection** is a translation followed by a reflection in a line parallel to the translation vector. Notice that the composition of isometries is another isometry.

Example **Triangle *ABC* has vertices *A*(3, 3), *B*(4, –2) and *C*(–1, –3). Graph △*ABC* and its image after a translation along⟨–2, –1⟩ and a reflection in the *x*-axis.**

Step 1 translation along ⟨–2, –1⟩

$(x, y) \rightarrow (x - 2, y - 1)$
$A(3, 3) \rightarrow A'(1, 2)$
$B(4, -2) \rightarrow B'(2, -3)$
$C(-1, -3) \rightarrow C'(-3, -4)$

Step 2 reflection in the *x*-axis.

$(x, y) \rightarrow (x, -y)$
$A'(1, 2) \rightarrow A''(1, -2)$
$B'(2, -3) \rightarrow B''(2, 3)$
$C'(-3, -4) \rightarrow C''(-3, 4)$

Step 3 Graph △*ABC* and its image △*A″B″C″*.

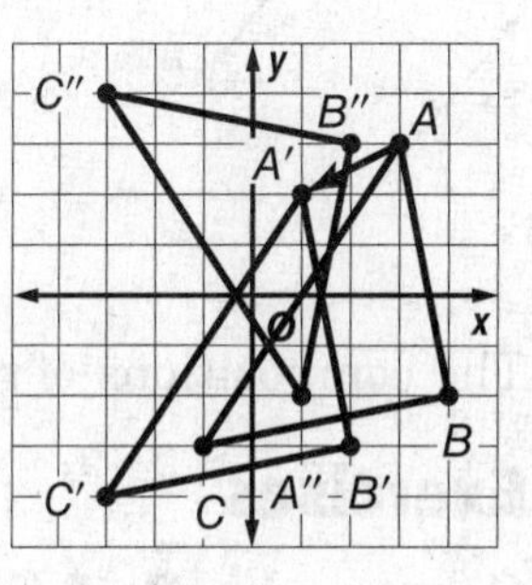

Exercises

Triangle *XYZ* has vertices *X*(6, 5), *Y*(7, –4) and *Z*(5, –5). Graph △*XYZ* and its image after the indicated glide reflection.

1. Translation: along ⟨1, 2⟩
Reflection: in *y*-axis

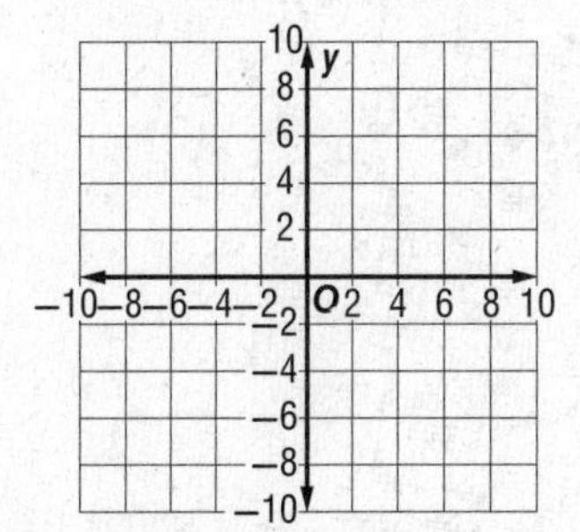

2. Translation: along ⟨–3, 4⟩
Reflection: in *x*-axis

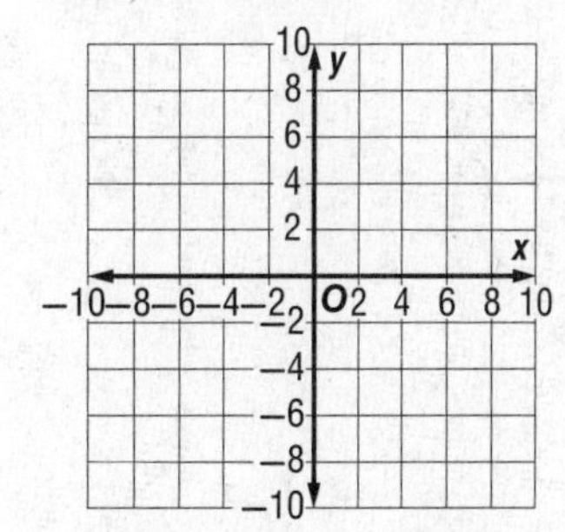

3. Translation: along ⟨2, 0⟩
Reflection: in $x = y$

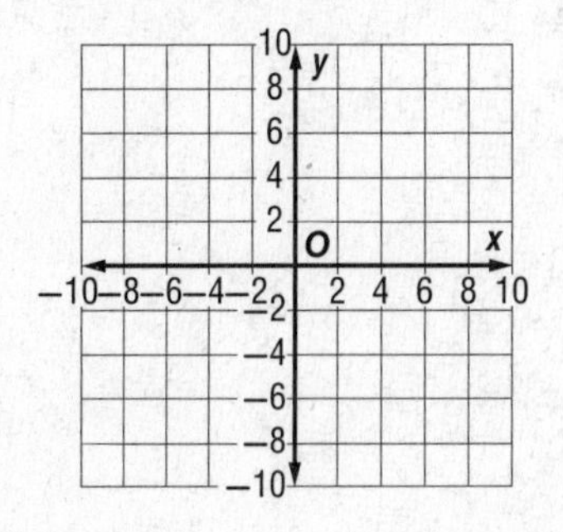

4. Translation: along ⟨–1, 3⟩
Reflection: in *x*-axis

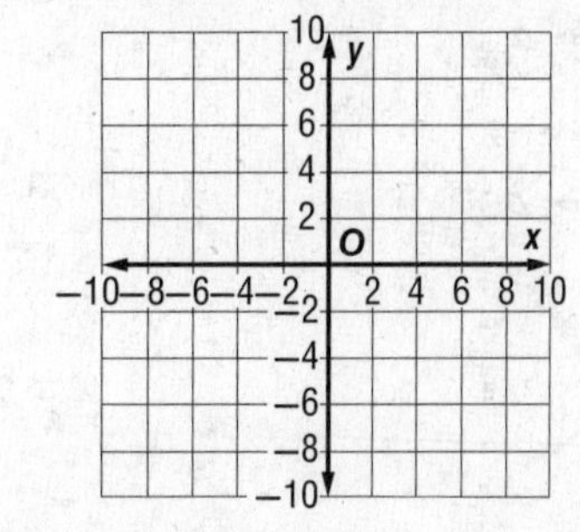

9-4 Study Guide and Intervention *(continued)*

Composition of Transformations

Compositions of Reflections The composition of two reflections in parallel lines is the same as a translation. The compositions of two reflections in intersecting lines is the same as a rotation.

Example **Copy and reflect figure A in line ℓ and then line m. Then describe a single transformation that maps A onto A''.**

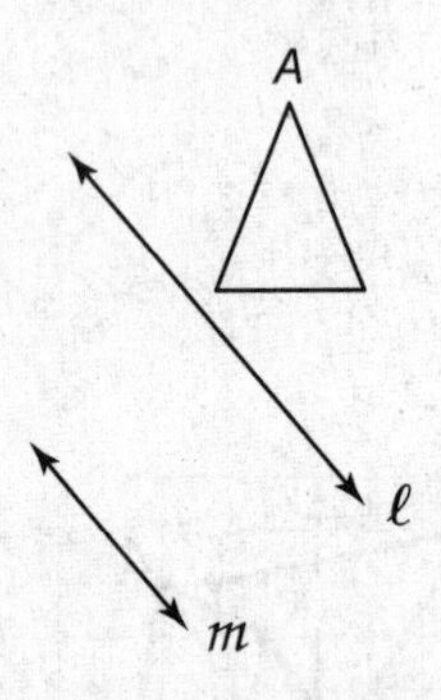

Step 1 Reflect A in line ℓ.

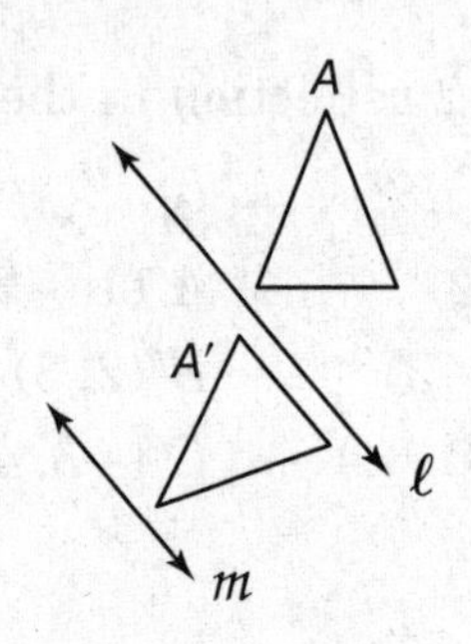

Step 2 Reflect A' in line m.

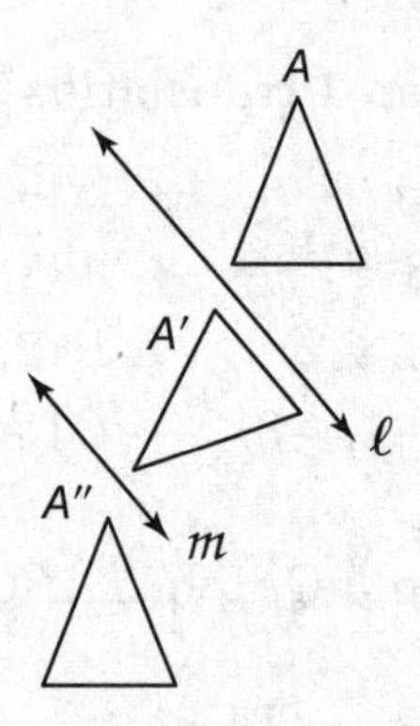

The compositions of two parallel lines is the same as a translation.

Exercises

Copy and reflect figure P in line a and then line b. Then describe a single transformation that maps P onto P''.

1.

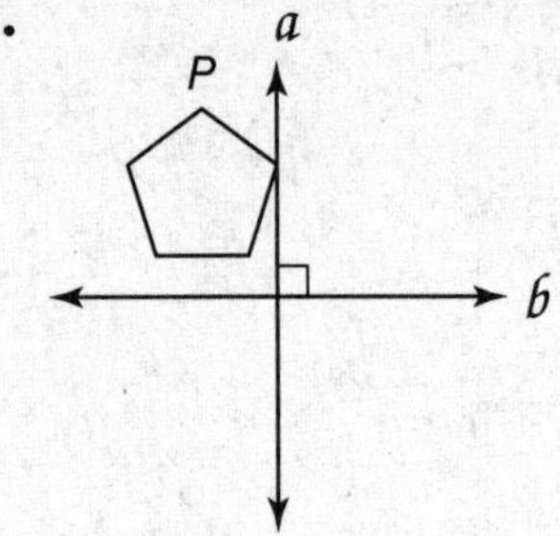

2.

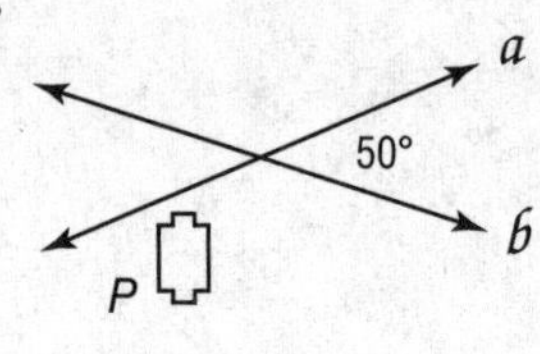

3.

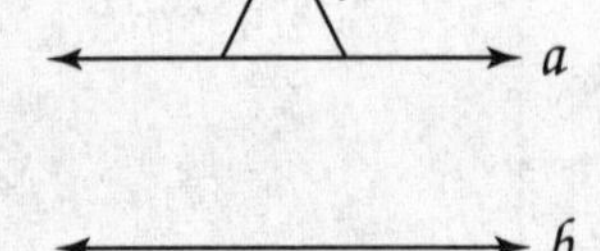

4.

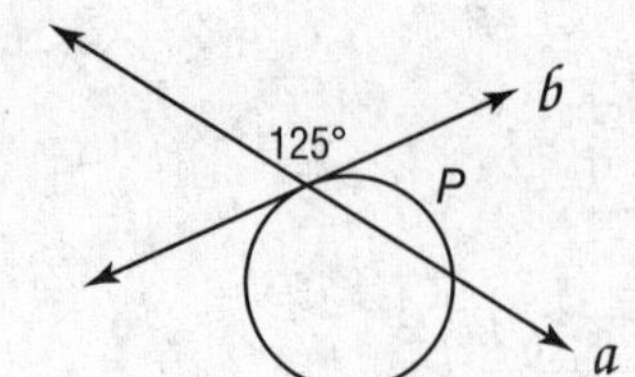

NAME ______________________ DATE __________ PERIOD __________

9-5 Study Guide and Intervention

Symmetry

Symmetry In Two-Dimensional Figures A two-dimensional figure has **line symmetry** if the figure can be mapped onto itself by a reflection in a line called the **line of symmetry.** A figure in a plane has **rotational symmetry** if the figure can be mapped onto itself by a rotation between 0° and 360° about the center of the figure, called the **center of symmetry.**

Example 1 **State whether the figure appears to have line symmetry. If so, draw all lines of symmetry, and state their number.**

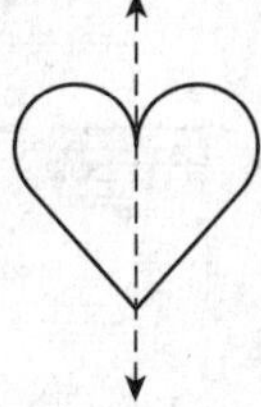

The heart has line symmetry. It has one line of symmetry.

Example 2 **State whether the figure appears to have rotational symmetry. If so, locate the center of symmetry and state the order and magnitude of symmetry.**

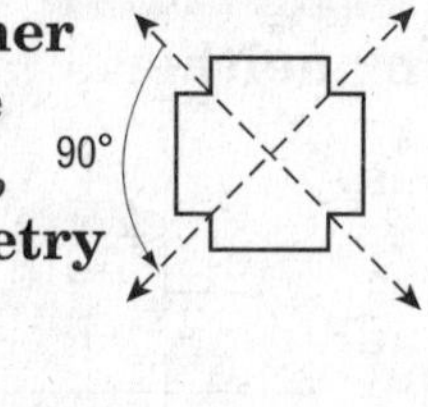

The figure has rotational symmetry. The figure has order 4 symmetry and magnitude of 360 ÷ 4 or 90°. The center of symmetry is the intersection of the diagonals.

Exercises

State whether the figure appears to have line symmetry. Write *yes* or *no*. If so, draw all lines of symmetry and state their number.

1.

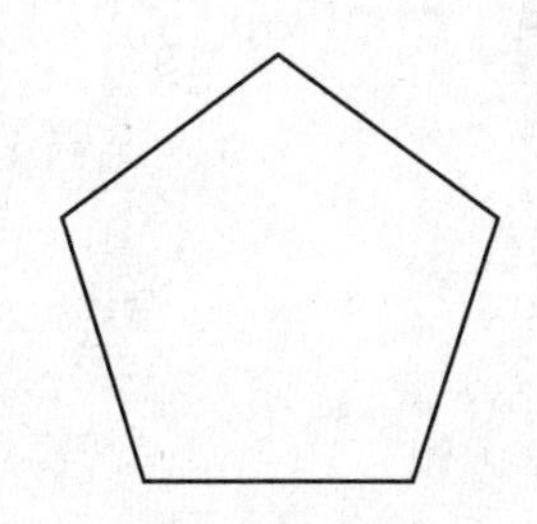

2.

3.

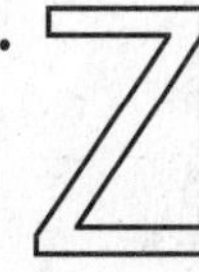

State whether the figure has rotational symmetry. Write *yes* or *no*. If so, locate the center of symmetry, and state the order and magnitude of symmetry.

4.

5.

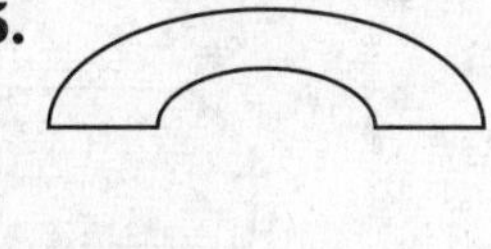

6.

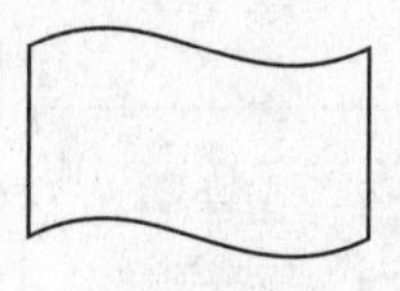

9-5 Study Guide and Intervention *(continued)*

Symmetry

Symmetry In Three-Dimensional Figures A three-dimensional figure has **plane symmetry** if the figure can be mapped onto itself by a reflection in a plane. A three-dimensional figure has **axis symmetry** if the figure can be mapped onto itself by a rotation between 0° and 360° in a line.

Example State whether the figure has *plane* symmetry, *axis* symmetry, *both*, or neither.

a.

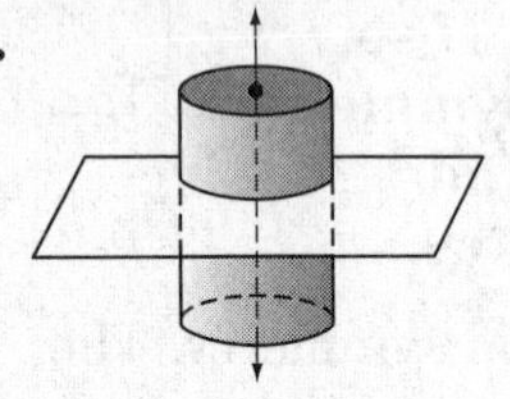

The cylinder has plane symmetry.
The cylinder has axis symmetry.

b.

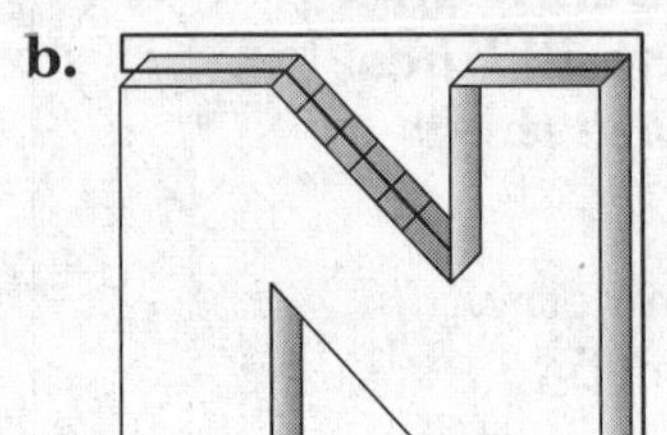

The letter N has plane symmetry.

Exercises

State whether the figure has *plane* symmetry, *axis* symmetry, *both*, or *neither*.

1.

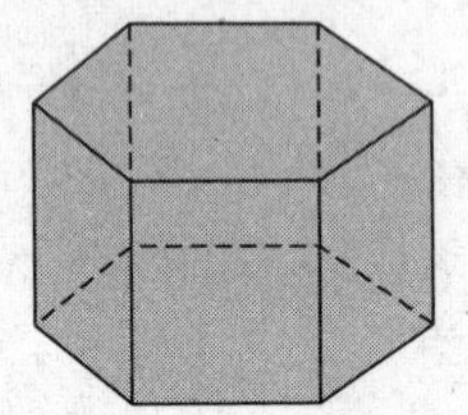

2.

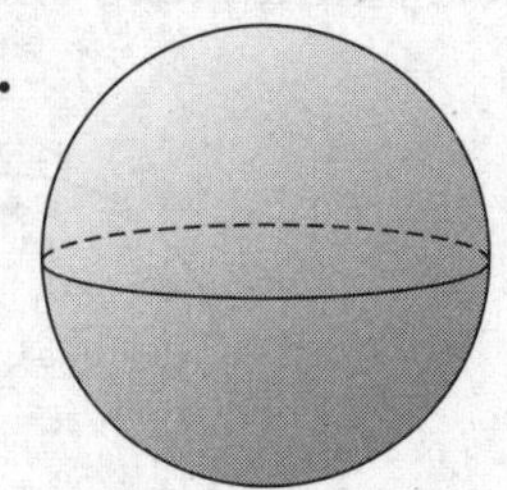

3.

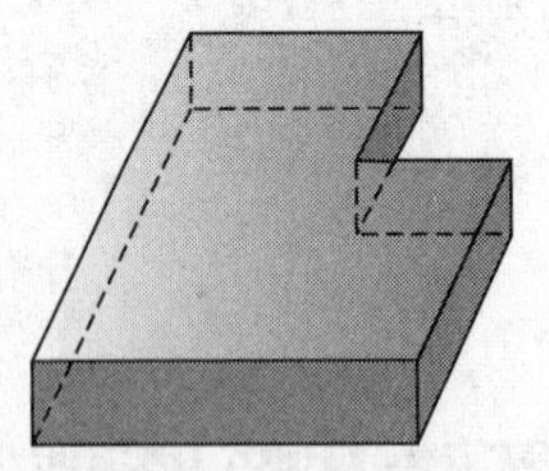

4.

5.

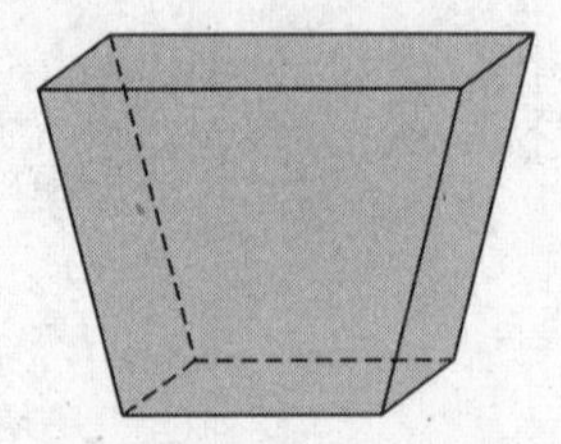

6.

NAME ______________________ DATE __________ PERIOD __________

9-6 Study Guide and Intervention

Dilations

Draw Dilations A dilation is a similarity transformation that enlarges or reduces a figure proportionally. Dilations are completed with respect to a center point and a scale factor.

Example **Draw the dilation image of $\triangle ABC$ with center O and $r = 2$.**

Draw $\overrightarrow{OA}$, $\overrightarrow{OB}$, and $\overrightarrow{OC}$. Label points A', B', and C' so that $OA' = 2(OA)$, $OB' = 2(OB)$ and $OC' = 2(OC)$. connect the points to draw $\triangle A'B'C'$. $\triangle A'B'C'$ is a dilation of $\triangle ABC$.

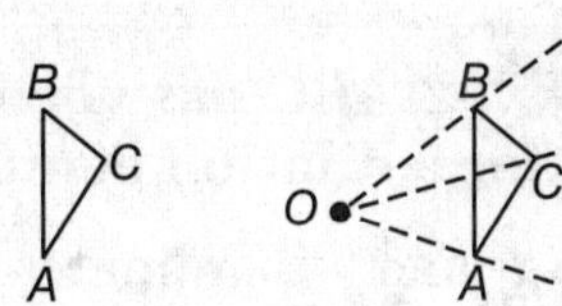

Exercises

Use a ruler to draw the image of the figure under a dilation with center S and the scale factor r indicated.

1. $r = 2$

2. $r = \frac{1}{2}$

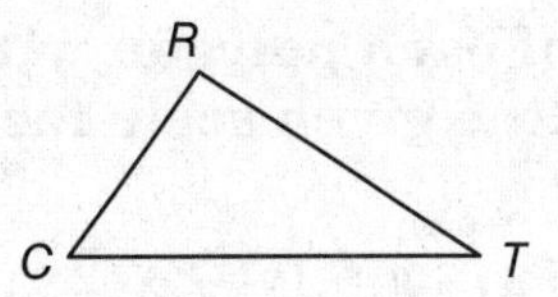

3. $r = 1$

4. $r = 3$

5. $r = \frac{2}{3}$

6. $r = 1$

9-6 Study Guide and Intervention *(continued)*

Dilations

Dilations In The Coordinate Plane To find the coordinates of an image after a dilation centered at the origin, multiply the x- and y-coordinates of each point on the preimage by the scale factor of the dilation, r.

$$(x, y) \rightarrow (rx, ry)$$

Example **$\triangle ABC$ has vertices $A(-2, -2)$, $B(1, -1)$, and $C(2, 0)$. Find the image of $\triangle ABC$ after a dilation centered at the origin with a scale factor of 2.**

Multiply the x- and y-coordinates of each vertex by the scale factor, 2.

(x, y)	$(2x, 2y)$
$A(-2, -2)$,	$A'(-4, -4)$
$B(1, -1)$	$B'(2, -2)$
$C(2, 0)$	$C'(4, 0)$

Graph $\triangle ABC$ and its image $\triangle A'B'C'$

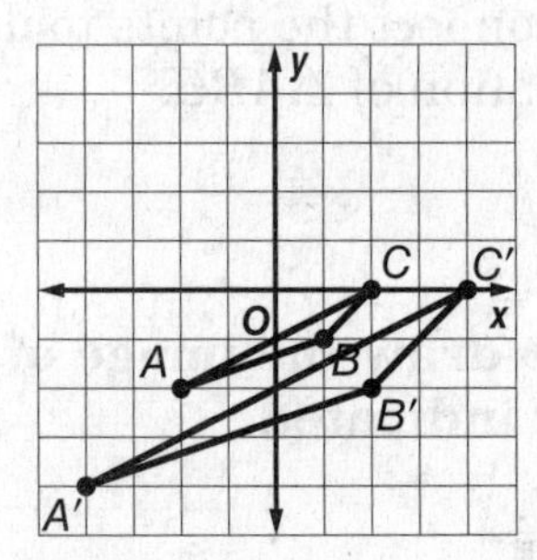

Exercises

Graph the image of each polygon with the given vertices after a dilation centered at the origin with the given scale factor.

1. $E(-2, -2)$, $F(-2, 4)$, $G(2, 4)$, $H(2, -2)$; $r = 0.5$

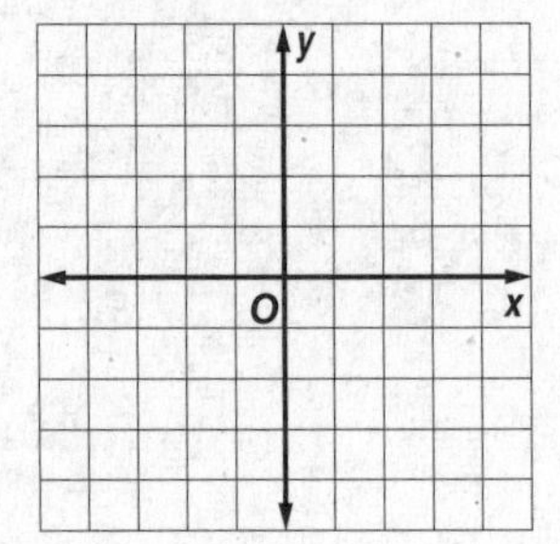

2. $A(0, 0)$, $B(3, 3)$, $C(6, 3)$, $D(6, -3)$, $E(3, -3)$; $r = \frac{1}{3}$

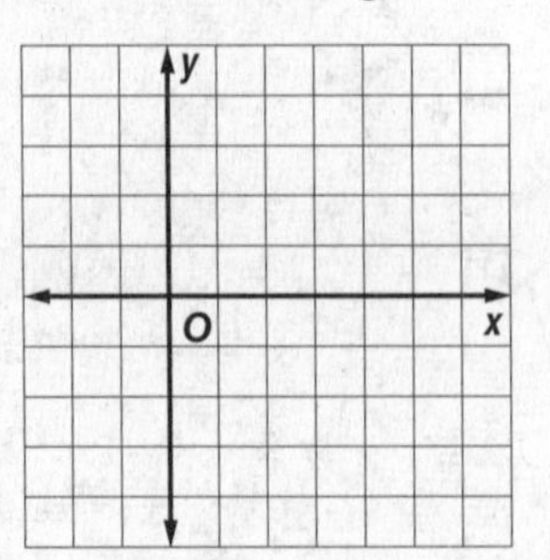

3. $A(-2, -2)$, $B(-1, 2)$, $C(2, 1)$; $r = 2$

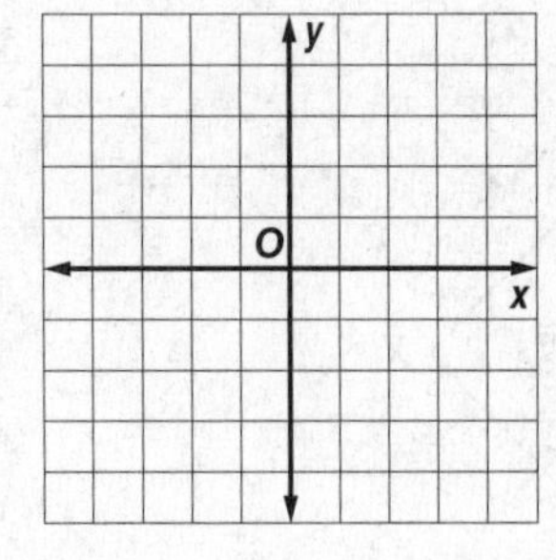

4. $A(2, 2)$, $B(3, 4)$, $C(5, 2)$; $r = 2.5$

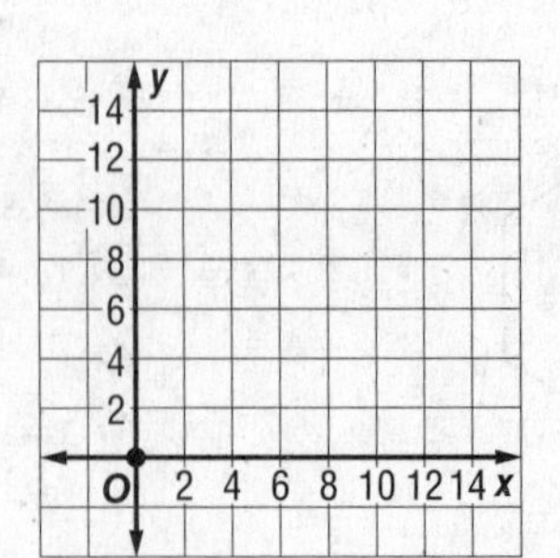

NAME ______________________ DATE __________ PERIOD __________

10-1 Study Guide and Intervention

Circles and Circumference

Segments in Circles A **circle** consists of all points in a plane that are a given distance, called the **radius**, from a given point called the **center**. A segment or line can intersect a circle in several ways.

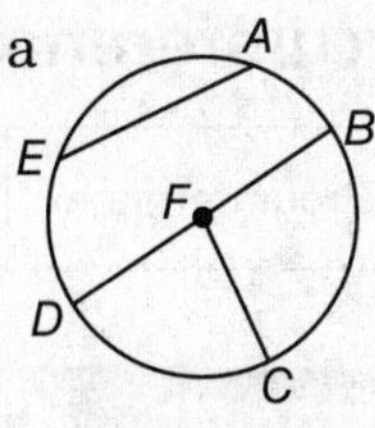

chord: $\overline{AE}$, $\overline{BD}$

radius: $\overline{FB}$, $\overline{FC}$, $\overline{FD}$

diameter: $\overline{BD}$

- A segment with endpoints that are at the center and on the circle is a **radius**.
- A segment with endpoints on the circle is a **chord**.

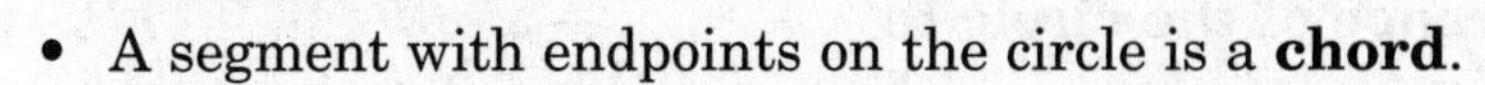

- A chord that passes through the circle's center and made up of collinear radii is a **diameter**.

For a circle that has radius r and diameter d, the following are true

$r = \frac{d}{2}$ $\qquad r = \frac{1}{2}d$ $\qquad d = 2r$

Example

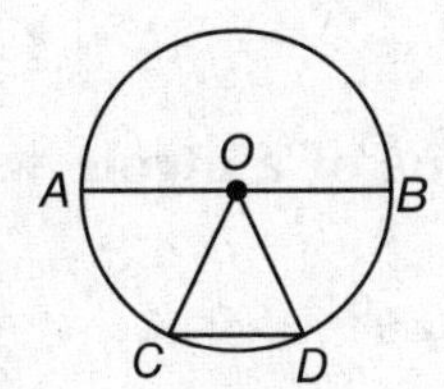

a. Name the circle.
The name of the circle is $\odot O$.

b. Name radii of the circle.
$\overline{AO}$, $\overline{BO}$, $\overline{CO}$, and $\overline{DO}$ are radii.

c. Name chords of the circle.
$\overline{AB}$ and $\overline{CD}$ are chords.

Exercises

For Exercises 1–7, refer to

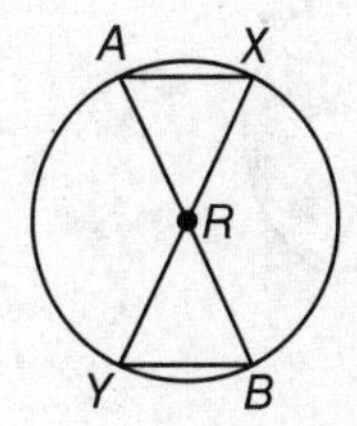

1. Name the circle.

2. Name radii of the circle.

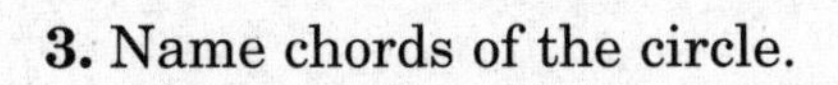

3. Name chords of the circle.

4. Name diameters of the circle.

5. If $AB = 18$ millimeters, find AR.

6. If $RY = 10$ inches, find AR and AB.

7. Is $\overline{AB} \cong \overline{XY}$? Explain.

NAME ______________________________ DATE ____________ PERIOD ________

10-1 Study Guide and Intervention *(continued)*

Circles and Circumference

Circumference The **circumference** of a circle is the distance around the circle.

Circumference	For a circumference of C units and a diameter of d units or a radius or r units, $C = \pi d$ or $C = 2\pi r$

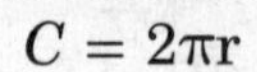

Find the circumference of the circle to the nearest hundredth.

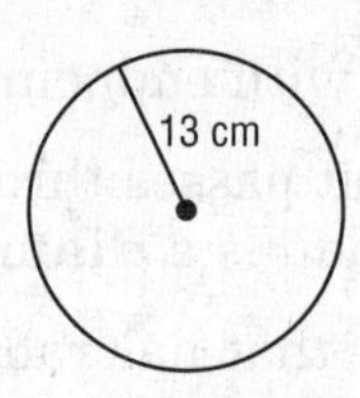

$C = 2\pi r$	Circumference formula
$= 2\pi(13)$	$r = 13$
$= 26\pi$	Simplify.
≈ 81.68	Use a calculator.

The circumference is 26π or about 81.68 centimeters.

Exercises

Find the diameter and radius of a circle with the given circumference. Round to the nearest hundredth.

1. $C = 40$ in.

2. $C = 256$ ft

3. $C = 15.62$ m

4. $C = 9$ cm

5. $C = 79.5$ yd

6. $C = 204.16$ m

Find the exact circumference of each circle using the given inscribed or circumscribed polygon.

7.

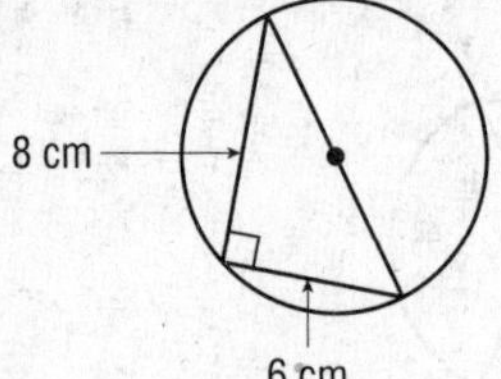

8.

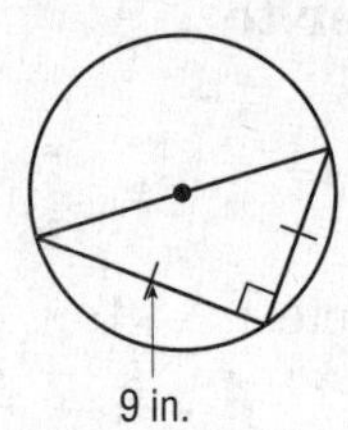

9.

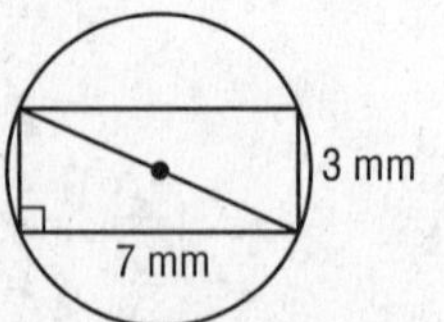

10.

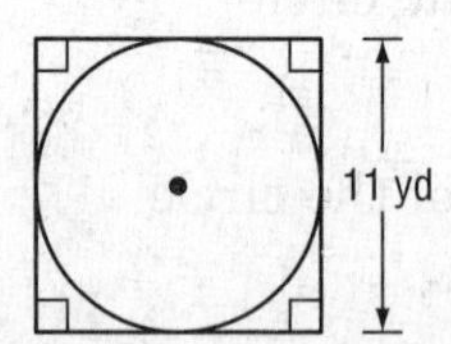

11.

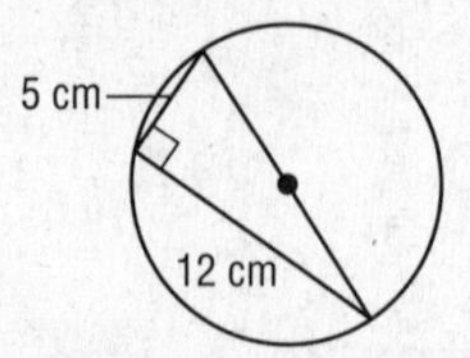

12.

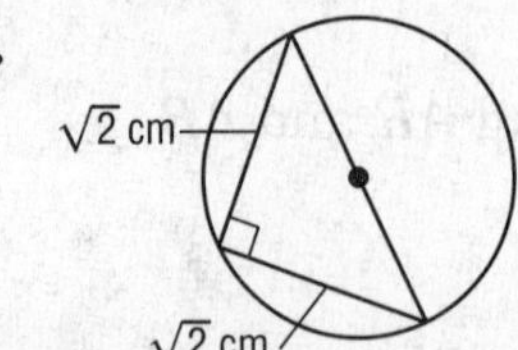

NAME ______________________ DATE __________ PERIOD __________

10-2 Study Guide and Intervention

Measuring Angles and Arcs

Angles and Arcs A **central angle** is an angle whose vertex is at the center of a circle and whose sides are radii. A central angle separates a circle into two arcs, a **major arc** and a **minor arc**.

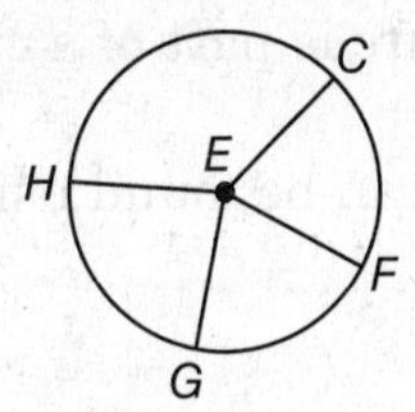

$\widehat{GF}$ is a minor arc.

$\widehat{CHG}$ is a major arc.

$\angle GEF$ is a central angle.

Here are some properties of central angles and arcs.

- The sum of the measures of the central angles of a circle with no interior points in common is 360. — $m\angle HEC + m\angle CEF + m\angle FEG + m\angle GEH = 360$
- The measure of a minor arc is less than 180 and equal to the measure of its central angle. — $m\widehat{CF} = m\angle CEF$
- The measure of a major arc is 360 minus the measure of the minor arc. — $m\widehat{CGF} = 360 - m\widehat{CF}$
- The measure of a semicircle is 180.
- Two minor arcs are congruent if and only if their corresponding central angles are congruent. — $\widehat{CF} \cong \widehat{FG}$ if and only if $\angle CEF \cong \angle FEG$.
- The measure of an arc formed by two adjacent arcs is the sum of the measures of the two arcs. **(Arc Addition Postulate)** — $m\widehat{CF} + m\widehat{FG} = m\widehat{CG}$

Example $\overline{AC}$ **is a diameter of** $\odot R$**. Find** $m\widehat{AB}$ **and** $m\widehat{ACB}$**.**

$\angle ARB$ is a central angle and $m\angle ARB = 42$, so $m\widehat{AB} = 42$.
Thus $m\widehat{ACB} = 360 - 42$ or 318.

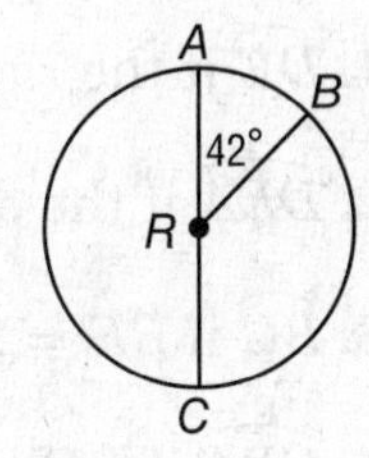

Exercises

Find the value of *x*.

1.

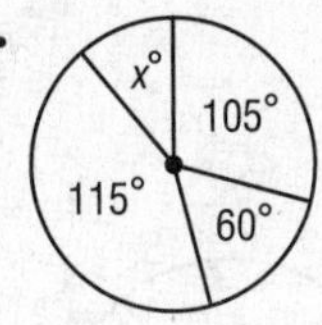

2.

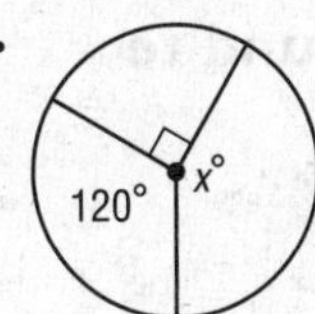

$\overline{BD}$ **and** $\overline{AC}$ **are diameters of** $\odot O$**. Identify each arc as a *major arc*, *minor arc*, or *semicircle* of the circle. Then find its measure.**

3. $m\widehat{BA}$

4. $m\widehat{BC}$

5. $m\widehat{CD}$

6. $m\widehat{ACB}$

7. $m\widehat{BCD}$

8. $m\widehat{AD}$

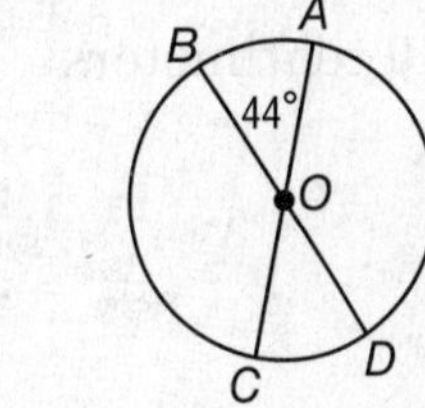

10-2 Study Guide and Intervention *(continued)*

Measuring Angles and Arcs

Arc Length An arc is part of a circle and its length is a part of the circumference of the circle.

The length of arc ℓ can be found using the following equations:
$\ell = \frac{x}{360} \cdot 2\pi r$

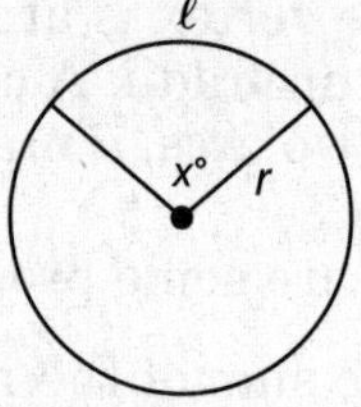

Example **Find the length of $\widehat{AB}$. Round to the nearest hundredth.**

The length of arc l, can be found using the following equation: $\widehat{AB} = \frac{x}{360} \cdot 2\pi r$

$\widehat{AB} = \frac{x}{360} \cdot 2\pi r$ Arc Length Equation

$\widehat{AB} = \frac{135}{360} \cdot 2\pi(8)$ Substitution

$\widehat{AB} \approx 18.85$ in. Use a calculator.

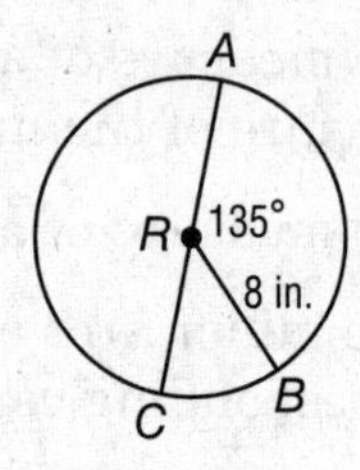

Exercises

Use ⊙O to find the length of each arc. Round to the nearest hundredth.

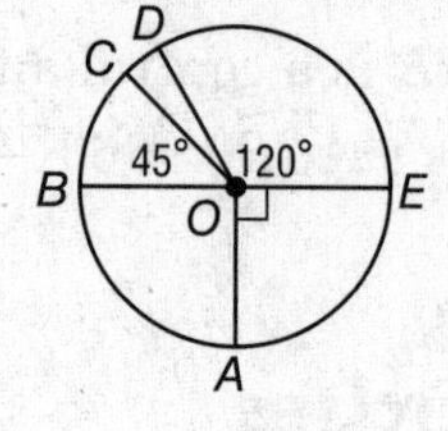

1. $\widehat{DE}$ if the radius is 2 meters

2. $\widehat{DEA}$ if the diameter is 7 inches

3. $\widehat{BC}$ if $BE = 24$ feet

4. $\widehat{CBA}$ if $DO = 3$ millimeters

Use ⊙P to find the length of each arc. Round to the nearest hundredth.

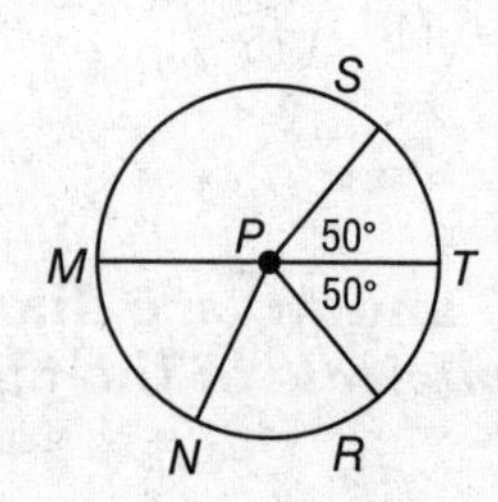

5. $\widehat{RT}$, if $MT = 7$ yards

6. $\widehat{MR}$, if $PR = 13$ feet

7. $\widehat{MST}$, if $MP = 2$ inches

8. $\widehat{MRS}$, if $PS = 10$ centimeters

NAME ______________________ DATE ____________ PERIOD ____________

10-3 Study Guide and Intervention

Arcs and Chords

Arcs and Chords Points on a circle determine both chords and arcs. Several properties are related to points on a circle. In a circle or in congruent circles, two minor arcs are congruent if and only if their corresponding chords are congruent.

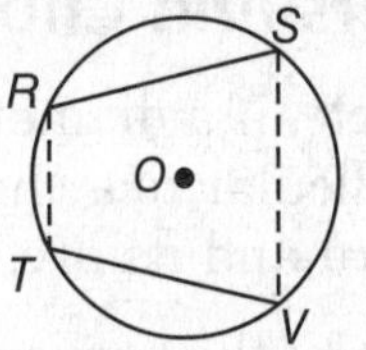

$\overparen{RS} \cong \overparen{TV}$ if and only if $\overline{RS} \cong \overline{TV}$.

Example **In $\odot K$, $\overparen{AB} \cong \overparen{CD}$. Find AB.**

$\overparen{AB}$ and $\overparen{CD}$ are congruent arcs, so the corresponding chords $\overline{AB}$ and $\overline{CD}$ are congruent.

$AB = CD$	Definition of congruent segments
$8x = 2x + 3$	Substitution
$x = \frac{1}{2}$	Simplify.

So, $AB = 8\left(\frac{1}{2}\right)$ or 4.

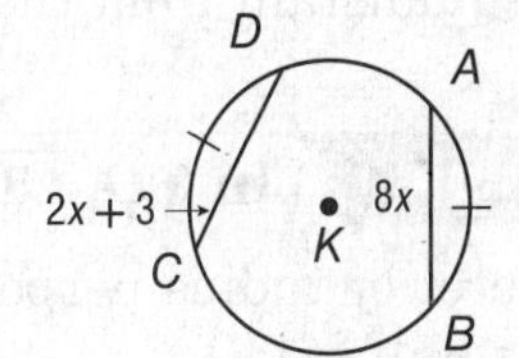

Exercises

ALGEBRA Find the value of x in each circle.

1.

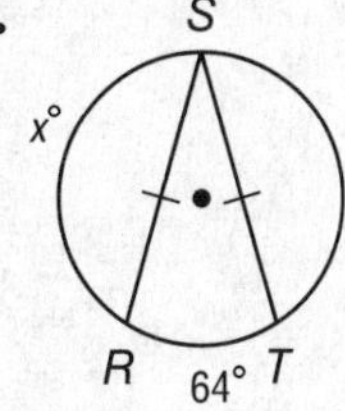

2.

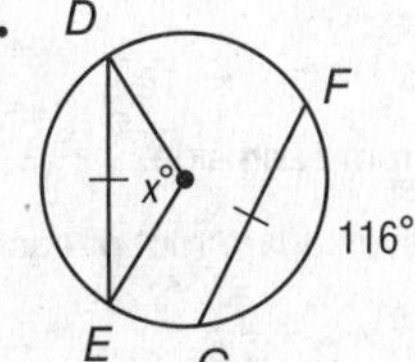

3.

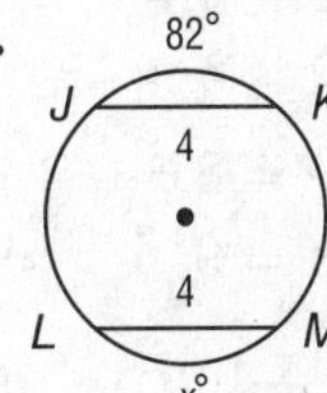

4.

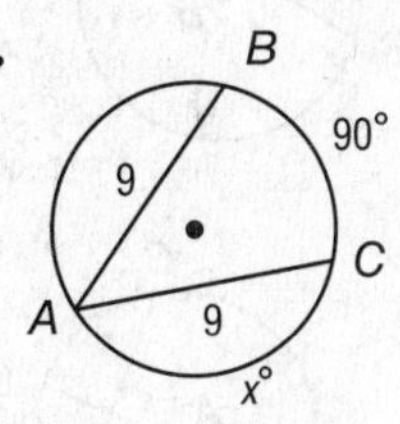

5.

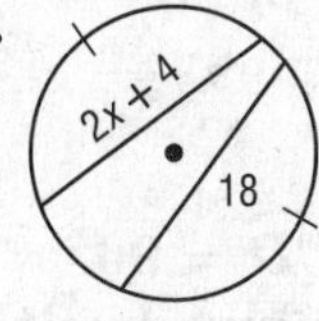

6.

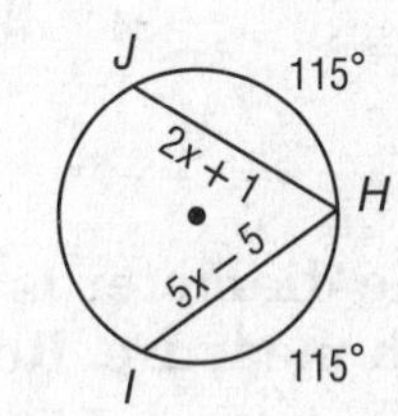

7.

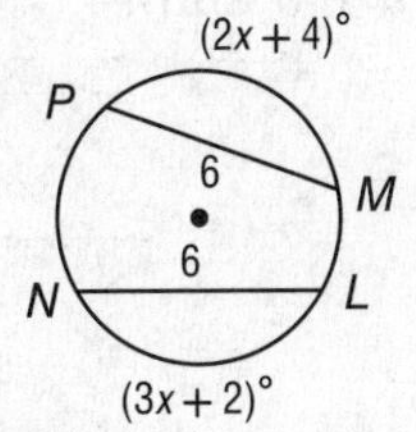

8. $\odot M \cong \odot P$

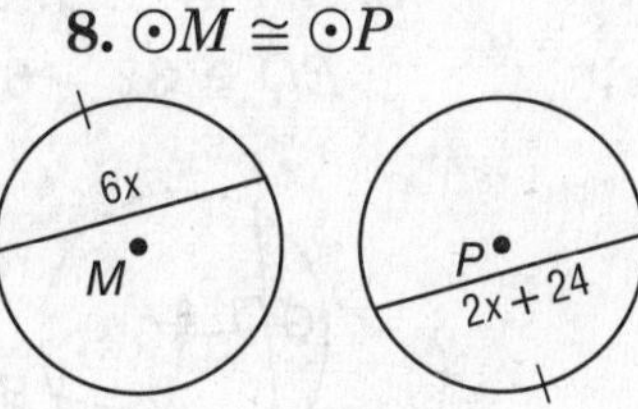

9. $\odot V \cong \odot W$

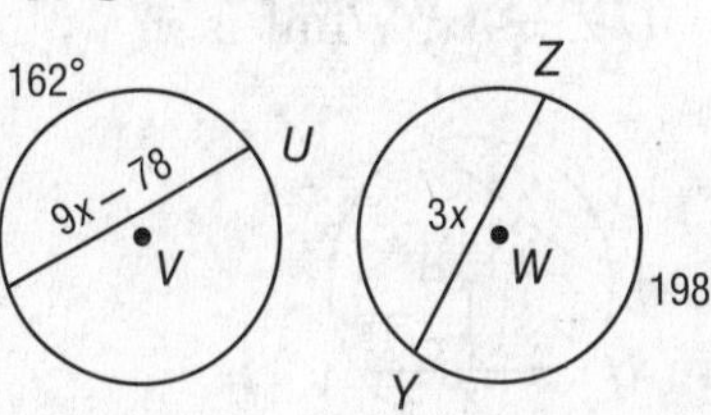

NAME ______________________________ DATE ____________ PERIOD ____________

10-3 Study Guide and Intervention *(continued)*

Arcs and Chords

Diameters and Chords

- In a circle, if a diameter (or radius) is perpendicular to a chord, then it bisects the chord and its arc.
- In a circle, the perpendicular bisector of a chord is the diameter (or radius).
- In a circle or in congruent circles, two chords are congruent if and only if they are equidistant from the center.

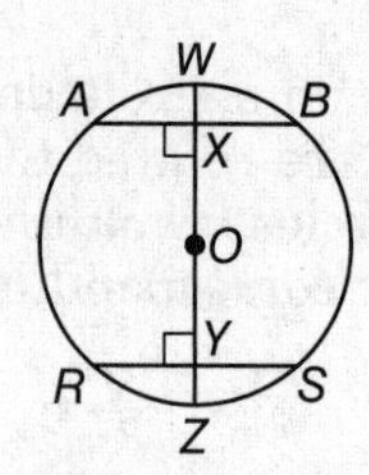

If $\overline{WZ} \perp \overline{AB}$, then $\overline{AX} \cong \overline{XB}$ and $\widehat{AW} \cong \widehat{WB}$.

If $OX = OY$, then $\overline{AB} \cong \overline{RS}$.

If $\overline{AB} \cong \overline{RS}$, then $\overline{AB}$ and $\overline{RS}$ are equidistant from point O.

Example **In $\odot O$, $\overline{CD} \perp \overline{OE}$, $OD = 15$, and $CD = 24$. Find OE.**

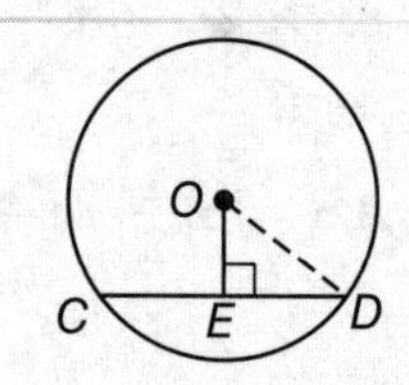

A diameter or radius perpendicular to a chord bisects the chord, so ED is half of CD.

$ED = \frac{1}{2}(24)$

$= 12$

Use the Pythagorean Theorem to find x in $\triangle OED$.

$(OE)^2 + (ED)^2 = (OD)^2$	Pythagorean Theorem
$(OE)^2 + 12^2 = 15^2$	Substitution
$(OE)^2 + 144 = 225$	Simplify.
$(OE)^2 = 81$	Subtract 144 from each side.
$OE = 9$	Take the positive square root of each side.

Exercises

In $\odot P$, the radius is 13 and $RS = 24$. Find each measure. Round to the nearest hundredth.

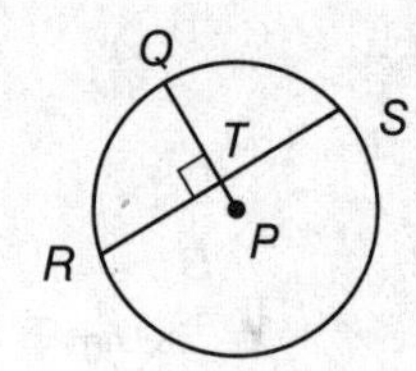

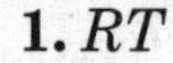

1. RT 2. PT 3. TQ

In $\odot A$, the diameter is 12, $CD = 8$, and $m\widehat{CD} = 90$. Find each measure. Round to the nearest hundredth.

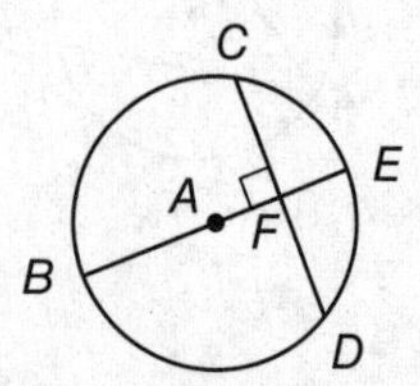

4. $m\widehat{DE}$ 5. FD 6. AF

7. In $\odot R$, $TS = 21$ and $UV = 3x$. What is x?

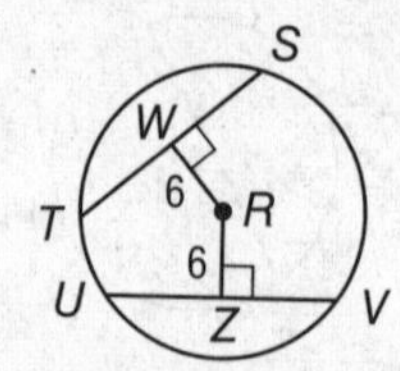

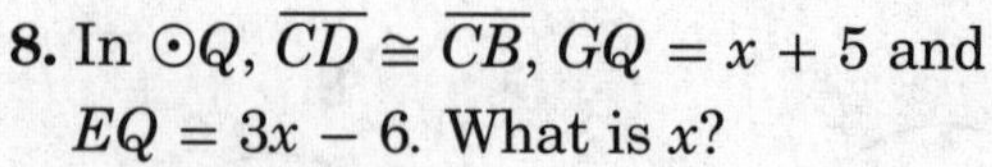
8. In $\odot Q$, $\overline{CD} \cong \overline{CB}$, $GQ = x + 5$ and $EQ = 3x - 6$. What is x?

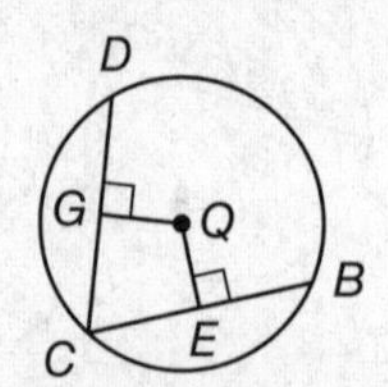

NAME ______________________________ DATE ____________ PERIOD ____________

10-4 Study Guide and Intervention

Inscribed Angles

Inscribed Angles An **inscribed angle** is an angle whose vertex is on a circle and whose sides contain chords of the circle. In ⊙G, minor arc $\widehat{DF}$ is the **intercepted arc** for inscribed angle $\angle DEF$.

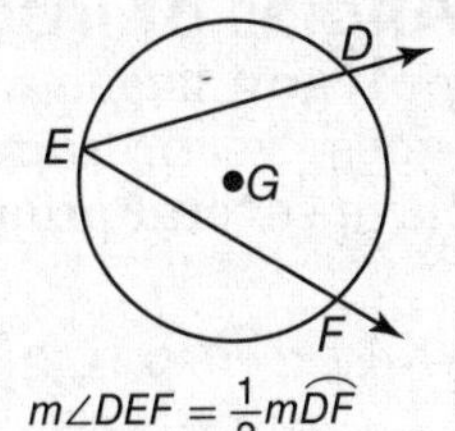

$m\angle DEF = \frac{1}{2}m\widehat{DF}$

Inscribed Angle Theorem	If an angle is inscribed in a circle, then the measure of the angle equals one-half the measure of its intercepted arc.

If two inscribed angles intercept the same arc or congruent arcs, then the angles are congruent.

Example **In ⊙G above, $m\widehat{DF} = 90$. Find $m\angle DEF$.**

$\angle DEF$ is an inscribed angle so its measure is half of the intercepted arc.

$m\angle DEF = \frac{1}{2}m\widehat{DF}$

$= \frac{1}{2}(90)$ or 45

Exercises

Find each measure.

1. $m\widehat{AC}$

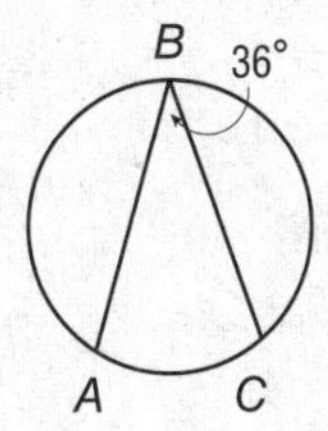

2. $m\angle N$

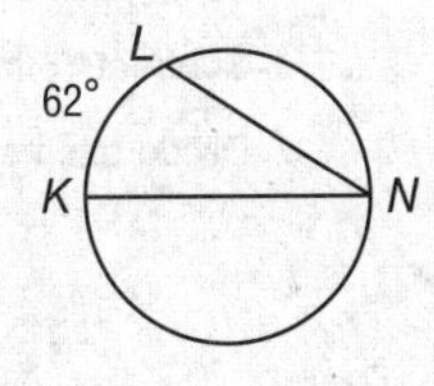

3. $m\widehat{QSR}$

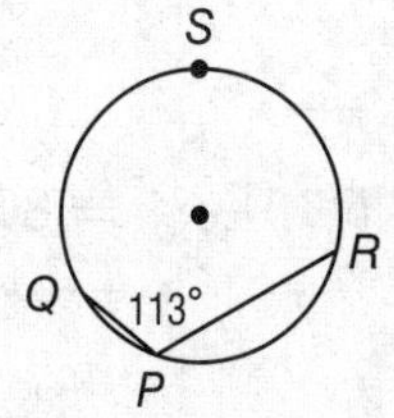

ALGEBRA Find each measure.

4. $m\angle U$

5. $m\angle T$

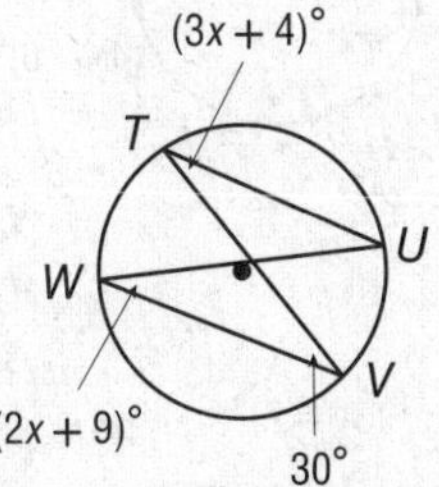

6. $m\angle A$

7. $m\angle C$

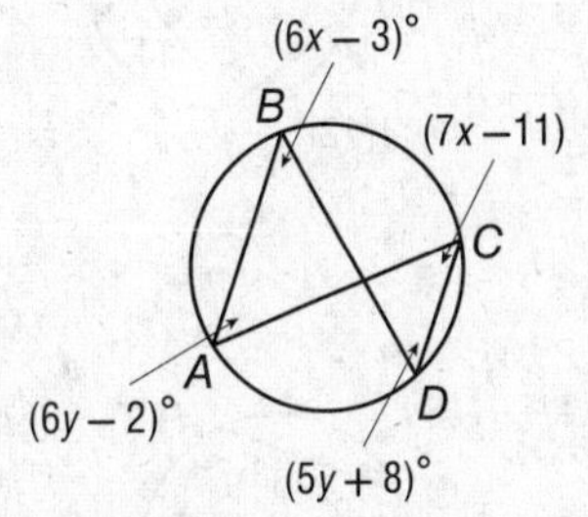

10-4 Study Guide and Intervention *(continued)*

Inscribed Angles

Angles of Inscribed Polygons An **inscribed polygon** is one whose sides are chords of a circle and whose vertices are points on the circle. Inscribed polygons have several properties.

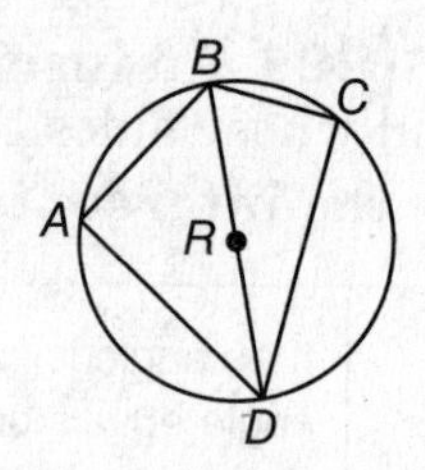

- An inscribed angle of a triangle intercepts a diameter or semicircle if and only if the angle is a right angle.

  If $\widehat{BCD}$ is a semicircle, then $m\angle BCD = 90$.

- If a quadrilateral is inscribed in a circle, then its opposite angles are supplementary.

  For inscribed quadrilateral $ABCD$, $m\angle A + m\angle C = 180$ and $m\angle ABC + m\angle ADC = 180$.

Example **Find $m\angle K$.**

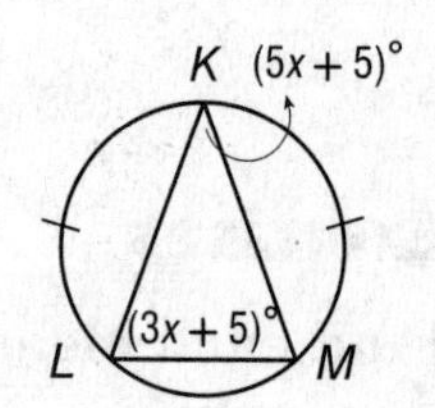

$\widehat{KL} \cong \widehat{KM}$, so $KL = KM$. The triangle is an isosceles triangle, therefore $m\angle L = m\angle M = 3x + 5$.

$m\angle L + m\angle M + m\angle K = 180$	Angle Sum Theorem
$(3x + 5) + (3x + 5) + (5x + 5) = 180$	Substitution
$11x + 15 = 180$	Simplify.
$11x = 165$	Subtract 15 from each side.
$x = 15$	Divide each side by 11.

So, $m\angle K = 5(15) + 5 = 80$.

Exercises

ALGEBRA Find each measure.

1. x

2. $m\angle W$

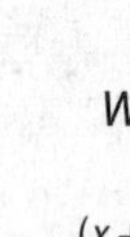
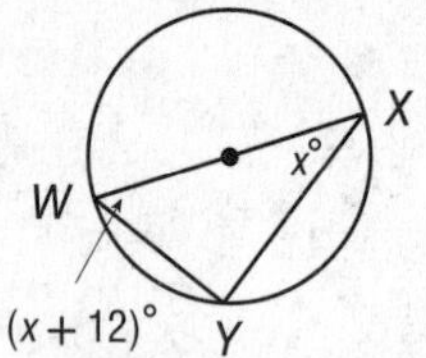

3. x

4. $m\angle T$

(2x)°
S
R
T
(4x – 6)°

5. $m\angle R$

6. $m\angle S$

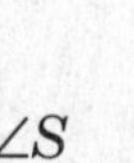

R
Q
(3x)°
110°
(2x)°
S
75°
P

7. $m\angle W$

8. $m\angle X$

(3y – 7)°
X
Y
(2x)°
(x + 18)°
(3y + 1)°
W
Z

NAME ______________________ DATE ____________ PERIOD ____________

10-5 Study Guide and Intervention

Tangents

Tangents A **tangent** to a circle intersects the circle in exactly one point, called the **point of tangency**. There are important relationships involving tangents. A **common tangent** is a line, ray, or segment that is tangent to two circles in the same plane.

- A line is tangent to a circle if and only if it is perpendicular to a radius at a point of tangency.
- If two segments from the same exterior point are tangent to a circle, then they are congruent.

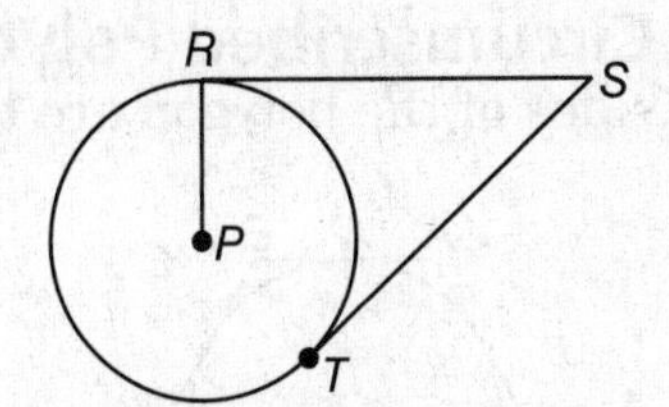

If $\overline{RS} \perp \overline{RP}$, then $\overline{SR}$ is tangent to $\odot P$. If $\overline{SR}$ is tangent to $\odot P$, then $\overline{RS} \perp \overline{RP}$.
If $\overline{SR}$ and $\overline{ST}$ are tangent to $\odot P$, then $\overline{SR} \cong \overline{ST}$.

Example $\overline{AB}$ **is tangent to** $\odot C$**. Find** x**.**

AB is tangent to $\odot C$, so $\overline{AB}$ is perpendicular to radius $\overline{BC}$. $\overline{CD}$ is a radius, so $CD = 8$ and $AC = 9 + 8$ or 17. Use the Pythagorean Theorem with right $\triangle ABC$.

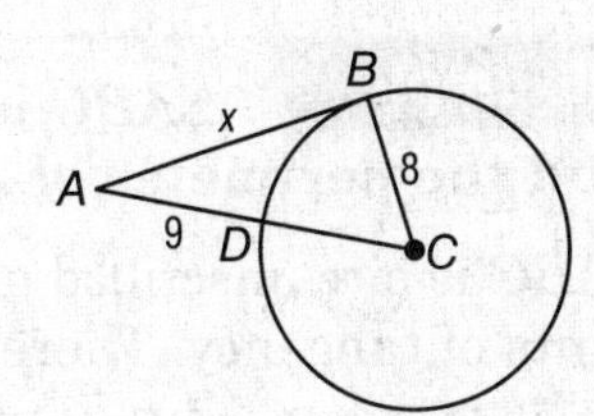

$(AB)^2 + (BC)^2 = (AC)^2$	Pythagorean Theorem
$x^2 + 8^2 = 17^2$	Substitution
$x^2 + 64 = 289$	Simplify.
$x^2 = 225$	Subtract 64 from each side.
$x = 15$	Take the positive square root of each side.

Exercises

Find x. Assume that segments that appear to be tangent are tangent.

1.

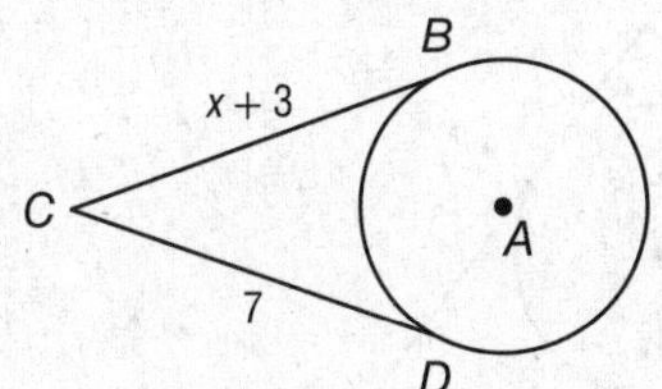

2.

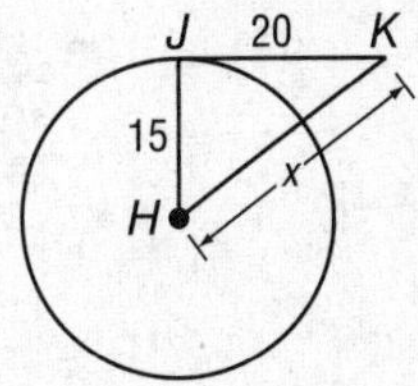

3.

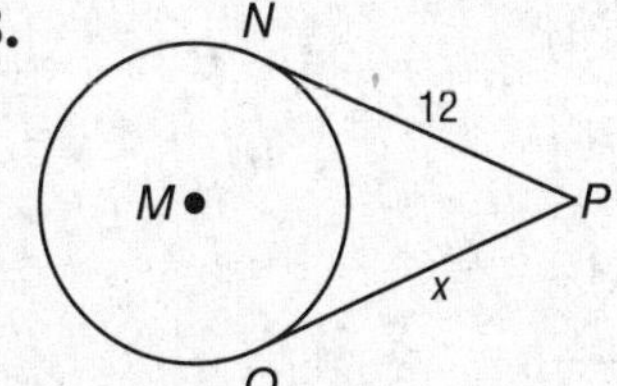

4.

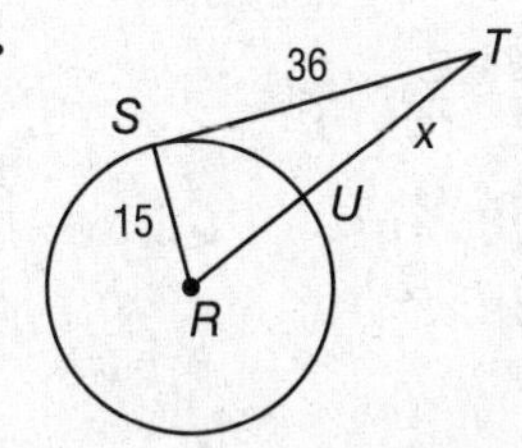

5.

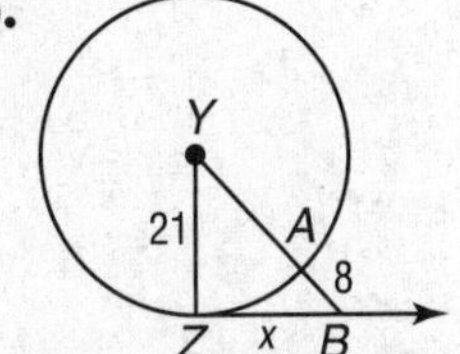

6.

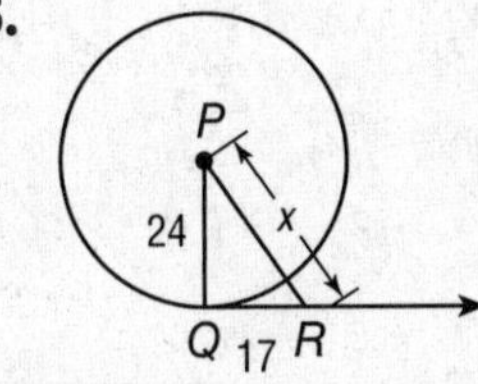

10-5 Study Guide and Intervention *(continued)*

Tangents

Circumscribed Polygons When a polygon is circumscribed about a circle, all of the sides of the polygon are tangent to the circle.

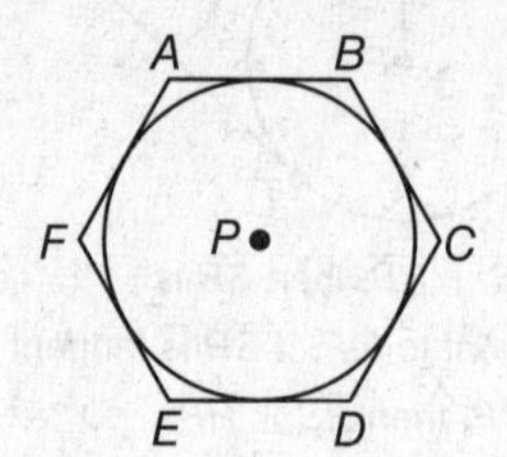

Hexagon *ABCDEF* is circumscribed about ⊙*P*.
$\overline{AB}$, $\overline{BC}$, $\overline{CD}$, $\overline{DE}$, $\overline{EF}$, and $\overline{FA}$ are tangent to ⊙*P*.

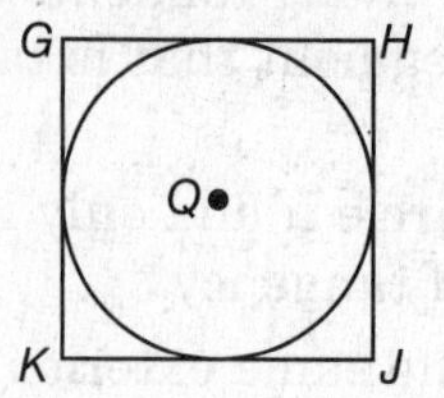

Square *GHJK* is circumscribed about ⊙*Q*.
$\overline{GH}$, $\overline{JH}$, $\overline{JK}$, and $\overline{KG}$ are tangent to ⊙*Q*.

Example **$\triangle ABC$ is circumscribed about ⊙*O*. Find the perimeter of $\triangle ABC$.**

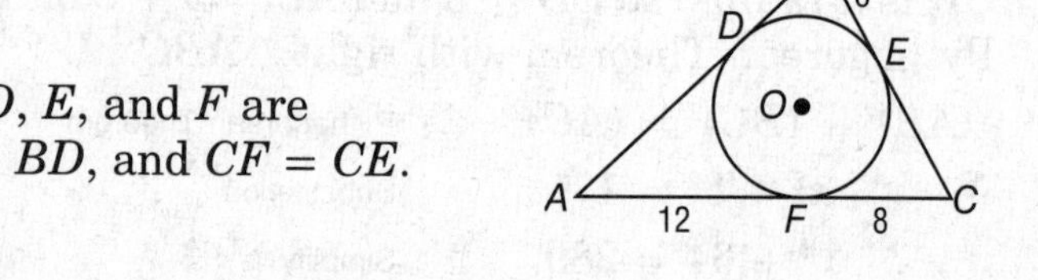

$\triangle ABC$ is circumscribed about ⊙*O*, so points *D*, *E*, and *F* are points of tangency. Therefore $AD = AF$, $BE = BD$, and $CF = CE$.

$$P = AD + AF + BE + BD + CF + CE$$
$$= 12 + 12 + 6 + 6 + 8 + 8$$
$$= 52$$

The perimeter is 52 units.

Exercises

For each figure, find *x*. Then find the perimeter.

1.

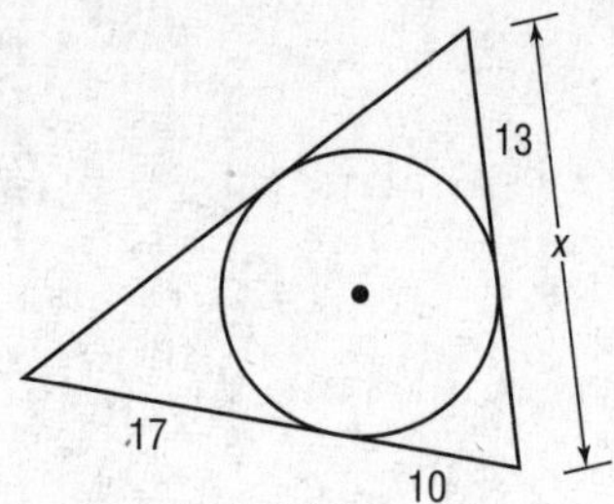

2.

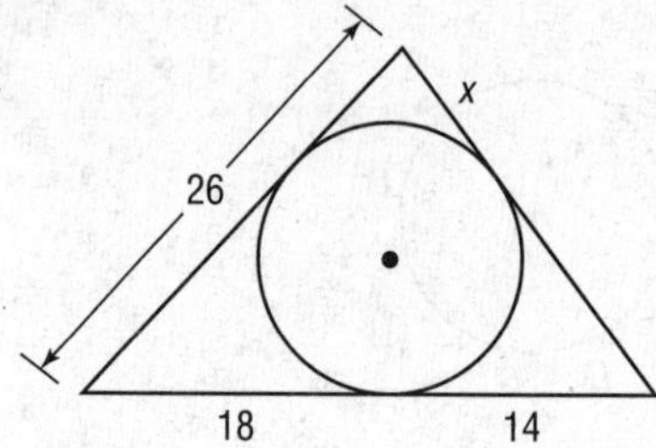

3.

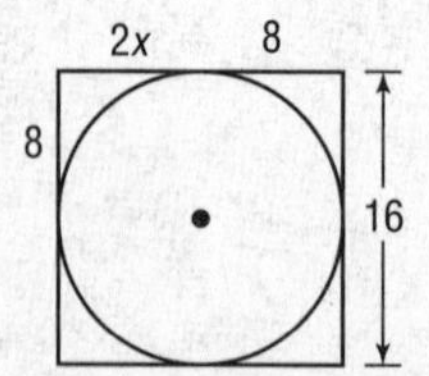

4.

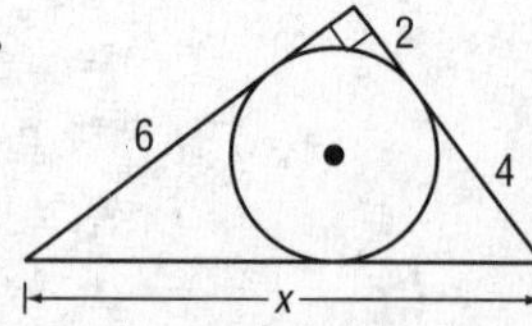

5.

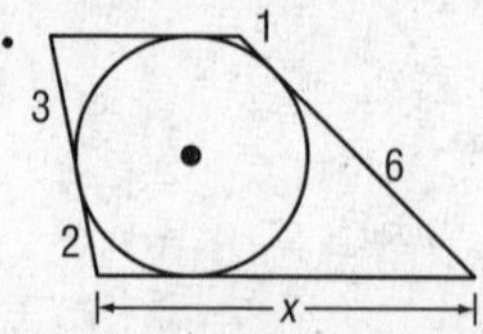

6.

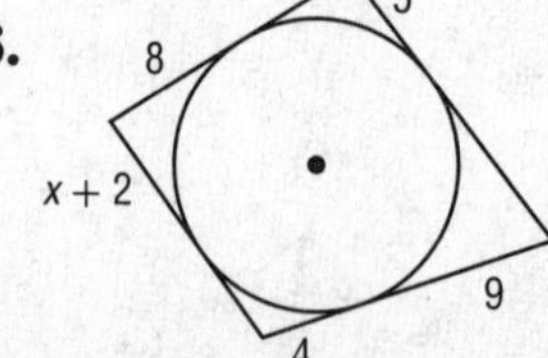

NAME ______________________ DATE ____________ PERIOD ________

10-6 Study Guide and Intervention

Secants, Tangents, and Angle Measures

Intersections On or Inside a Circle A line that intersects a circle in exactly two points is called a **secant**. The measures of angles formed by secants and tangents are related to intercepted arcs.

- If two secants or chords intersect in the interior of a circle, then the measure of the angle formed is one half the sum of the measure of the arcs intercepted by the angle and its vertical angle.

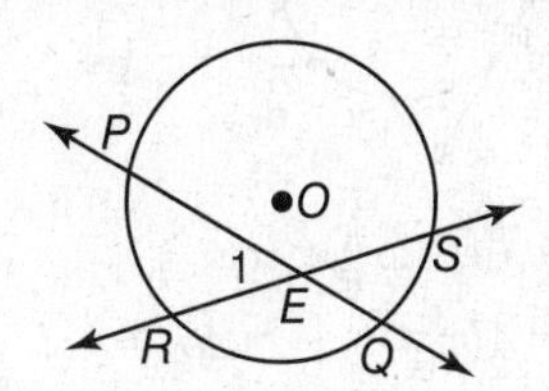

$m\angle 1 = \frac{1}{2}(m\widehat{PR} + m\widehat{QS})$

- If a secant (or chord) and a tangent intersect at the point of tangency, then the measure of each angle formed is one half the measure of its intercepted arc.

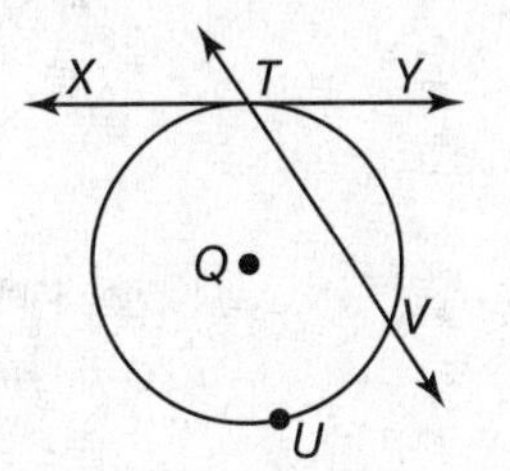

$m\angle XTV = \frac{1}{2}m\widehat{TUV}$

$m\angle YTV = \frac{1}{2}m\widehat{TV}$

Example 1 **Find x.**

The two chords intersect inside the circle, so x is equal to one half the sum of the measures of the arcs intercepted by the angle and its vertical angle.

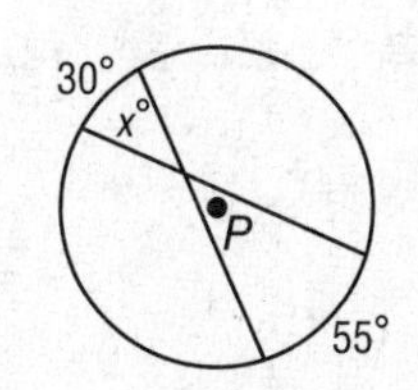

$x = \frac{1}{2}(30 + 55)$

$= \frac{1}{2}(85)$

$= 42.5$

Example 2 **Find y.**

The chord and the tangent intersect at the point of tangency, so the measure of the angle is one half the measure of its intercepted arc.

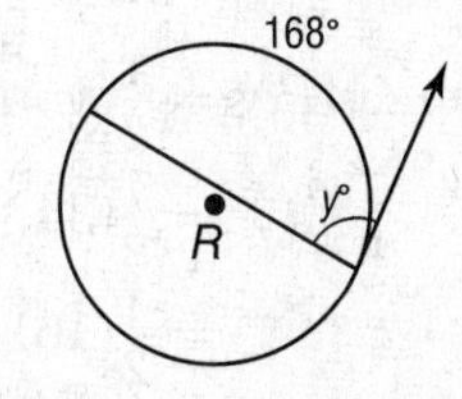

$y = \frac{1}{2}(168)$

$= 84$

Exercises

Find each measure. Assume that segments that appear to be tangent are tangent.

1. $m\angle 1$

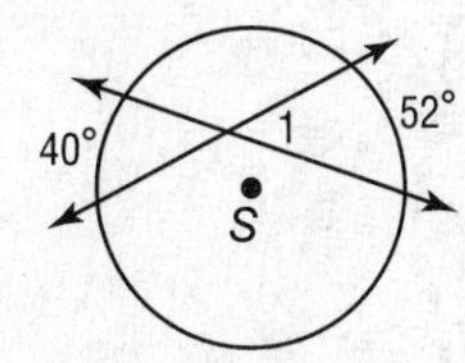

2. $m\widehat{GH}$

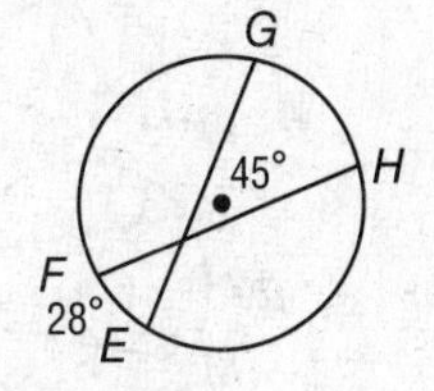

3. $m\angle 3$

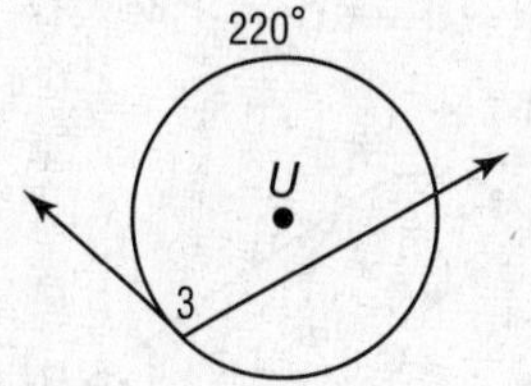

4. $m\widehat{RT}$

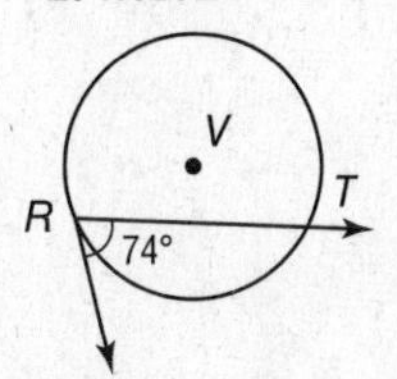

5. $m\angle 5$

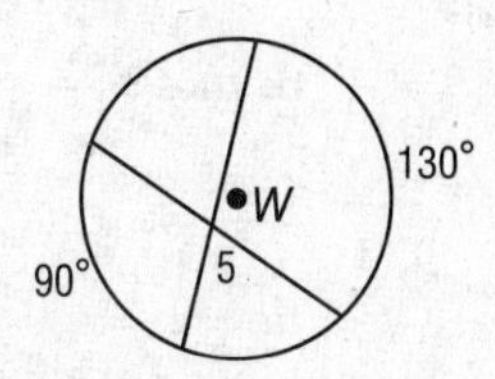

6. $m\angle 6$

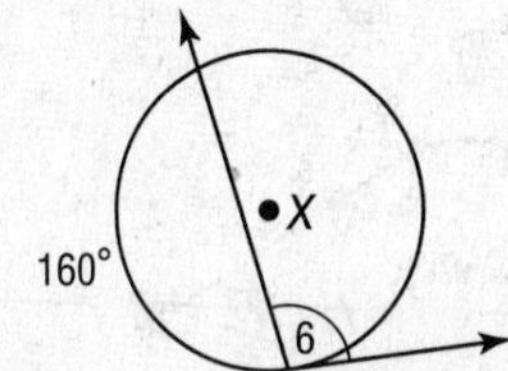

10-6 Study Guide and Intervention *(continued)*

Secants, Tangents, and Angle Measures

Intersections Outside a Circle If secants and tangents intersect outside a circle, they form an angle whose measure is related to the intercepted arcs.

If two secants, a secant and a tangent, or two tangents intersect in the exterior of a circle, then the measure of the angle formed is one half the difference of the measures of the intercepted arcs.

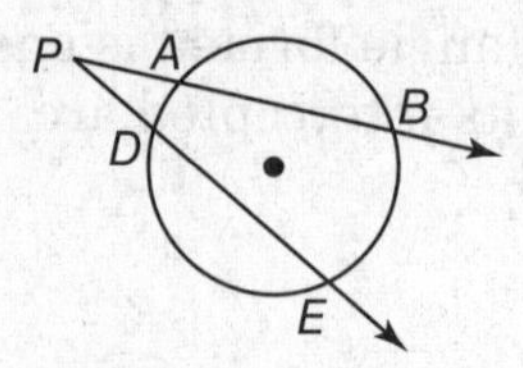

$\overrightarrow{PB}$ and $\overrightarrow{PE}$ are secants.

$m\angle P = \frac{1}{2}(m\widehat{BE} - m\widehat{AD})$

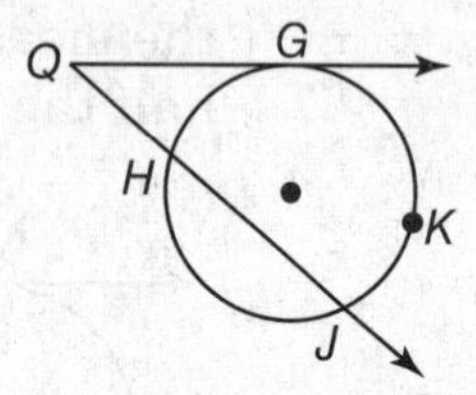

$\overrightarrow{QG}$ is a tangent. $\overrightarrow{QJ}$ is a secant.

$m\angle Q = \frac{1}{2}(m\widehat{GKJ} - m\widehat{GH})$

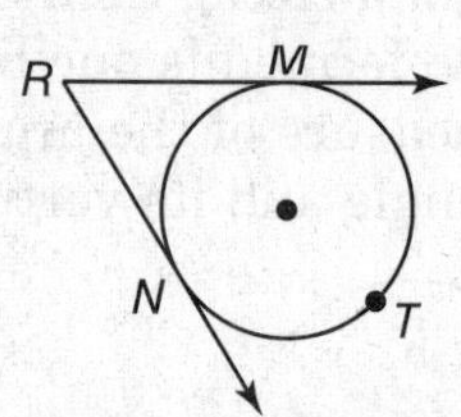

$\overrightarrow{RM}$ and $\overrightarrow{RN}$ are tangents.

$m\angle R = \frac{1}{2}(m\widehat{MTN} - m\widehat{MN})$

Example **Find $m\angle MPN$.**

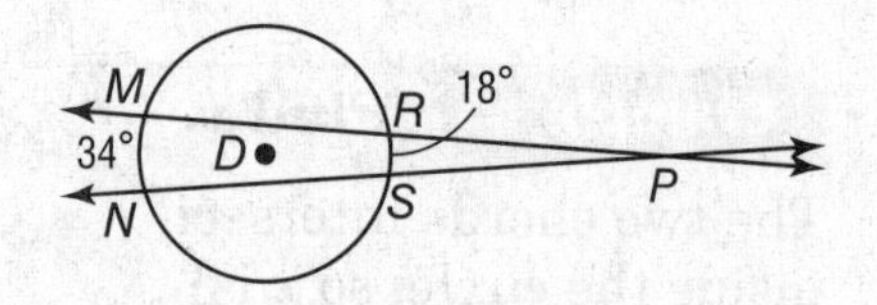

$\angle MPN$ is formed by two secants that intersect in the exterior of a circle.

$$m\angle MPN = \frac{1}{2}(m\widehat{MN} - m\widehat{RS})$$
$$= \frac{1}{2}(34 - 18)$$
$$= \frac{1}{2}(16) \text{ or } 8$$

The measure of the angle is 8.

Exercises

Find each measure. Assume that segments that appear to be tangent are tangent.

1. $m\angle 1$

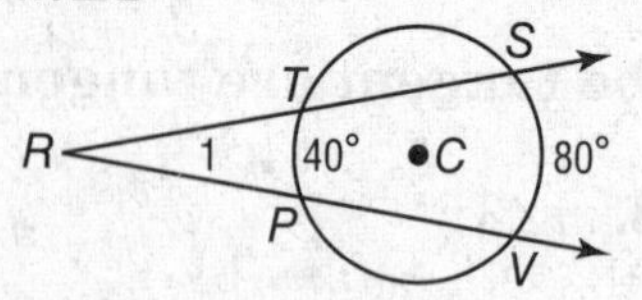

2. $m\angle 2$

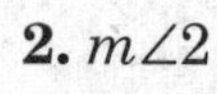
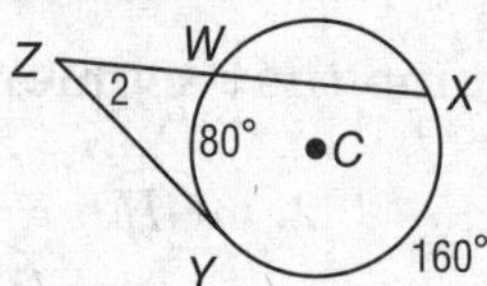

3. $m\angle 3$

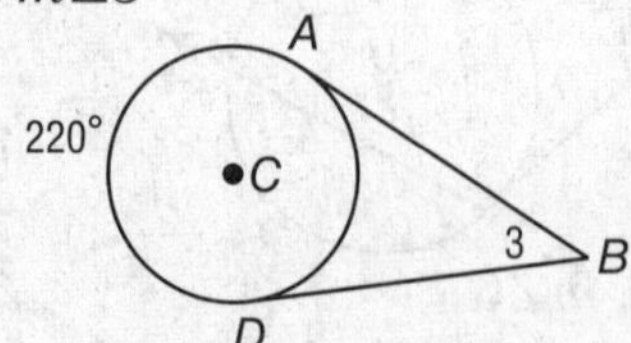

4. $m\widehat{JP}$

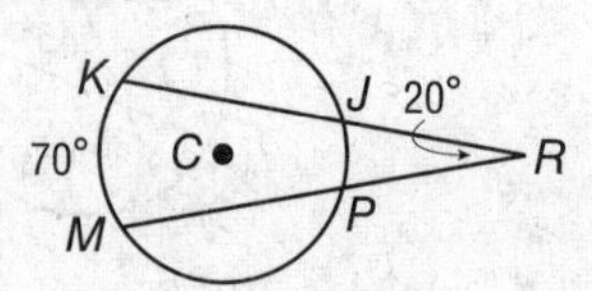

5. $m\widehat{LN}$

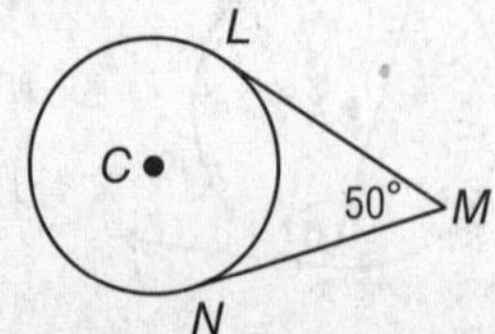

6. $m\angle V$

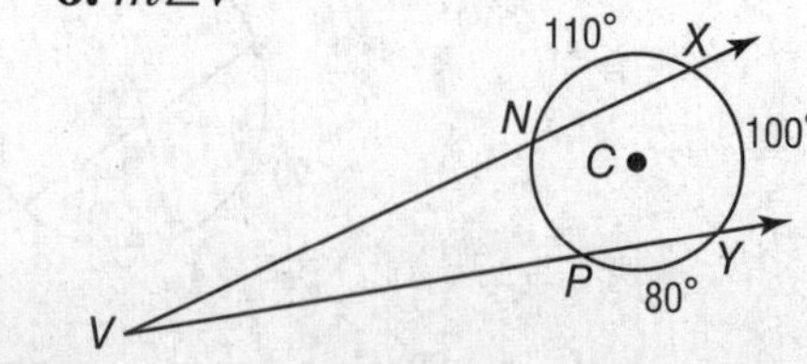

NAME ______________________ DATE ____________ PERIOD ____________

10-7 Study Guide and Intervention

Special Segments in a Circle

Segments Intersecting Inside a Circle If two chords intersect in a circle, then the products of the lengths of the chord segments are equal.

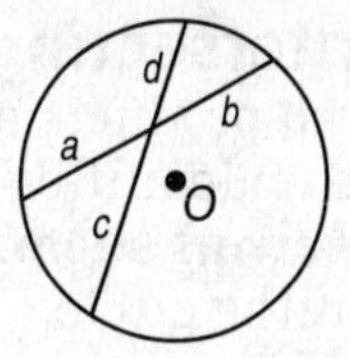

$a \cdot b = c \cdot d$

Example **Find x.**

The two chords intersect inside the circle, so the products $AB \cdot BC$ and $EB \cdot BD$ are equal.

$AB \cdot BC = EB \cdot BD$

$6 \cdot x = 8 \cdot 3$ Substitution

$6x = 24$ Multiply.

$x = 4$ Divide each side by 6.

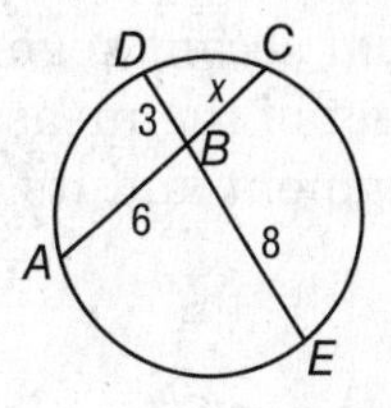

$AB \cdot BC = EB \cdot BD$

Exercises

Find x. Assume that segments that appear to be tangent are tangent. Round to the nearest tenth.

1.

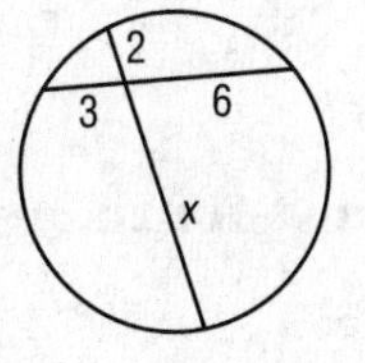

2.

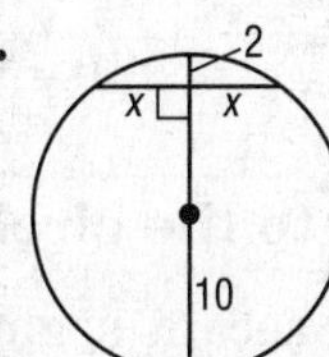

3.

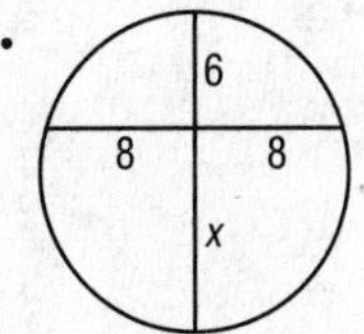

4.

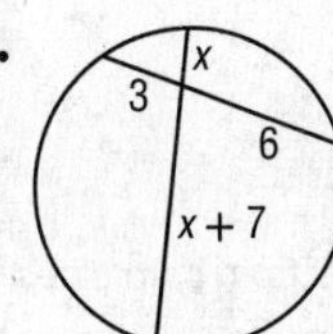

5.

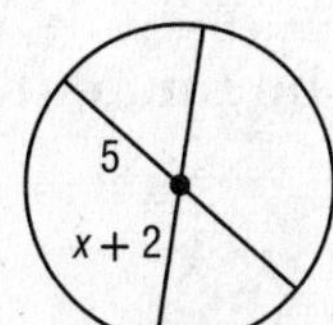

6.

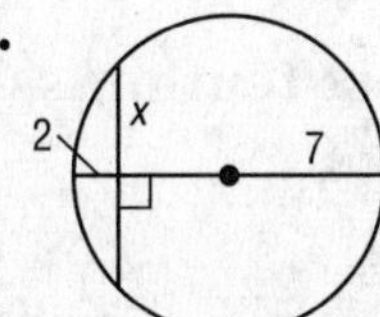

7.

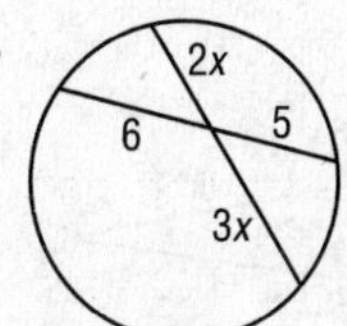

8.

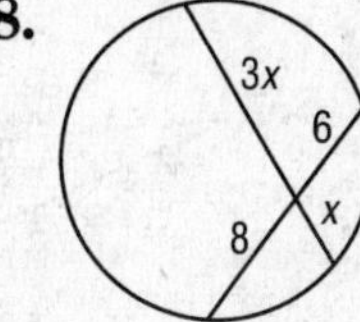

NAME ______________________________ DATE ____________ PERIOD ____________

10-7 Study Guide and Intervention *(continued)*

Special Segments in a Circle

Segments Intersecting Outside a Circle If secants and tangents intersect outside a circle, then two products are equal. A **secant segment** is a segment of a secant line that has exactly one endpoint on the circle. A secant segment that lies in the exterior of the circle is called an **external secant segment**. A **tangent segment** is a segment of a tangent with one endpoint on the circle.

- If two secants are drawn to a circle from an exterior point, then the product of the measures of one secant segment and its external secant segment is equal to the product of the measures of the other secant segment and its external secant segment.

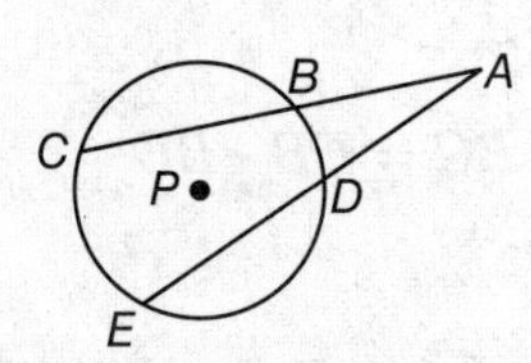

$\overline{AC}$ and $\overline{AE}$ are secant segments.

$\overline{AB}$ and $\overline{AD}$ are external secant segments.

$AC \cdot AB = AE \cdot AD$

- If a tangent segment and a secant segment are drawn to a circle from an exterior point, then the square of the measure of the tangent segment is equal to the product of the measures of the secant segment and its external secant segment.

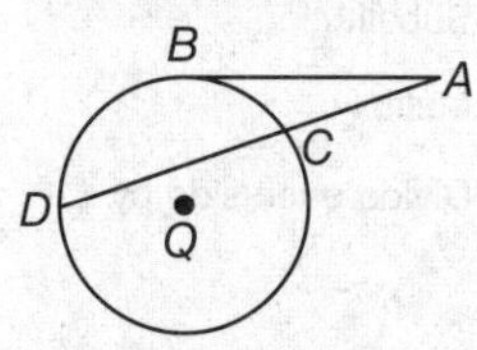

$\overline{AB}$ is a tangent segment.

$\overline{AD}$ is a secant segment.

$\overline{AC}$ is an external secant segment.

$(AB)^2 = AD \cdot AC$

Example $\overline{AB}$ **is tangent to the circle. Find** x**. Round to the nearest tenth.**

The tangent segment is $\overline{AB}$, the secant segment is $\overline{BD}$, and the external secant segment is $\overline{BC}$.

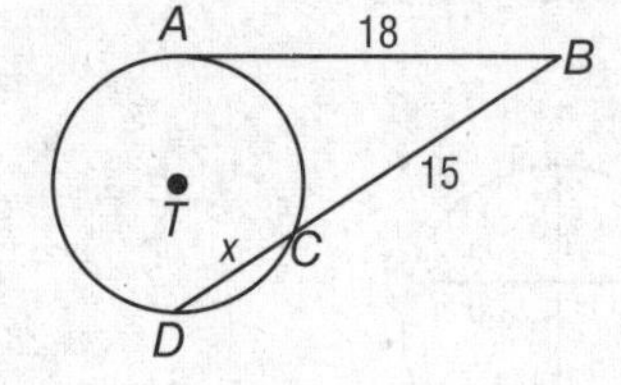

$(AB)^2 = BC \cdot BD$

$(18)^2 = 15(15 + x)$ Substitution.

$324 = 225 + 15x$ Multiply.

$99 = 15x$ Subtract 225 from both sides.

$6.6 = x$ Divide both sides by 15.

Exercises

Find x. Round to the nearest tenth. Assume segments that appear to be tangent are tangent.

1.

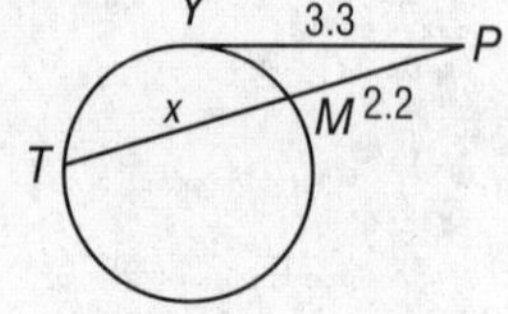

2.

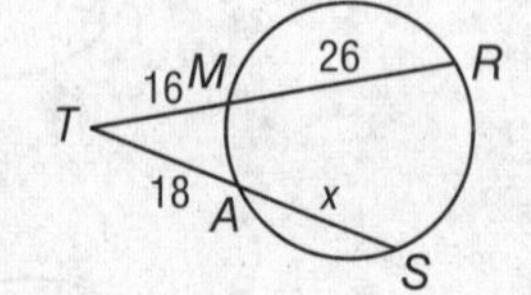

3.

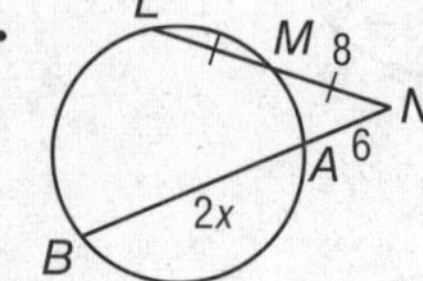

4.

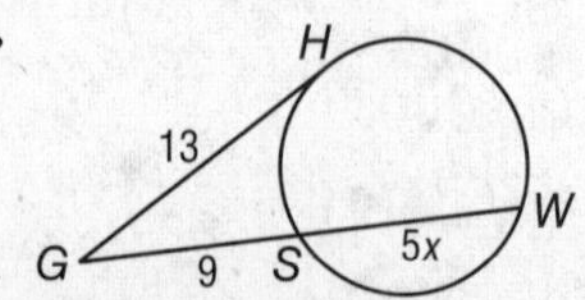

5.

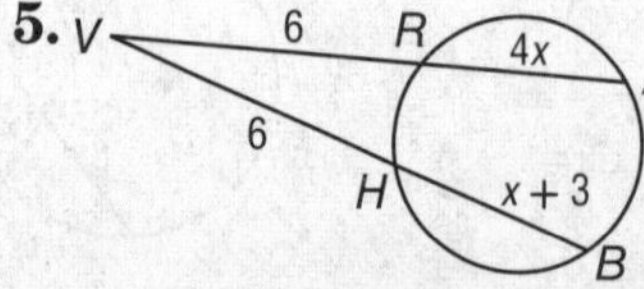

6.

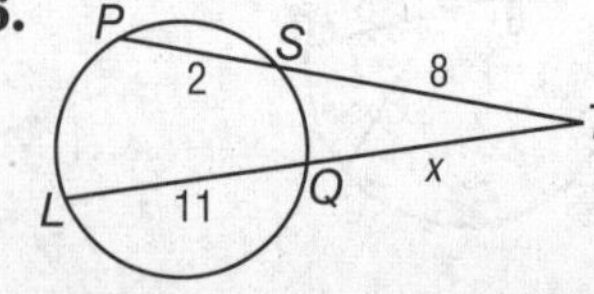

10-8 Study Guide and Intervention

Equations of Circles

Equation of a Circle A **circle** is the locus of points in a plane equidistant from a given point. You can use this definition to write an equation of a circle.

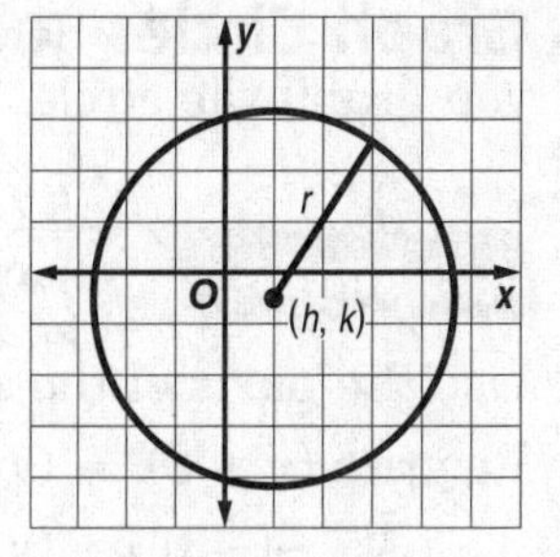

Standard Equation of a Circle	An equation for a circle with center at (h, k) and a radius of r units is $(x - h)^2 + (y - k)^2 = r^2$.

Example **Write an equation for a circle with center (–1, 3) and radius 6.**

Use the formula $(x - h)^2 + (y - k)^2 = r^2$ with $h = -1$, $k = 3$, and $r = 6$.

$(x - h)^2 + (y - k)^2 = r^2$ Equation of a circle

$(x - (-1))^2 + (y - 3)^2 = 6^2$ Substitution

$(x + 1)^2 + (y - 3)^2 = 36$ Simplify.

Exercises

Write the equation of each circle.

1. center at (0, 0), radius 8

2. center at (–2, 3), radius 5

3. center at (2, –4), radius 1

4. center at (–1, –4), radius 2

5. center at (–2, –6), diameter 8

6. center at origin, diameter 4

7. center at (3, –4), passes through (–1, –4)

8. center at (0, 3), passes through (2, 0)

9.

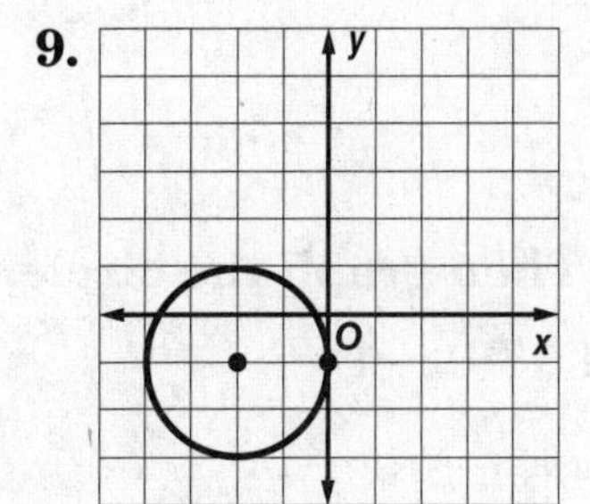

10.

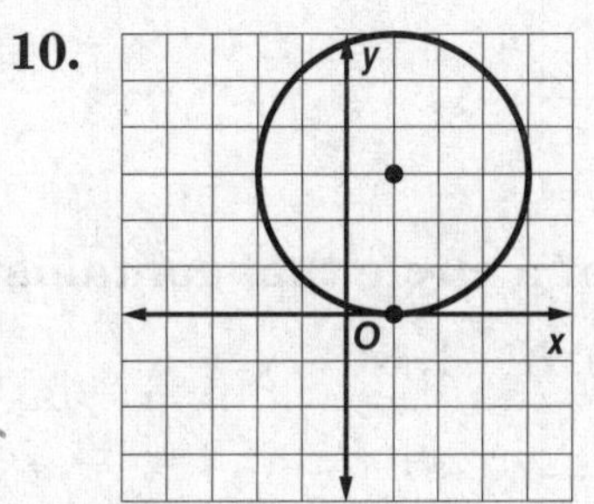

NAME ______________________ DATE ____________ PERIOD ________

10-8 Study Guide and Intervention *(continued)*

Equations of Circles

Graph Circles If you are given an equation of a circle, you can find information to help you graph the circle.

Example **Graph $(x + 3)^2 + (y + 1)^2 = 9$.**

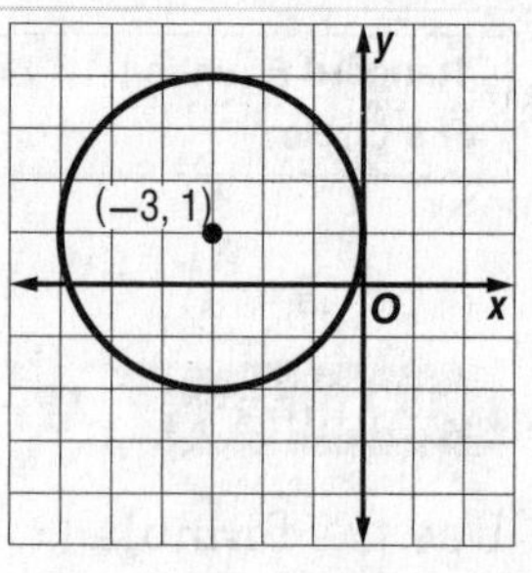

Use the parts of the equation to find (h, k) and r.

Rewrite $(x + 3)^2 + (y - 1)^2 = 9$ to find the center and the radius.

$$[x - (-3)]^2 + (y - 1)^2 = 3^2$$
$$\uparrow \quad \uparrow \quad \uparrow$$
$$(x - h)^2 + (y - k)^2 = r^2$$

So $h = -3$, $k = 1$, and $r = 3$. The center is at $(-3, 1)$ and the radius is 3.

Exercises

For each circle with the given equation, state the coordinates of the center and the measure of the radius. Then graph the equation.

1. $x^2 + y^2 = 16$

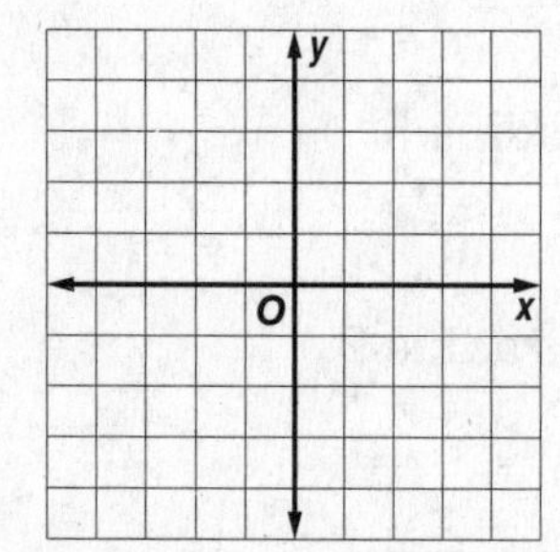

2. $(x - 2)^2 + (y - 1)^2 = 9$

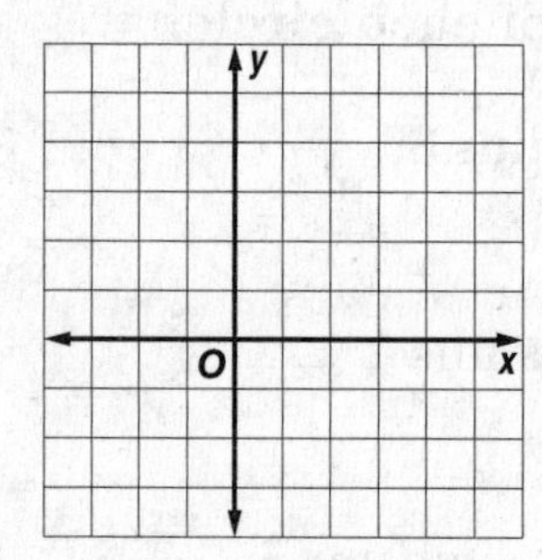

3. $(x + 2)^2 + y^2 = 16$

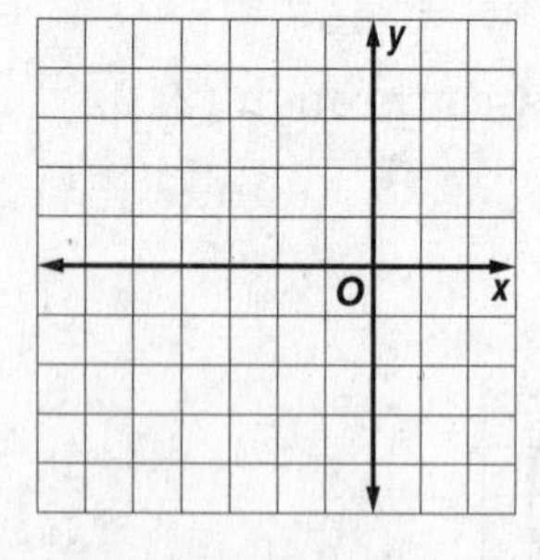

4. $x^2 + (y - 1)^2 = 9$

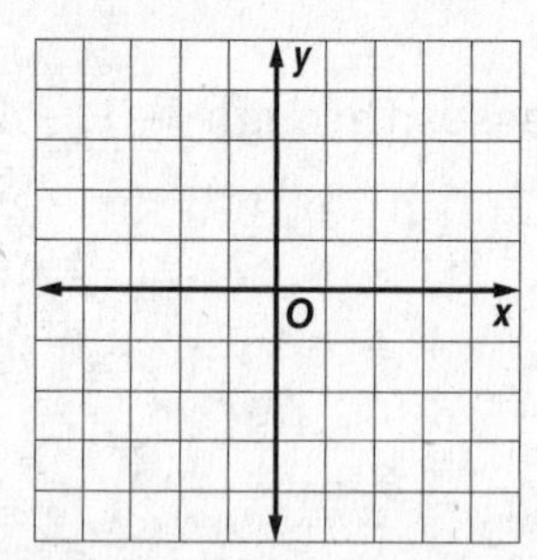

Write an equation of a circle that contains each set of points. Then graph the circle.

5. $F(-2, 2)$, $G(-1, 1)$, $H(-1, 3)$

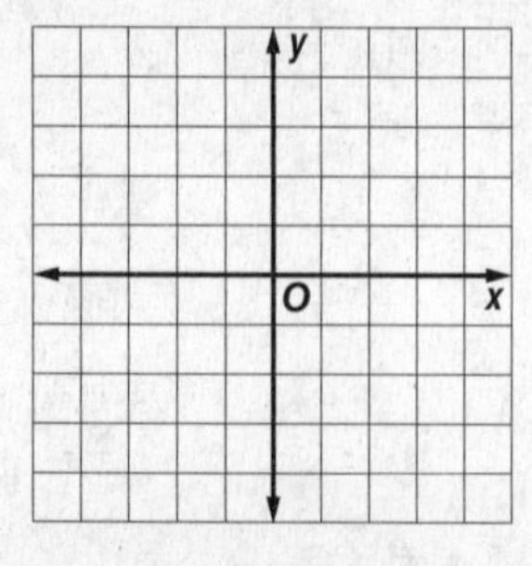

6. $R(-2, 1)$, $S(-4, -1)$, $T(0, -1)$

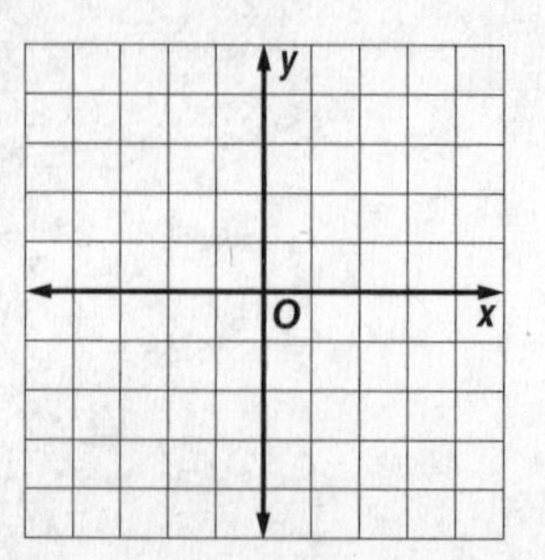

NAME ______________________________ DATE ____________ PERIOD ____________

11-1 Study Guide and Intervention

Areas of Parallelograms and Triangles

Areas of Parallelograms Any side of a parallelogram can be called a **base**. The **height** of a parallelogram is the perpendicular distance between any two parallel bases. The area of a parallelogram is the product of the base and the height.

Area of a Parallelogram	If a parallelogram has an area of A square units, a base of b units, and a height of h units, then $A = bh$.

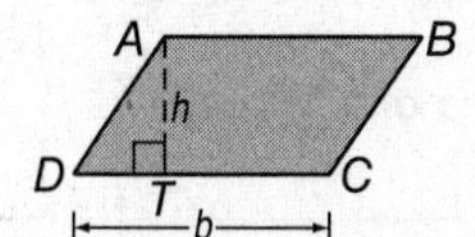

Example **Find the area of parallelogram *EFGH*.**

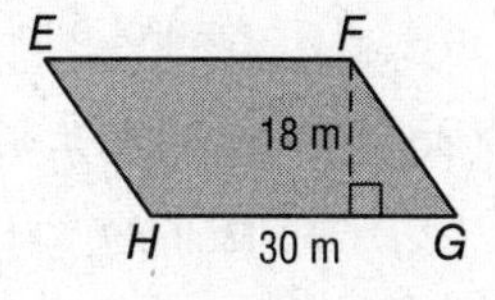

$A = bh$ — Area of a parallelogram

$= 30(18)$ — $b = 30$, $h = 18$

$= 540$ — Multiply.

The area is 540 square meters.

Exercises

Find the perimeter and area of each parallelogram. Round to the nearest tenth if necessary.

1.

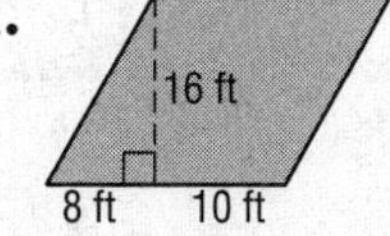

2.

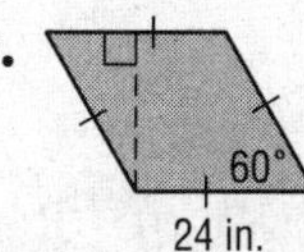

3.

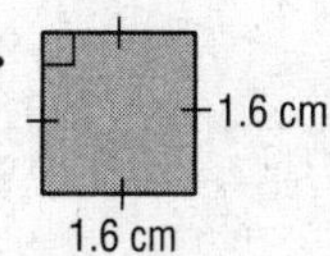

4.

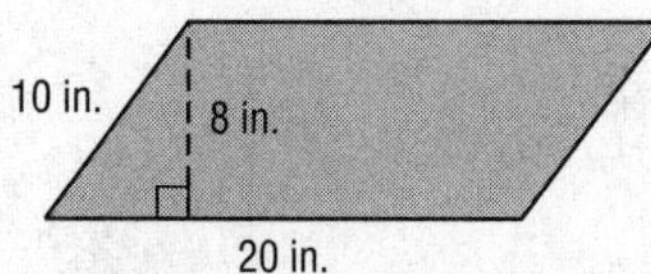

5.

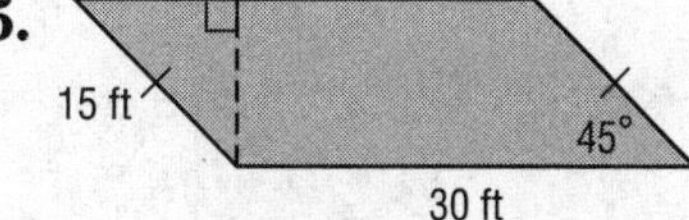

6.

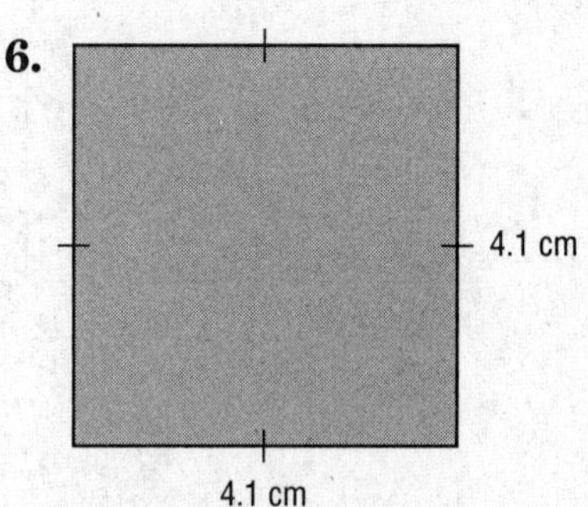

7. TILE FLOOR A bathroom tile floor is made of black-and-white parallelograms. Each parallelogram is made of two triangles with dimensions as shown. Find the perimeter and area of one parallelogram.

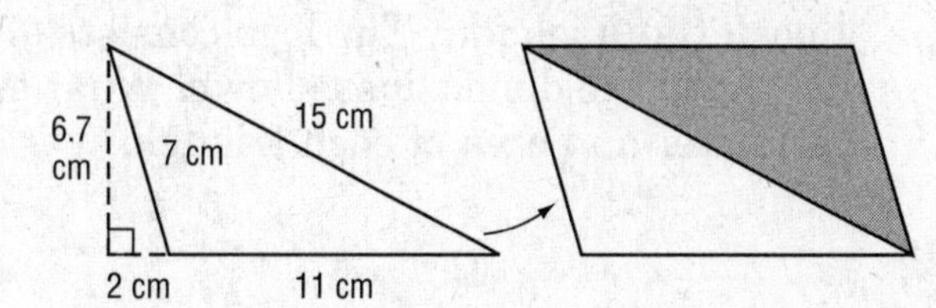

NAME ______________________ DATE __________ PERIOD __________

11-1 Study Guide and Intervention *(continued)*

Areas of Parallelograms and Triangles

Areas Of Triangles The area of a triangle is one half the product of the base and its corresponding height. Like a parallelogram, the base can be any side, and the height is the length of an altitude drawn to a given base.

Area of a Triangle	If a triangle has an area of A square units, a base of b units, and a corresponding height of h units, then $A = \frac{1}{2}bh$.

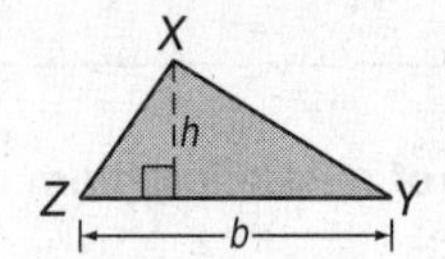

Example **Find the area of the triangle.**

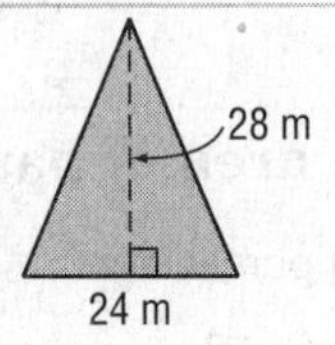

$A = \frac{1}{2}bh$ Area of a triangle

$= \frac{1}{2}(24)(28)$ $b = 24$, $h = 28$

$= 336$ Multiply.

The area is 336 square meters.

Exercises

Find the perimeter and area of each triangle. Round to the nearest tenth if necessary.

1.

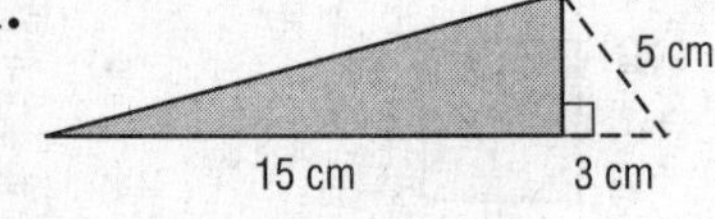

2.

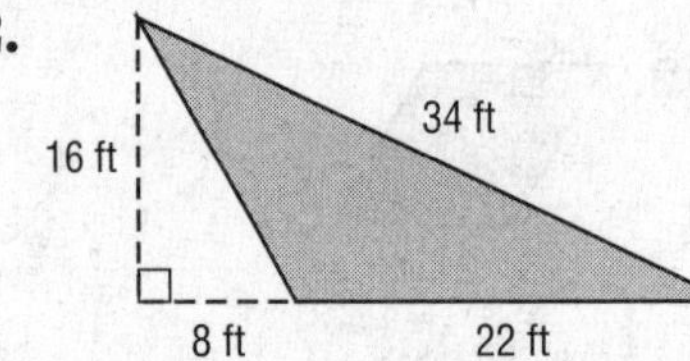

3.

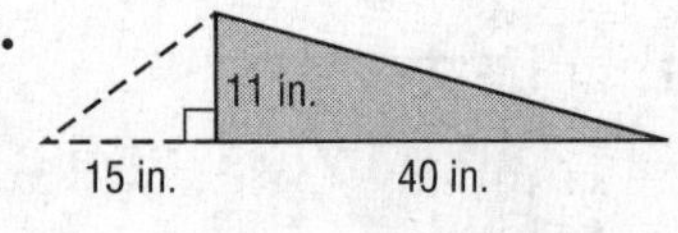

4.

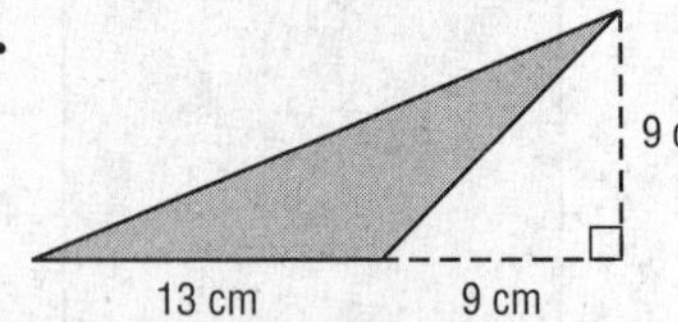

5.

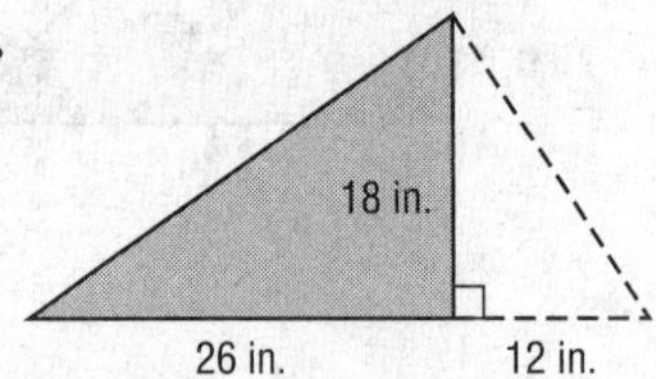

6.

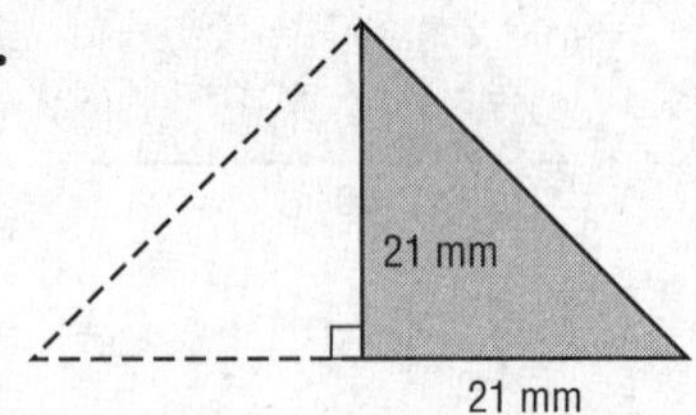

7. LOGO The logo for an engineering company is on a poster at a job fair. The logo consists of two triangles that have the dimensions shown. What are the perimeter and area of each triangle?

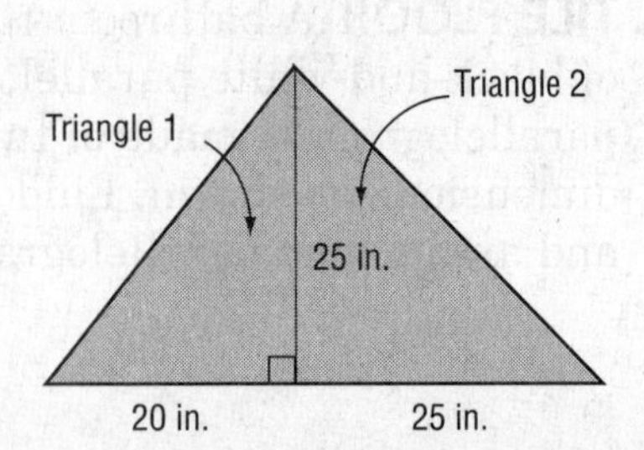

NAME ______________________________ DATE ____________ PERIOD ________

11-2 Study Guide and Intervention

Areas of Trapezoids, Rhombi, and Kites

Areas of Trapezoids A trapezoid is a quadrilateral with exactly one pair of parallel sides, called bases. The **height of a trapezoid** is the perpendicular distance between the bases. The area of a trapezoid is the product of one half the height and the sum of the lengths of the bases.

Area of a Trapezoid	If a trapezoid has an area of A square units, bases of b_1 and b_2 units, and a height of h units, then $A = \frac{1}{2}h(b_1 + b_2)$

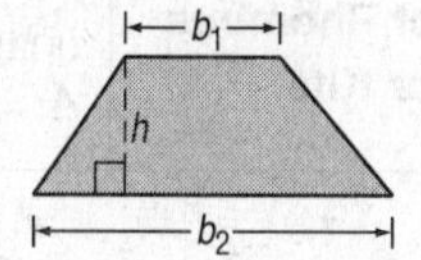

Example **Find the area of the trapezoid.**

$A = \frac{1}{2}h(b_1 + b_2)$ Area of a trapezoid

$= \frac{1}{2}(15)(18 + 40)$ $h = 15$, $b_1 = 18$, and $b_2 = 40$

$= 435$ Simplify.

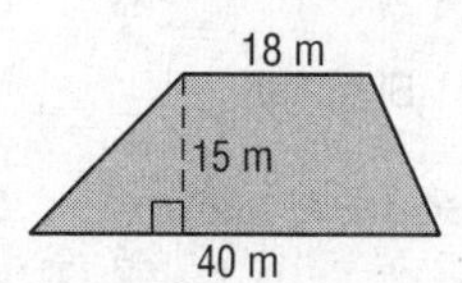

The area of the trapezoid is 435 square meters.

Exercises

Find the area of each trapezoid.

1.

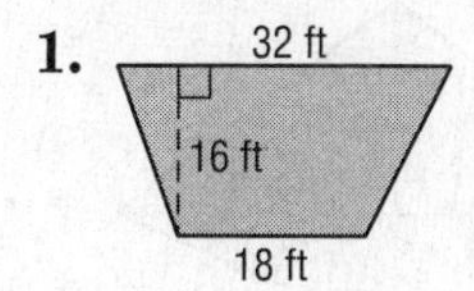

2.

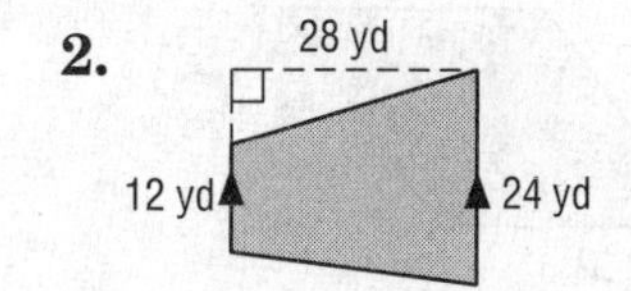

3.

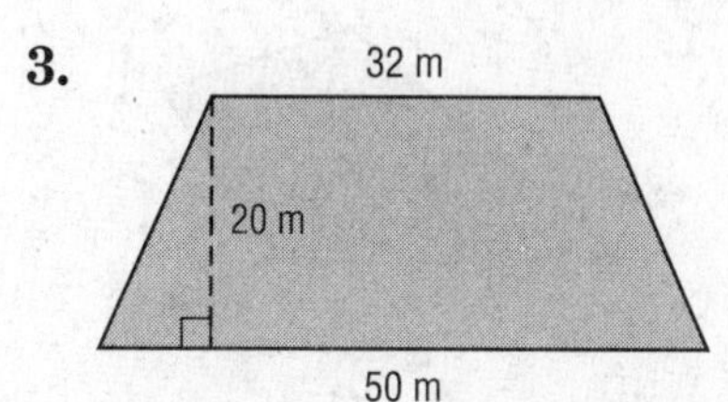

4.

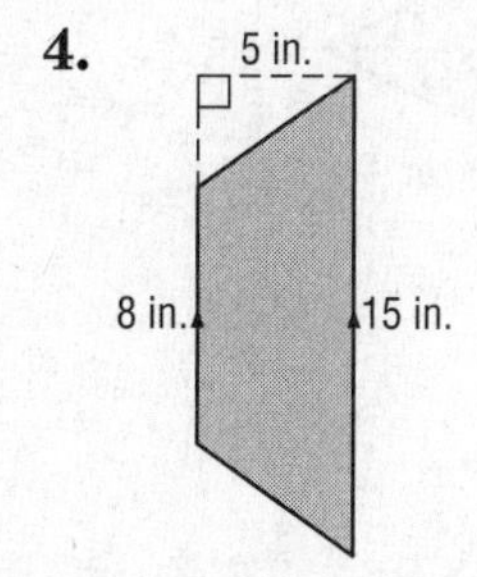

5.

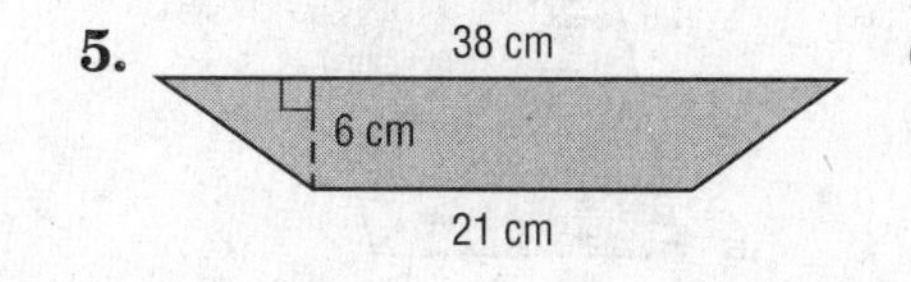

6.

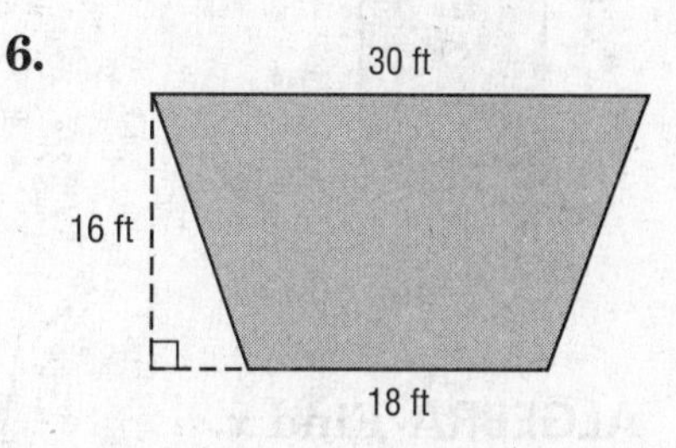

7. OPEN ENDED Ryan runs a landscaping business. A new customer has a trapezodial shaped backyard, shown at the right. How many square feet of grass will Ryan have to mow?

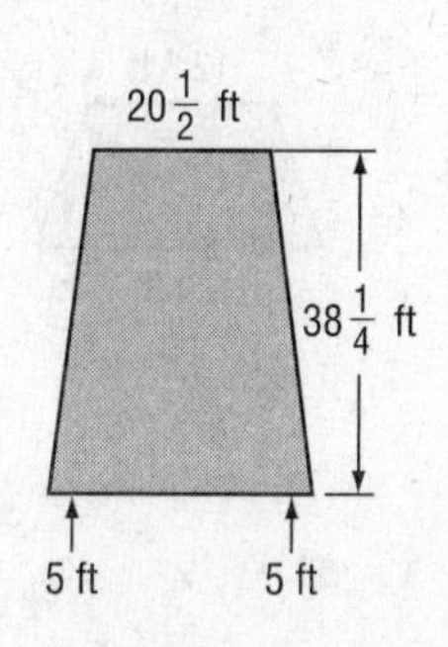

NAME ______________________ DATE ____________ PERIOD ____________

11-2 Study Guide and Intervention *(continued)*

Areas of Trapezoids, Rhombi, and Kites

Areas of Rhombi and Kites A rhombus is a parallelogram with all four sides congruent. A kite is a quadrilateral with exactly two pairs of consecutive sides congruent.

Area of Rhombus or Kite	If a rhombus or kite has an area of A square units, and diagonals of d_1 and d_2 units, then $A = \frac{1}{2} d_1 \cdot d_2$.

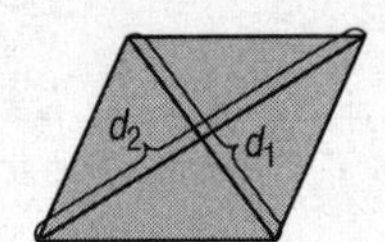

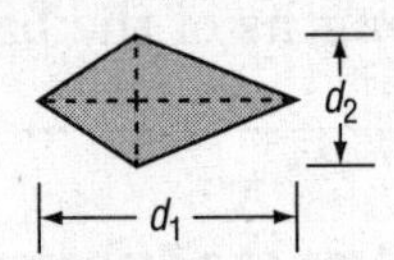

Example **Find the area of the rhombus.**

$A = \frac{1}{2} d_1 d_2$ Area of rhombus

$= \frac{1}{2}(7)(9)$ $d_1 = 7$, and $d_2 = 9$

$= 31.5$ Simplify.

The area is 31.5 square meters.

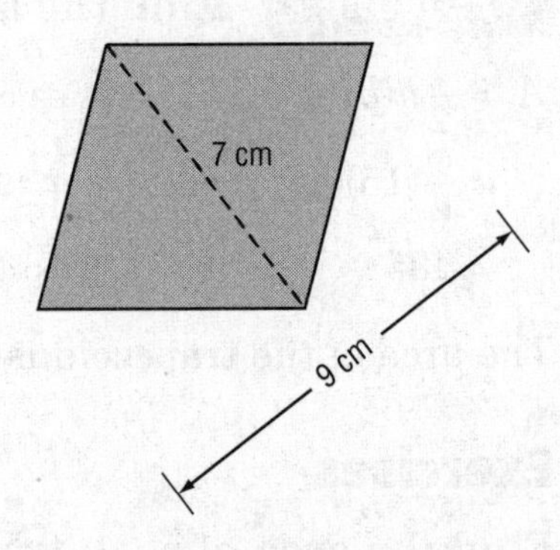

Exercises

Find the area of each rhombus or kite.

1.

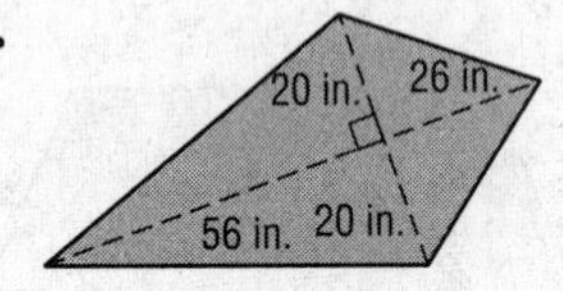

2.

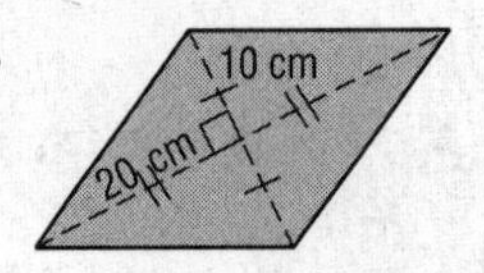

3.

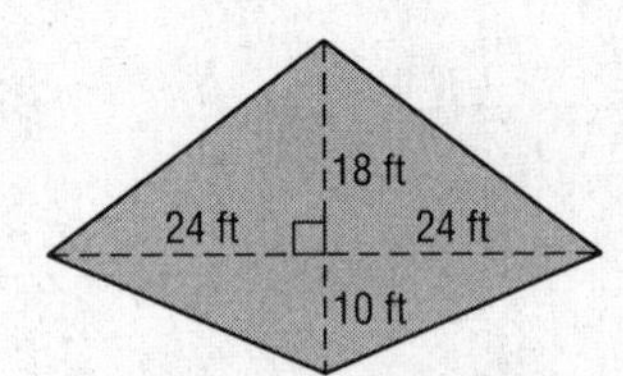

4.

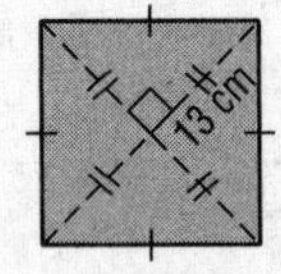

5.

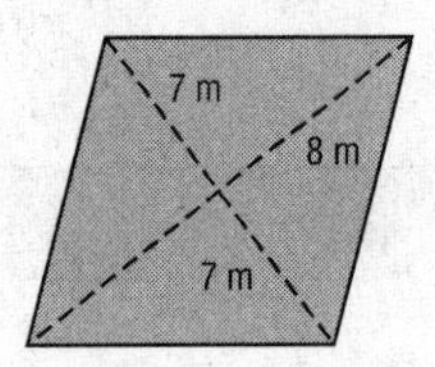

6.

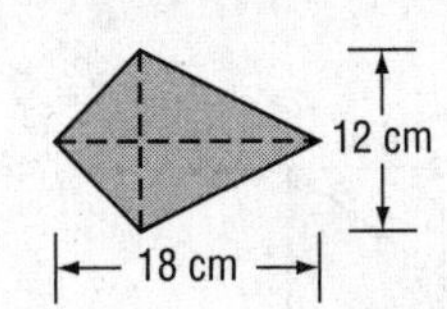

ALGEBRA Find x.

7. $A = 164$ ft^2

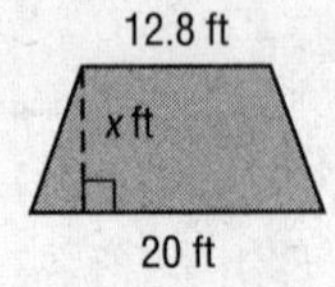

8. $A = 340$ cm^2

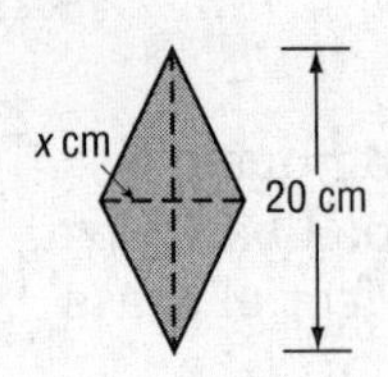

9. $A = 247.5$ mm^2

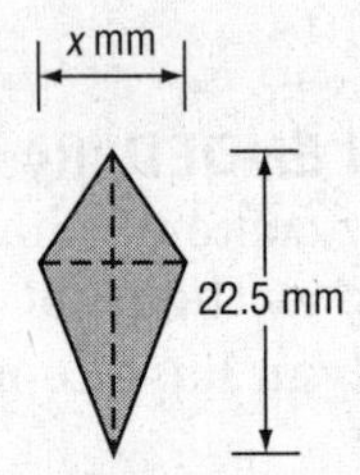

11-3 Study Guide and Intervention

Areas of Circles and Sectors

Areas Of Circles The area of a circle is equal to π times the square of radius.

Area of a Circle	If a circle has an area of A square units and a radius of r units, then $A = \pi r^2$.

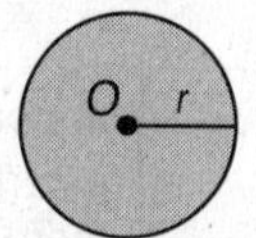

Example **Find the area of the circle *p*.**

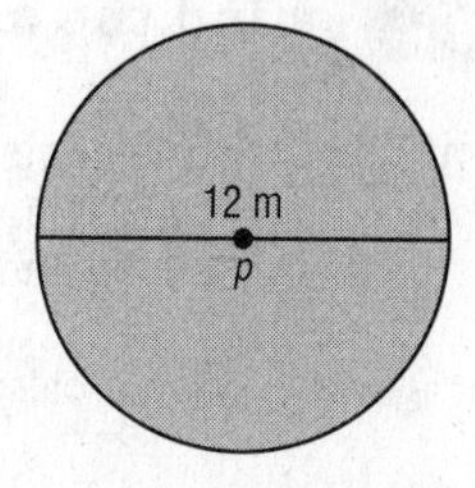

$A = \pi r^2$ Area of a circle

$= \pi(6)^2$ $r = 6$

≈ 113.1 Use a calculator.

The area of the circle is about 113.1 square meters. If $d = 12$ m, then $r = 6$ m.

Exercises

Find the area of each circle. Round to the nearest tenth.

1.

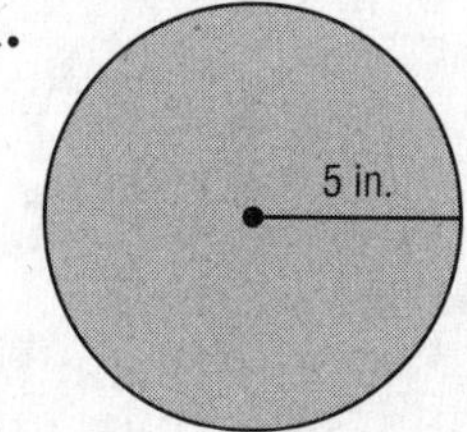

2.

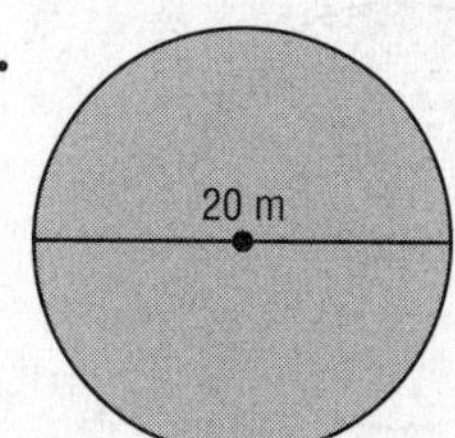

3.

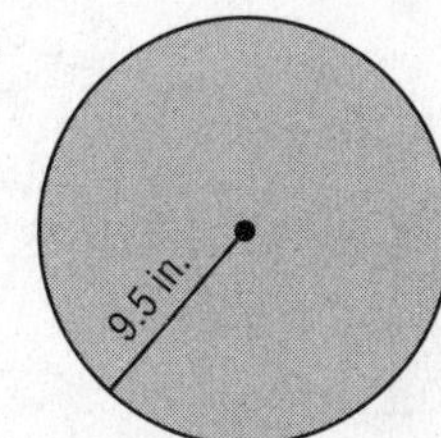

4.

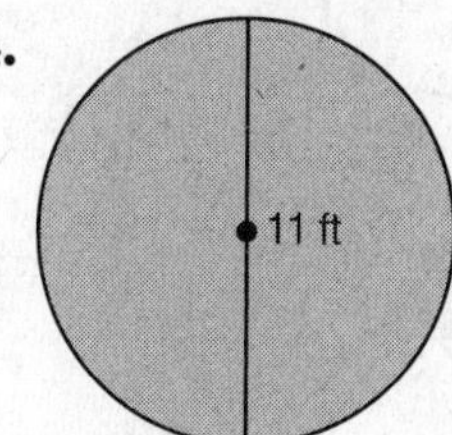

5.

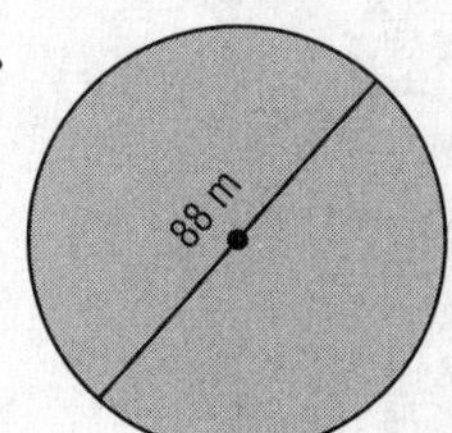

6.

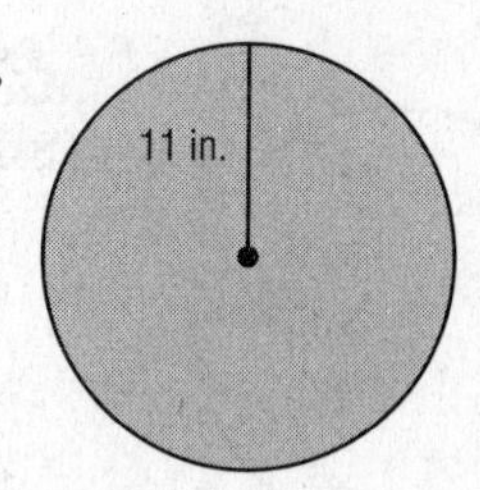

Find the indicated measure. Round to the nearest tenth.

7. The area of a circle is 153.9 square centimeters. Find the diameter.

8. Find the diameter of a circle with an area of 490.9 square millimeters.

9. The area of a circle is 907.9 square inches. Find the radius.

10. Find the radius of a circle with an area of 63.6 square feet.

11-3 Study Guide and Intervention *(continued)*

Areas of Circles and Sectors

Areas of Sectors A sector of a circle is a region bounded by a central angle and its intercepted arc.

Area of a Sector	If a sector of a circle has an area of A square units, a central angle measuring $x°$, and a radius of r units, then $A = \frac{x}{360}\pi r^2$.

Example **Find the area of the shaded sector.**

$A = \frac{x}{360} \cdot \pi r^2$	Area of a sector
$= \frac{36}{360} \cdot \pi(5)^2$	$x = 36$ and $r = 5$
≈ 7.85	Use a calculator.

36°
5 in.

The area of the sector is about 7.85 square inches.

Exercises

Find the area of each shaded sector. Round to the nearest tenth.

1.

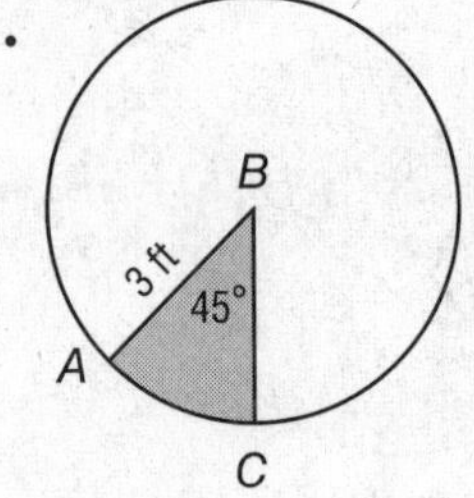

2.

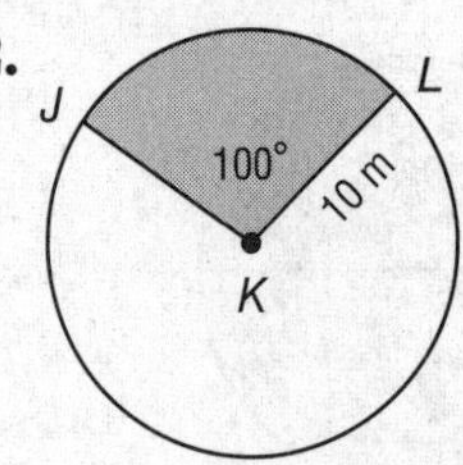

3.

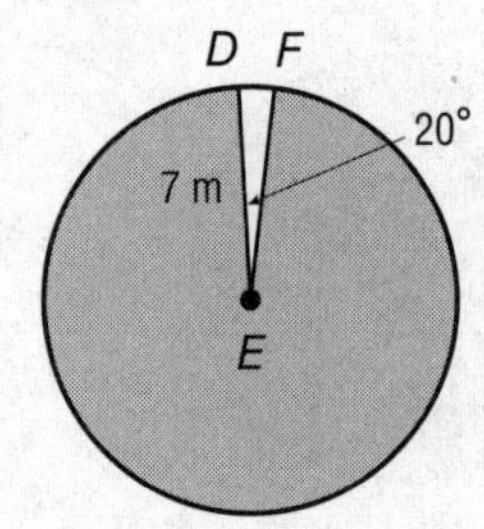

4.

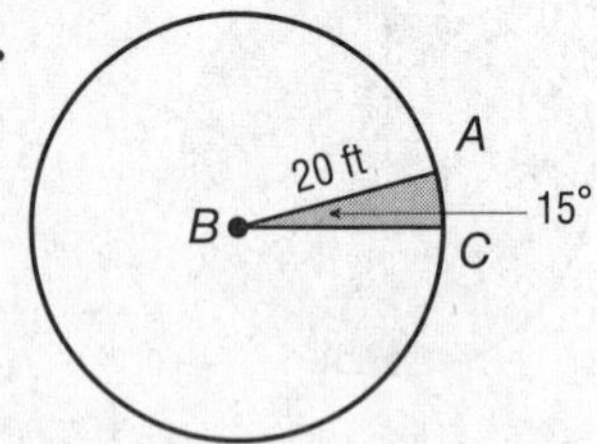

5.

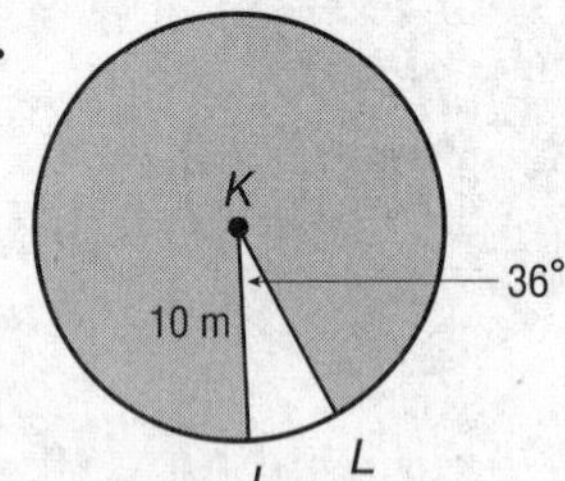

6.

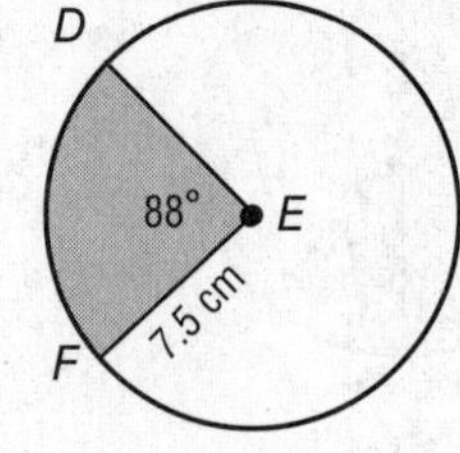

7. SANDWICHES For a party, Samantha wants to have finger sandwiches. She cuts sandwiches into circles. If she cuts each circle into three congruent pieces, what is the area of each piece?

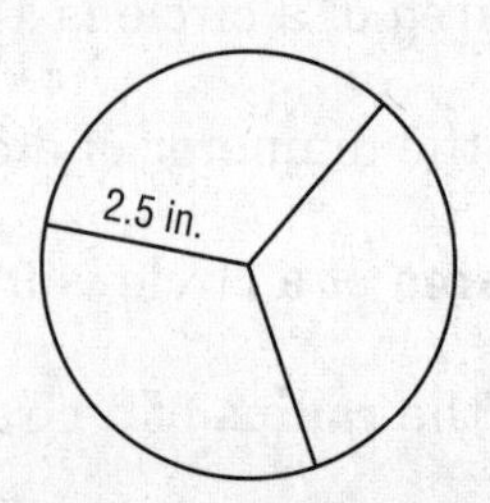

NAME ______________________ DATE __________ PERIOD __________

11-4 Study Guide and Intervention

Areas of Regular Polygons and Composite Figures

Areas of Regular Polygons In a regular polygon, the segment drawn from the center of the polygon perpendicular to the opposite side is called the **apothem**. In the figure at the right, $\overline{AP}$ is the apothem and $\overline{AR}$ is the radius of the circumscribed circle.

Area of a Regular Polygon	If a regular polygon has an area of A square units, a perimeter of P units, and an apothem of a units, then $A = \frac{1}{2}aP$.

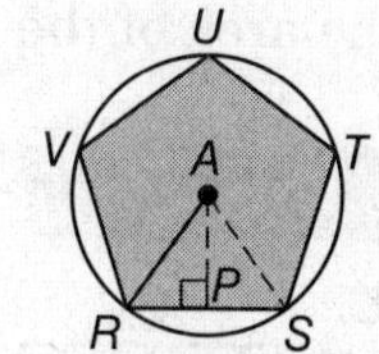

Example 1 **Verify the formula $A = \frac{1}{2}aP$ for the regular pentagon above.**

For $\triangle RAS$, the area is $A = \frac{1}{2}bh = \frac{1}{2}(RS)(AP)$. So the area of the pentagon is $A = 5\left(\frac{1}{2}\right)(RS)(AP)$. Substituting P for $5RS$ and substituting a for AP, then $A = \frac{1}{2}aP$.

Example 2 **Find the area of regular pentagon *RSTUV* above if its perimeter is 60 centimeters.**

First find the apothem.

The measure of central angle RAS is $\frac{360°}{5}$ or 72°. Therefore, $m\angle RAP = 36$. The perimeter is 60, so $RS = 12$ and $RP = 6$.

$$\tan m\angle RAP = \frac{RP}{AP}$$

$$\tan 36° = \frac{6}{AP}$$

$$AP = \frac{6}{\tan 36°}$$

$$\approx 8.26$$

So, $A = \frac{1}{2}aP = \frac{1}{2}(60)(8.26)$ or 247.8.

The area is about 248 square centimeters.

Exercises

Find the area of each regular polygon. Round to the nearest tenth.

1.

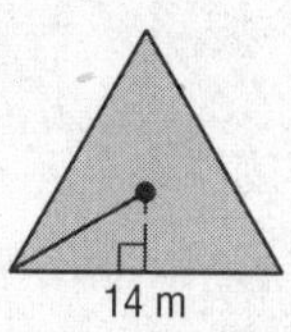

2.

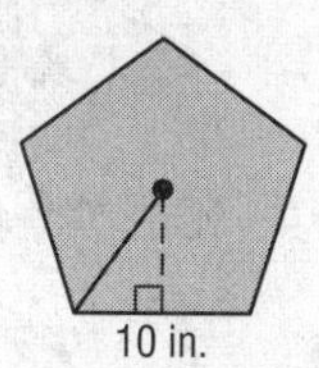

3.

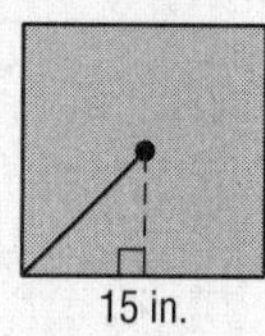

4.

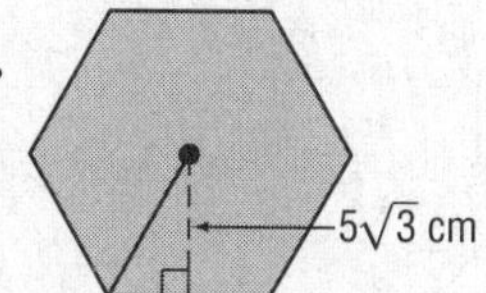

5.

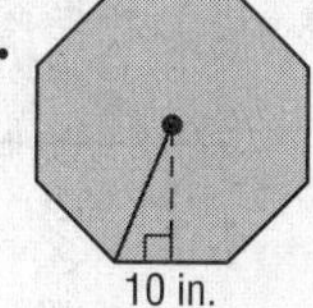

6.

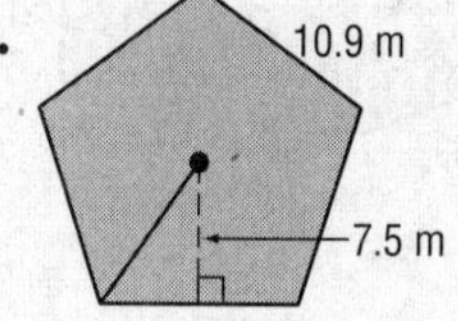

NAME ______________________ DATE __________ PERIOD __________

11-4 Study Guide and Intervention *(continued)*

Areas of Regular Polygons and Composite Figures

Areas of Composite Figures A composite figure is a figure that can be seprated into regions that are basic figures. To find the area of a composite figure, separate the figure into basic figures of which we can find the area. The sum of the areas of the basic figures is the area of the figure.

Example **Find the area of the shaded region.**

a.

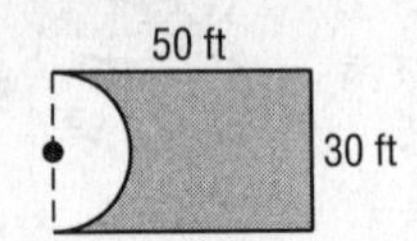

The figure is a rectangle minus one half of a circle. The radius of the circle is one half of 30 or 15.

$A = \ell w - \frac{1}{2}\pi r^2$

$= 50(30) - 0.5\pi(15)^2$

≈ 1146.6 or about 1147 ft^2

b.

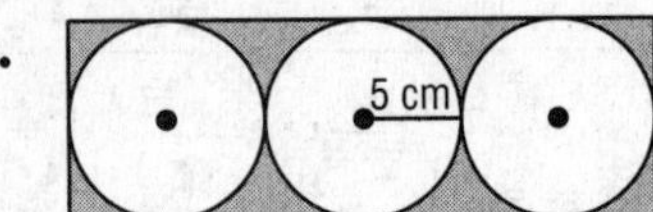

The dimensions of the rectangle are 10 centimeters and 30 centimeters. The area of the shaded region is

$(10)(30) - 3\pi(5^2) = 300 - 75\pi$

≈ 64.4 cm^2

Exercises

Find the area of each figure. Round to the nearest tenth if necessary.

1.

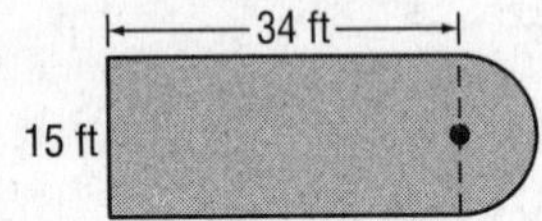

2.

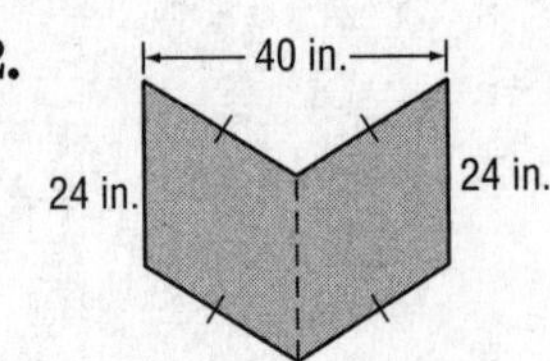

3.

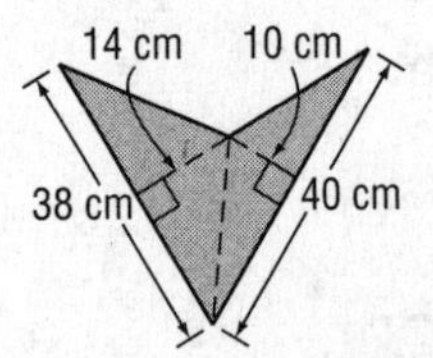

4.

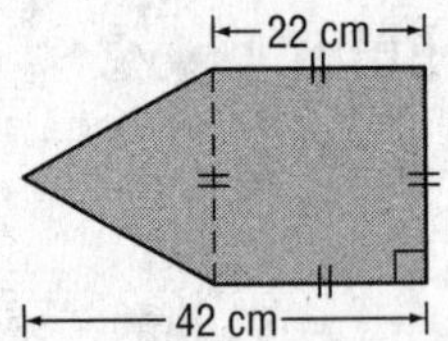

5.

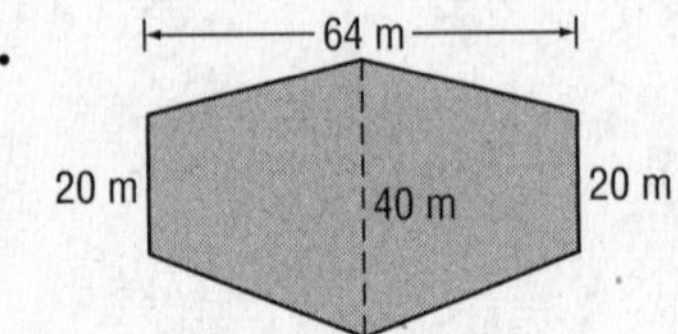

6.

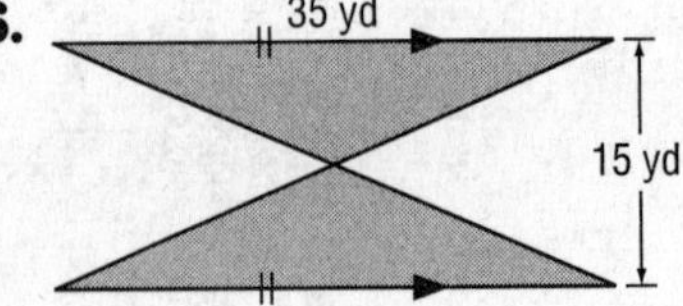

NAME ______________________ DATE __________ PERIOD __________

11-5 Study Guide and Intervention

Areas of Similar Figures

Areas of Similar Figures If two polygons are similar, then their areas are proportional to the square of the scale factor between them.

Example $\triangle JKL \cong \triangle PQR$.
The area of $\triangle JKL$ is 40 square inches.
Find the area of the shaded triangle.

Find the scale factor: $\frac{12}{10}$ or $\frac{6}{5}$.

The ratio of their areas is $\left(\frac{6}{5}\right)^2$.

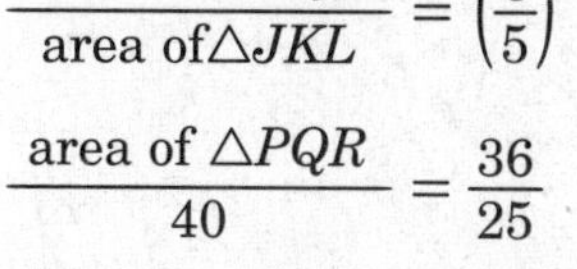

$\frac{\text{area of } \triangle PQR}{\text{area of} \triangle JKL} = \left(\frac{6}{5}\right)^2$ Write a proportion.

$\frac{\text{area of } \triangle PQR}{40} = \frac{36}{25}$ Area of $\triangle JKL = 40$; $\left(\frac{6}{5}\right)^2 = \frac{36}{25}$

$\text{area of } \triangle PQR = \frac{36}{25} \cdot 40$ Multiply each side by 40.

$\text{area of } \triangle PQR = 57.6$ Simplify.

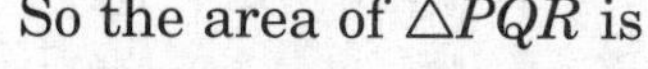

So the area of $\triangle PQR$ is 57.6 square inches.

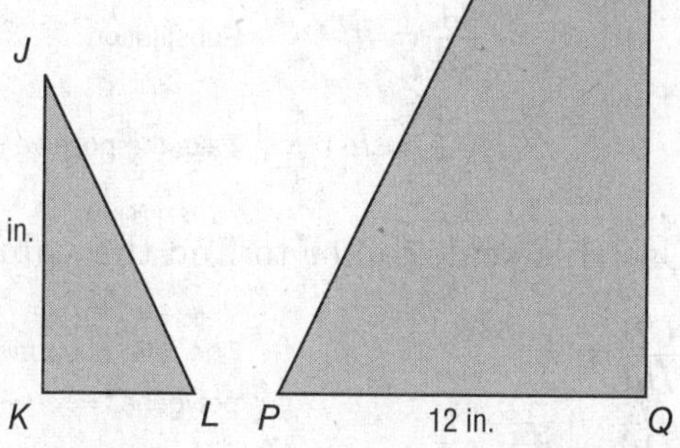

Exercises

For each pair of similar figures, find the area of the shaded figure. Round to the nearest tenth if necessary.

1.

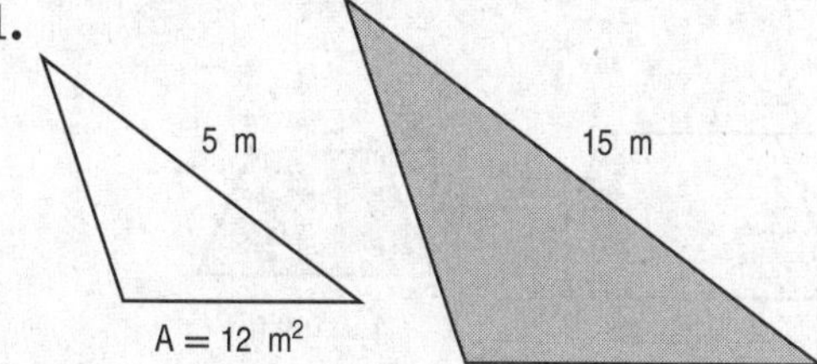

2.

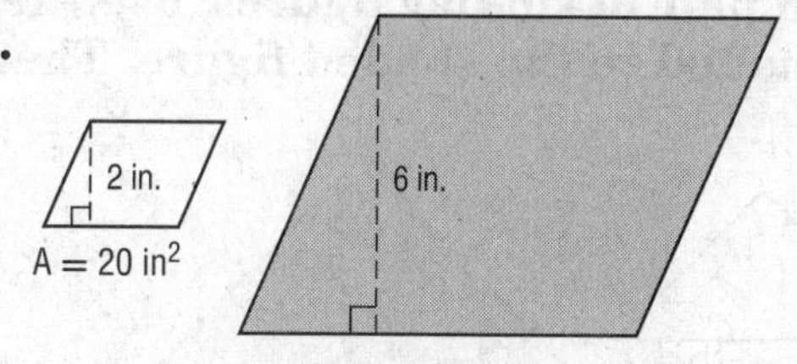

3.

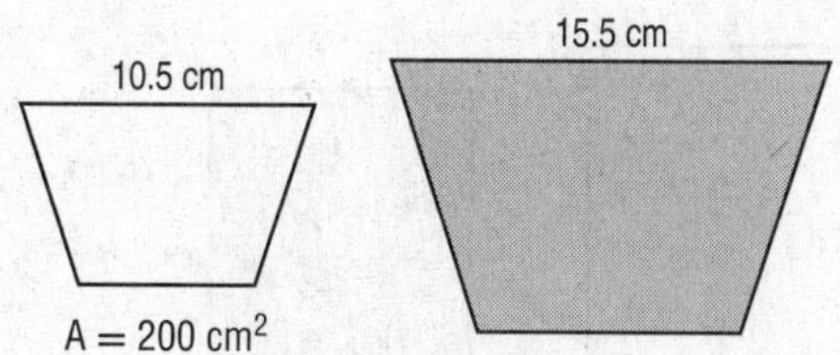

4.

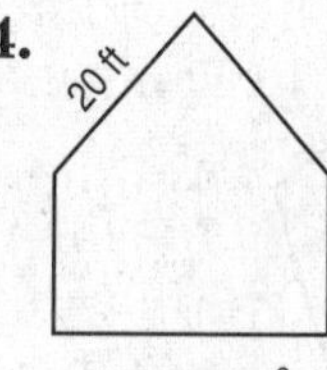

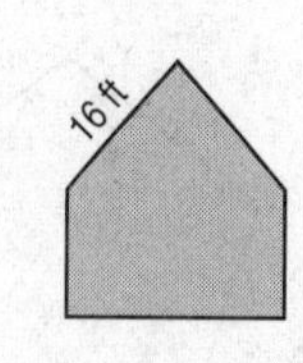

11-5 Study Guide and Intervention *(continued)*

Areas of Similar Figures

Scale Factors and Missing Measures in Similar Figures You can use the areas of similar figures to find the scale factor between them or a missing measure.

Example **If $\square ABDC$ is similar to $\square FGJH$, find the value of x.**

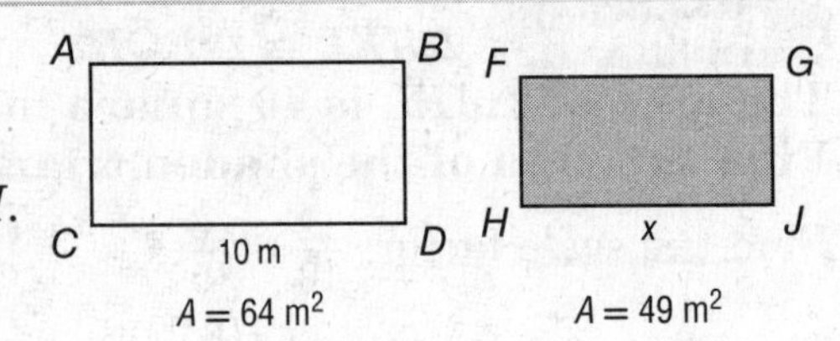

Let k be the scale factor between $\square ABDC$ and $\square FGJH$.

$\frac{\text{area } \square ABCD}{\text{area } \square FGJH} = k^2$ Theorem 11.1

$\frac{64}{49} = k^2$ Substitution

$\frac{8}{7} = k$ Take the positive square root of each side.

Use this scale factor to find the value of x.

$\frac{CD}{HJ} = k$ The ratio of corresponding lengths of similar polygons is equal to the scale factor between the polygons.

$\frac{10}{x} = \frac{8}{7}$ Substitution

$x = \frac{7}{8} \cdot 10 \text{ or } 8.75$ Multiply each side by 10.

Exercises

For each pair of similar figures, use the given areas to find the scale factor from the unshaded to the shaded figure. Then find x.

1.

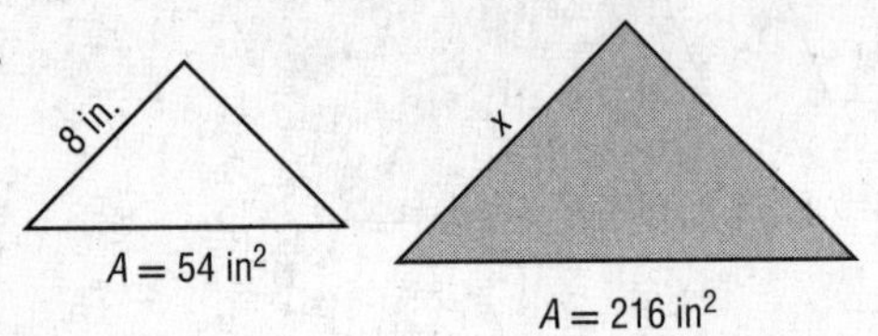

2.

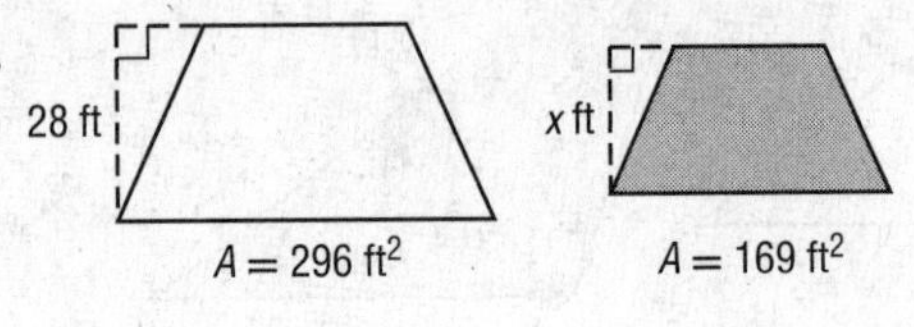

3.

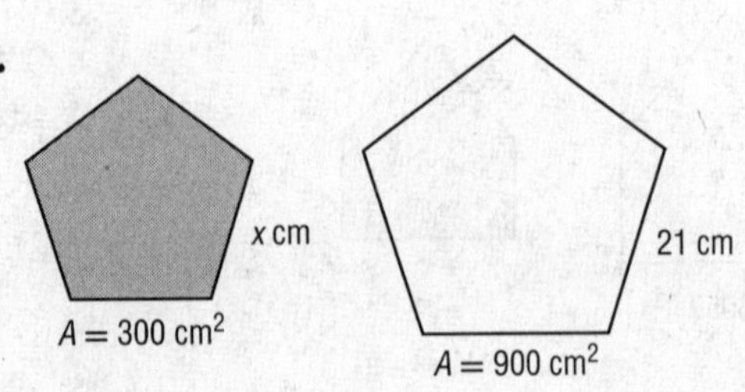

4.

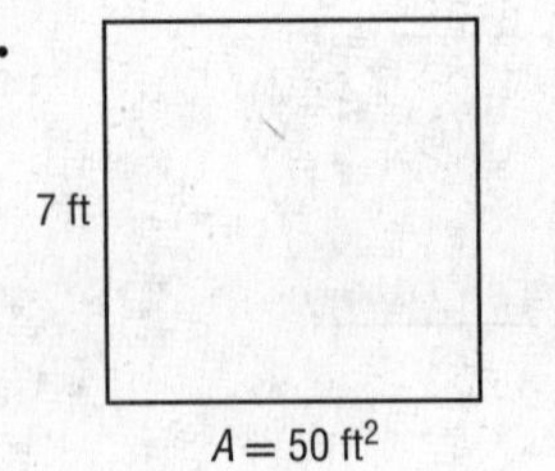

NAME ______________________ DATE ____________ PERIOD ____________

12-1 Study Guide and Intervention

Representations of Three-Dimensional Figures

Draw Isometric Views Isometric dot paper can be used to draw isometric views, or corner views, of a three-dimensional object on two-dimensional paper.

Example 1 **Use isometric dot paper to sketch a triangular prism 3 units high, with two sides of the base that are 3 units long and 4 units long.**

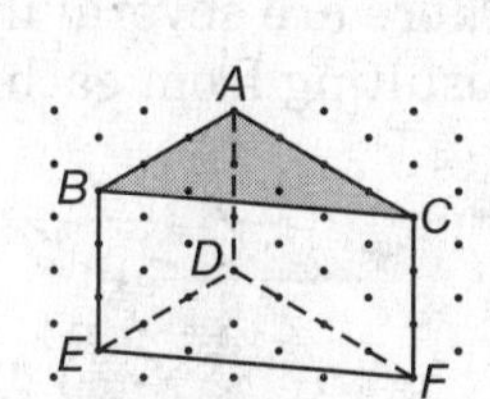

Step 1 Draw $\overline{AB}$ at 3 units and draw $\overline{AC}$ at 4 units.

Step 2 Draw $\overline{AD}$, $\overline{BE}$, and $\overline{CF}$, each at 3 units.

Step 3 Draw $\overline{BC}$ and $\triangle DEF$.

Example 2 **Use isometric dot paper and the orthographic drawing to sketch a solid.**

top view   left view   front view   right view

- The top view indicates two columns.
- The right and left views indicate that the height of figure is three blocks.
- The front view indicates that the columns have heights 2 and 3 blocks.

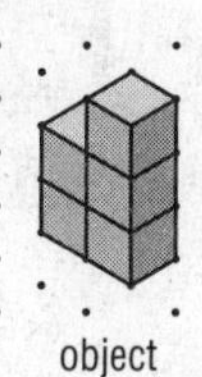

object

Connect the dots on the isometric dot paper to represent the edges of the solid. Shade the tops of each column.

Exercises

Sketch each solid using isometric dot paper.

1. cube with 4 units on each side

2. rectangular prism 1 unit high, 5 units long, and 4 units wide

Use isometric dot paper and each orthographic drawing to sketch a solid.

3.

top view   left view   front view   right view

4.

top view   left view   front view   right view

12-1 Study Guide and Intervention *(continued)*

Representations of Three-Dimensional Figures

Cross Sections The intersection of a solid and a plane is called a **cross section** of the solid. The shape of a cross section depends upon the angle of the plane.

Example

There are several interesting shapes that are cross sections of a cone. Determine the shape resulting from each cross section of the cone.

a. 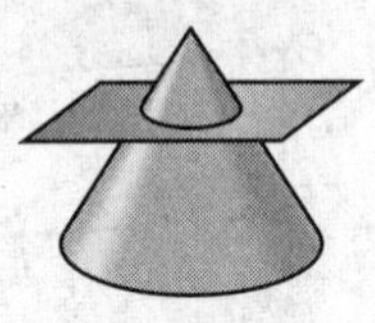

If the plane is parallel to the base of the cone, then the resulting cross section will be a circle.

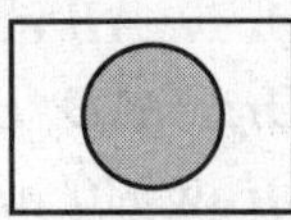

b. 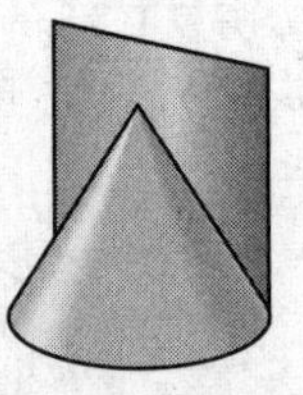

If the plane cuts through the cone perpendicular to the base and through the center of the cone, then the resulting cross section will be a triangle.

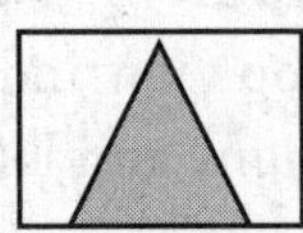

c. 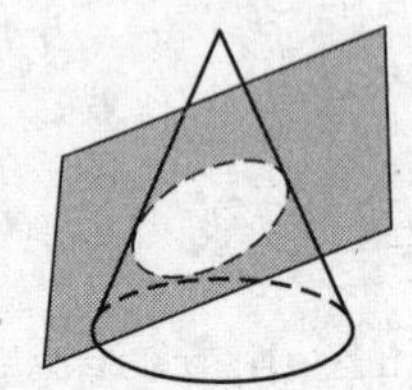

If the plane cuts across the entire cone, then the resulting cross section will be an ellipse.

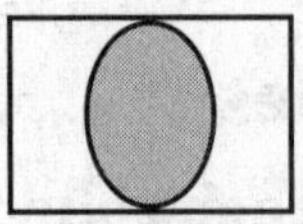

Exercises

Describe each cross section.

1. 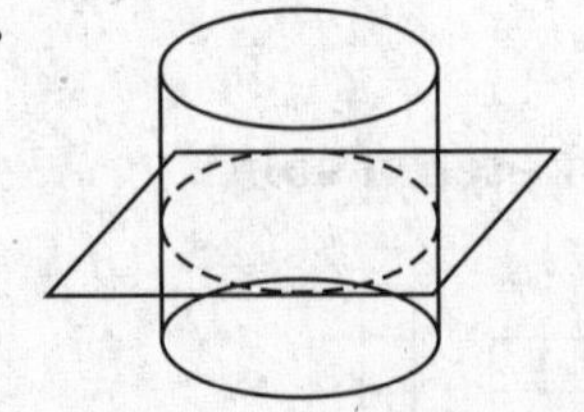

2. 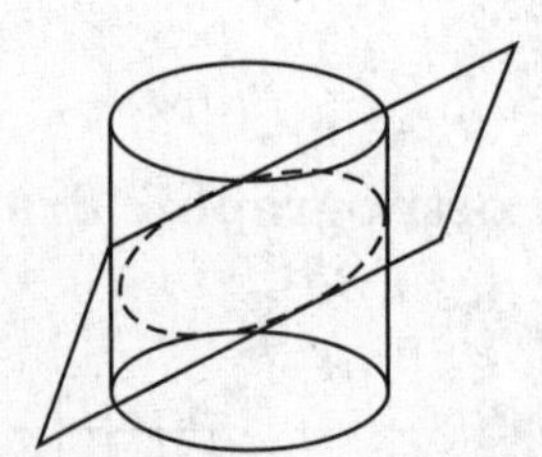

3. 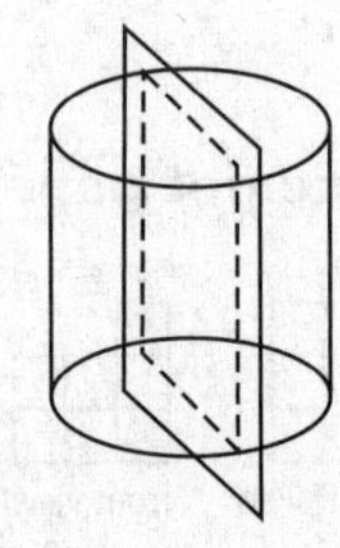

NAME ______________________ DATE ____________ PERIOD ____________

12-2 Study Guide and Intervention

Surface Areas of Prisms and Cylinders

Lateral and Surface Areas of Prisms In a solid figure, faces that are not bases are **lateral faces.** The **lateral area** is the sum of the area of the lateral faces. The **surface area** is the sum of the lateral area and the area of the bases.

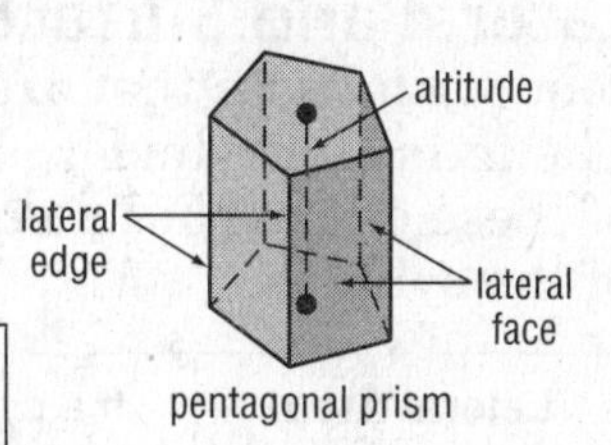

Lateral Area of a Prism	If a prism has a lateral area of L square units, a height of h units, and each base has a perimeter of P units, then $L = Ph$.
Surface Area of a Prism	If a prism has a surface area of S square units, a lateral area of L square units, and each base has an area of B square units, then $S = L + 2B$ or $S = Ph + 2B$

Example **Find the lateral and surface area of the regular pentagonal prism above if each base has a perimeter of 75 centimeters and the height is 10 centimeters.**

$L = Ph$ Lateral area of a prism

$= 75(10)$ $P = 75, h = 10$

$= 750$ Multiply.

The lateral area is 750 square centimeters and the surface area is about 1524.2 square centimeters.

$S = L + 2B$

$= 750 + 2\left(\frac{1}{2}aP\right)$

$= 750 + \left(\frac{7.5}{\tan 30°}\right)(75)$

≈ 1524.2

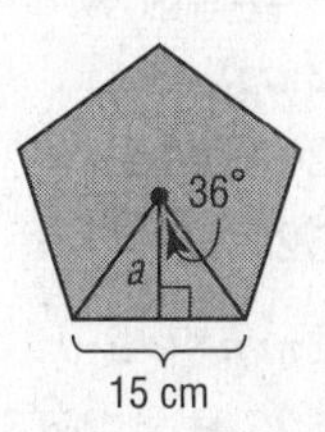

$\tan 36° = \frac{7.5}{a}$

$a = \frac{7.5}{\tan 36°}$

Exercises

Find the lateral area and surface area of each prism. Round to the nearest tenth if necessary.

1.

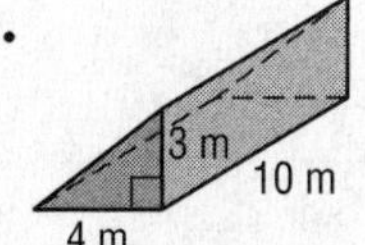

2.

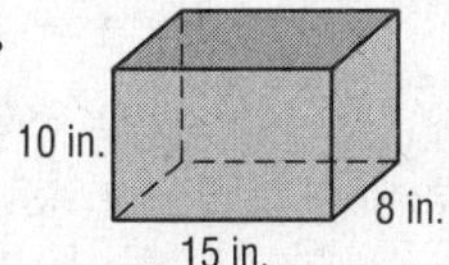

3.

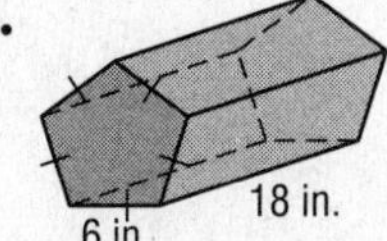

4.

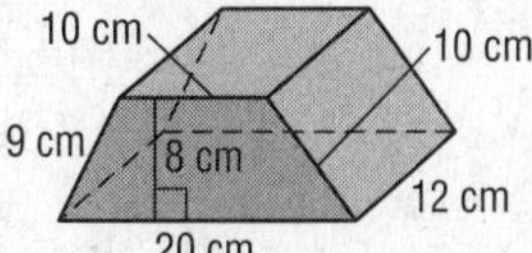

5.

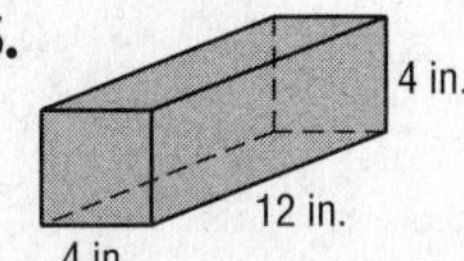

6.

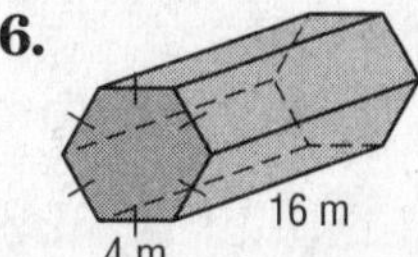

NAME ______________________________ DATE ____________ PERIOD ____________

12-2 Study Guide and Intervention *(continued)*

Surface Areas of Prisms and Cylinders

Lateral and Surface Areas of Cylinders A **cylinder** is a solid with bases that are congruent circles lying in parallel planes. The **axis** of a cylinder is the segment with endpoints at the centers of these circles. For a **right cylinder**, the axis is also the altitude of the cylinder.

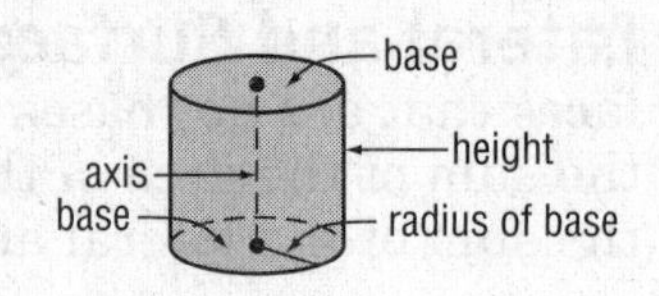

Lateral Area of a Cylinder	If a cylinder has a lateral area of L square units, a height of h units, and a base has a radius of r units, then $L = 2\pi rh$.
Surface Area of a Cylinder	If a cylinder has a surface area of S square units, a height of h units, and a base has a radius of r units, then $S = L + 2B$ or $2\pi rh + 2\pi r^2$.

Example **Find the lateral and surface area of the cylinder. Round to the nearest tenth.**

12 cm
14 cm

If $d = 12$ cm, then $r = 6$ cm.

$L = 2\pi rh$ — Lateral area of a cylinder

$= 2\pi(6)(14)$ — $r = 6, h = 14$

≈ 527.8 — Use a calculator.

$S = 2\pi rh + 2\pi r^2$ — Surface area of a cylinder

$\approx 527.8 + 2\pi(6)^2$ — $2\pi rh \approx 527.8, r = 6$

≈ 754.0 — Use a calculator.

The lateral area is about 527.5 square centimeters and the surface area is about 754.0 square centimeters.

Exercises

Find the lateral area and surface area of each cylinder. Round to the nearest tenth.

1.

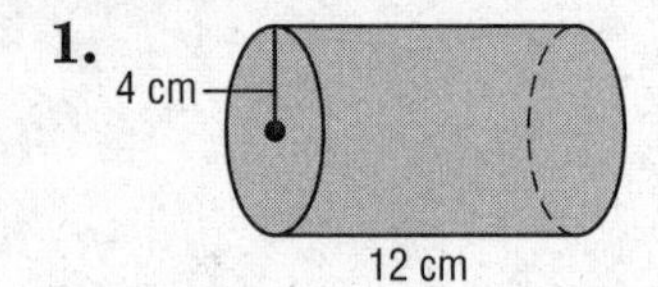

2.

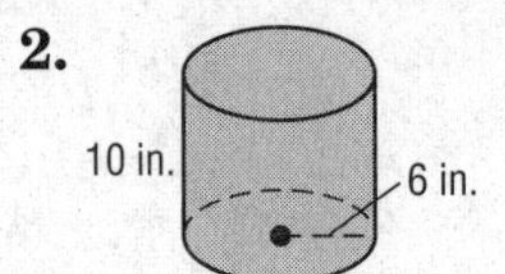

3.

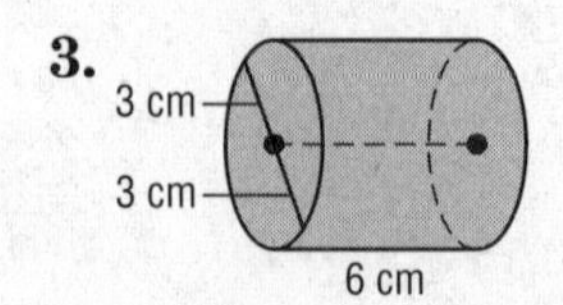

4.

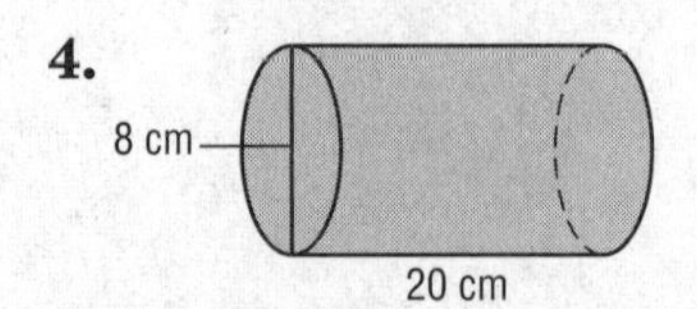

5.

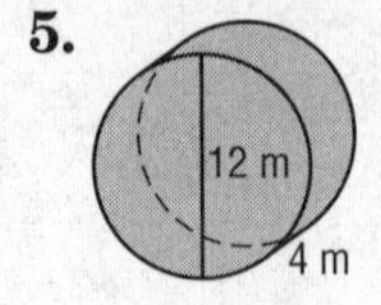

6.

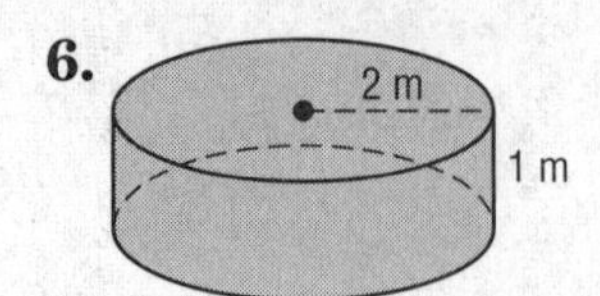

NAME ______________________ DATE ____________ PERIOD ____________

12-3 Study Guide and Intervention

Surface Areas of Pyramids and Cones

Lateral and Surface Areas of Pyramids A **pyramid** is a solid with a polygon base. The lateral faces intersect in a common point known as the vertex. The altitude is the segment from the vertex that is perpendicular to the base. For a **regular pyramid,** the base is a regular polygon and the altitude has an endpoint at the center of the base. All the lateral edges are congruent and all the lateral faces are congruent isosceles triangles. The height of each lateral face is called the **slant height.**

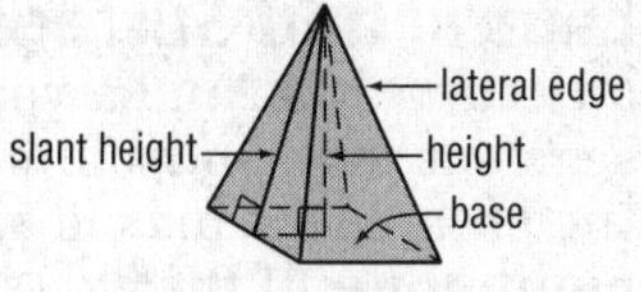

Lateral Area of a Regular Pyramid	The lateral area L of a regular pyramid is $L = \frac{1}{2}P\ell$, where ℓ is the slant height and P is the perimeter of the base.
Surface Area of a Regular Pyramid	The surface area S of a regular pyramid is $S = \frac{1}{2}P\ell + B$, where ℓ is the slant height, P is the perimeter of the base, and B is the area of the base.

Example **For the regular square pyramid above, find the lateral area and surface area if the length of a side of the base is 12 centimeters and the height is 8 centimeters. Round to the nearest tenth if necessary.**

Find the slant height.

$\ell^2 = 6^2 + 8^2$ Pythagorean Theorem

$\ell^2 = 100$ Simplify.

$\ell = 10$ Take the positive square root of each side.

$L = \frac{1}{2}P\ell$ Lateral area of a regular pyramid

$= \frac{1}{2}(48)(10)$ $P = 4 \cdot 12$ or 48, $\ell = 10$

$= 240$ Simplify.

$S = \frac{1}{2}P\ell + B$ Surface area of a regular pyramid

$= (240) + 144$ $P\ell = 240$, $B = 12 \cdot 12$ or 144

$= 384$

The lateral area is 240 square centimeters, and the surface area is 384 square centimeters.

Exercises

Find the lateral area and surface area of each regular pyramid. Round to the nearest tenth if necessary.

1.

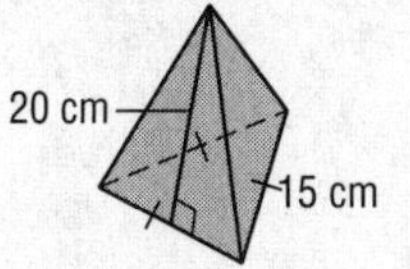

2.

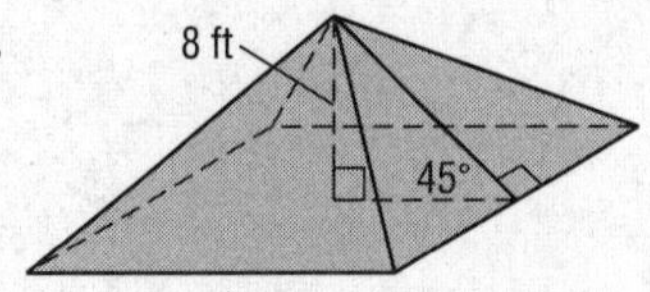

3.

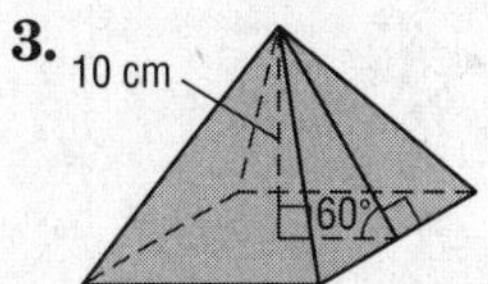

4.

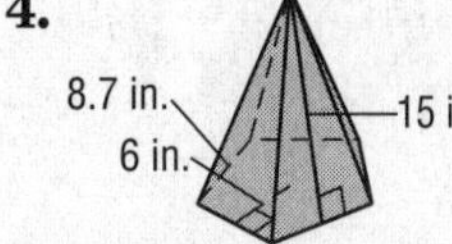

12-3 Study Guide and Intervention *(continued)*

Surface Areas of Pyramids and Cones

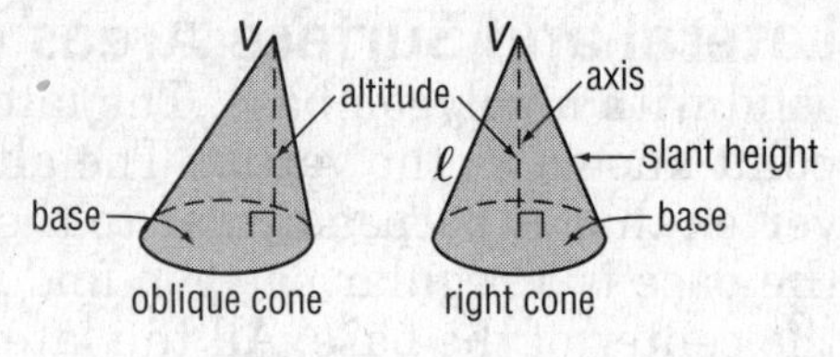

Lateral and Surface Areas of Cones A cone has a circular base and a vertex. The axis of the cone is the segment with endpoints at the vertex and the center of the base. If the axis is also the altitude, then the cone is a **right cone**. If the axis is not the altitude, then the cone is an **oblique cone**.

Lateral Area of a Cone	The lateral area L of a right circular cone is $L = \pi r\ell$, where r is the radius and ℓ is the slant height.
Surface Area of a Cone	The surface area S of a right cone is $S = \pi r\ell + \pi r^2$, where r is the radius and ℓ is the slant height.

Example **For the right cone above, find the lateral area and surface area if the radius is 6 centimeters and the height is 8 centimeters. Round to the nearest tenth if necessary.**

Find the slant height.

$\ell^2 = 6^2 + 8^2$ Pythagorean Theorem

$\ell^2 = 100$ Simplify.

$\ell = 10$ Take the positive square root of each side.

$L = \pi r\ell$ Lateral area of a right cone

$= \pi(6)(10)$ $r = 6, \ell = 10$

≈ 188.5 Simplify.

$S = \pi r\ell + \pi r^2$ Surface area of a right cone

$\approx 188.5 + \pi(6^2)$ $\pi r\ell \approx 188.4, r = 6$

≈ 301.6 Simplify.

The lateral area is about 188.5 square centimeters and the surface area is about 301.6 square centimeters.

Exercises

Find the lateral area and surface area of each cone. Round to the nearest tenth if necessary.

1.

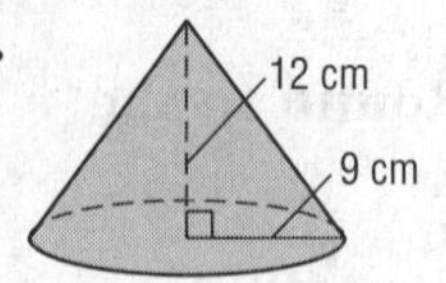

2.

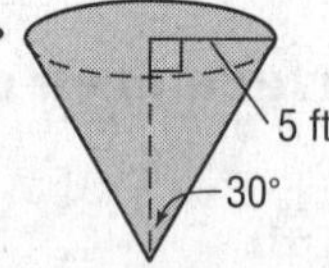

3.

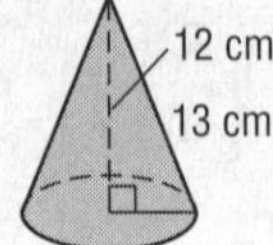

4.

12-4 Study Guide and Intervention

Volumes of Prisms and Cylinders

Volumes of Prisms The measure of the amount of space that a three-dimensional figure encloses is the **volume** of the figure. Volume is measured in units such as cubic feet, cubic yards, or cubic meters. One cubic unit is the volume of a cube that measures one unit on each edge.

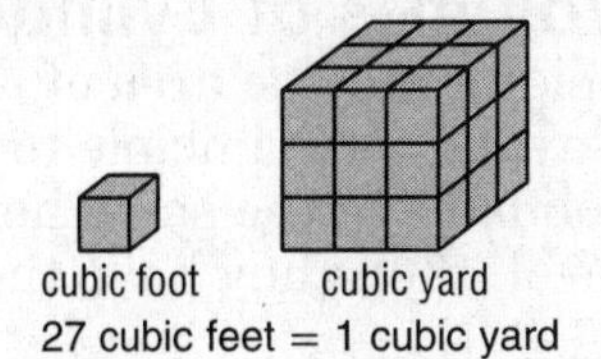

Volume of a Prism	If a prism has a volume of V cubic units, a height of h units, and each base has an area of B square units, then $V = Bh$.

Example 1 **Find the volume of the prism.**

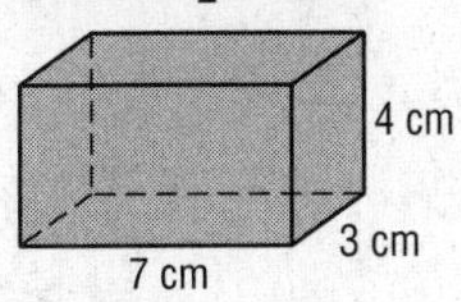

$V = Bh$ Volume of a prism

$= (7)(3)(4)$ $B = (7)(3)$, $h = 4$

$= 84$ Multiply.

The volume of the prism is 84 cubic centimeters.

Example 2 **Find the volume of the prism if the area of each base is 6.3 square feet.**

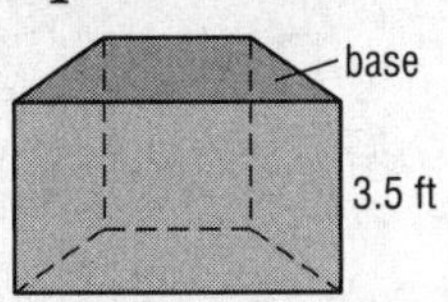

$V = Bh$ Volume of a prism

$= (6.3)(3.5)$ $B = 6.3$, $h = 3.5$

$= 22.05$ Multiply.

The volume is 22.05 cubic feet.

Exercises

Find the volume of each prism.

1.

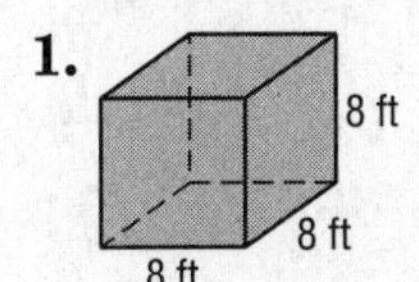

2.

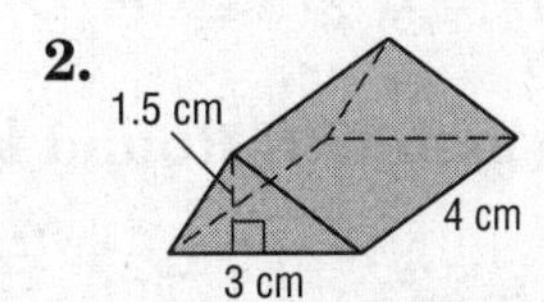

3.

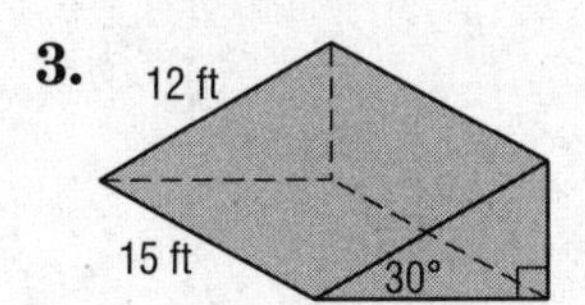

4.

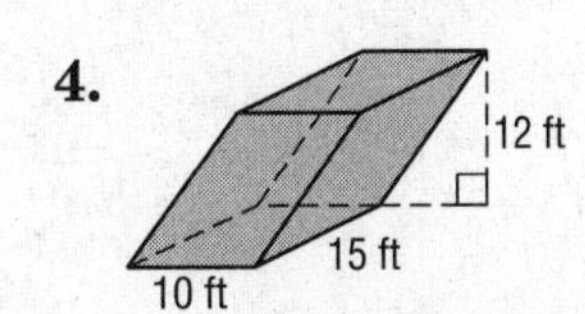

5.

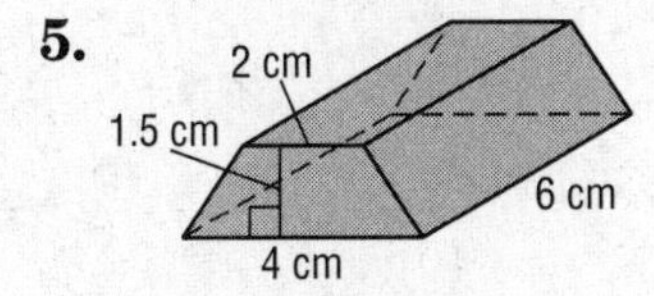

6.

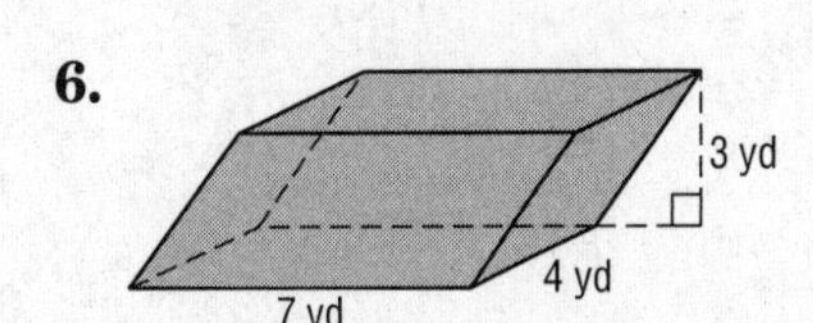

NAME _______________ DATE _______ PERIOD _______

12-4 Study Guide and Intervention *(continued)*

Volumes of Prisms and Cylinders

Volumes of Cylinders The volume of a cylinder is the product of the height and the area of the base. When a solid is not a right solid, use Cavalieri's Priniciple to find the volume. The principle states that if two solids have the same height and the same cross sectional area at every level, then they have the same volume.

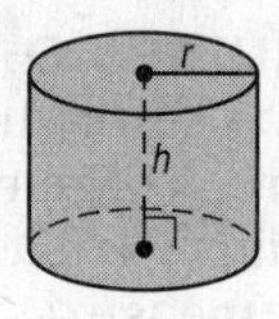

Volume of a Cylinder	If a cylinder has a volume of V cubic units, a height of h units, and the bases have a radius of r units, then $V = \pi r^2 h$.

Example 1 **Find the volume of the cylinder.**

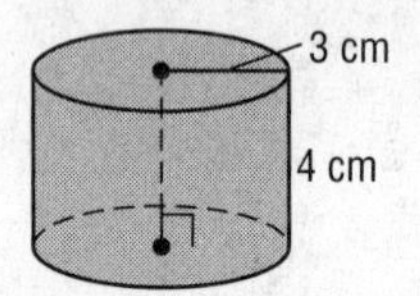

$V = \pi r^2 h$ Volume of a cylinder

$= \pi(3)^2(4)$ $r = 3, h = 4$

≈ 113.1 Simplify.

The volume is about 113.1 cubic centimeters.

Example 2 **Find the volume of the oblique cylinder.**

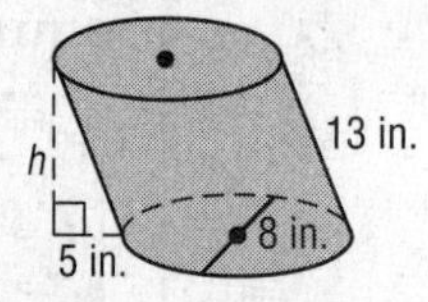

Use the Pythagorean Theorem to find the height of the cylinder.

$h^2 + 5^2 = 13^2$ Pythagorean Theorem

$h^2 = 144$ Simplify.

$h = 12$ Take the positive square root of each side.

$V = \pi r^2 h$ Volume of a cylinder

$= \pi(4)^2(12)$ $r = 4, h = 12$

≈ 603.2 Simplify.

The Volume is about 603.2 cubic inches.

Exercises

Find the volume of each cylinder. Round to the nearest tenth.

1.

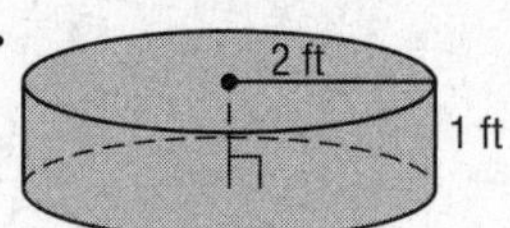

2.

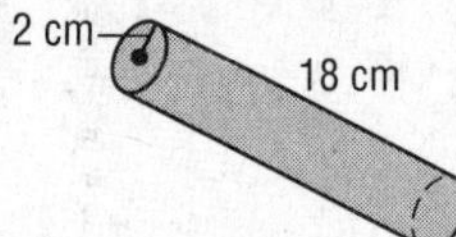

3.

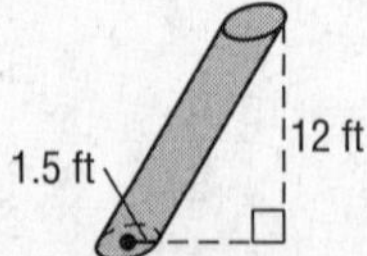

4.

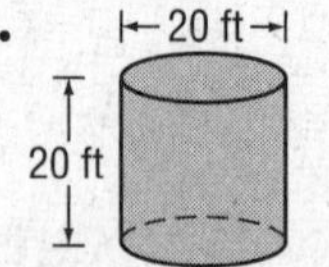

5.

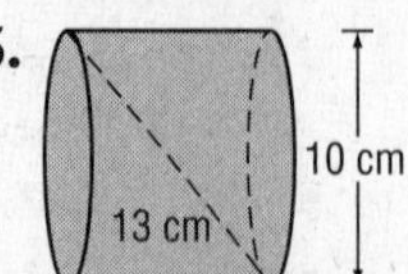

6.

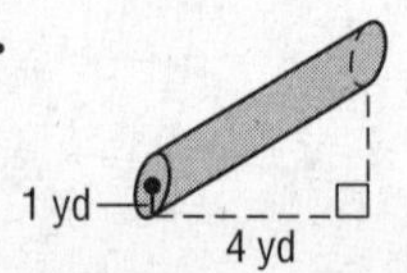

NAME ______________________ DATE ____________ PERIOD ____________

12-5 Study Guide and Intervention

Volumes of Pyramids and Cones

Volumes of Pyramids This figure shows a prism and a pyramid that have the same base and the same height. It is clear that the volume of the pyramid is less than the volume of the prism. More specifically, the volume of the pyramid is one-third of the volume of the prism.

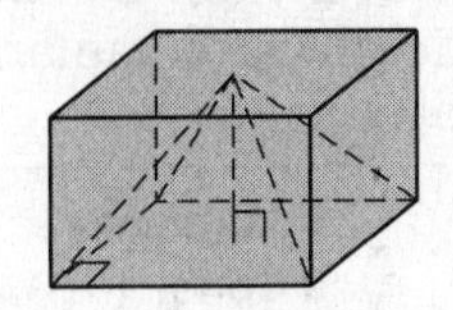

Volume of a Pyramid	If a pyramid has a volume of V cubic units, a height of h units, and a base with an area of B square units, then $V = \frac{1}{3}Bh$.

Example **Find the volume of the square pyramid.**

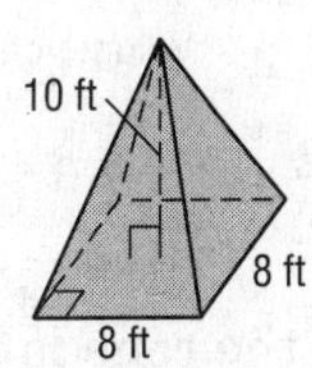

$V = \frac{1}{3}Bh$ Volume of a pyramid

$= \frac{1}{3}(8)(8)10$ $B = (8)(8)$, $h = 10$

≈ 213.3 Multiply.

The volume is about 213.3 cubic feet.

Exercises

Find the volume of each pyramid. Round to the nearest tenth if necessary.

1.

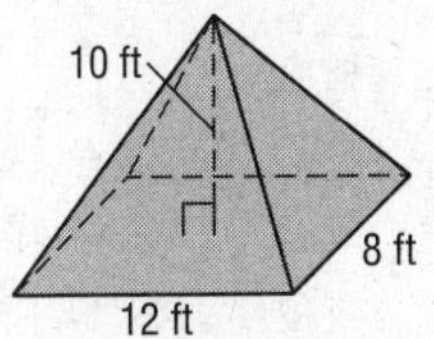

2.

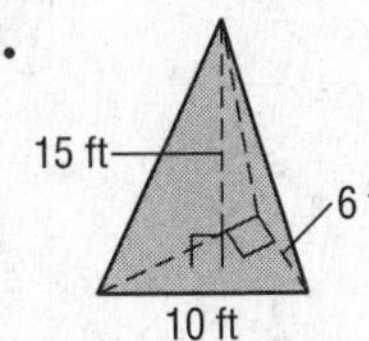

3.

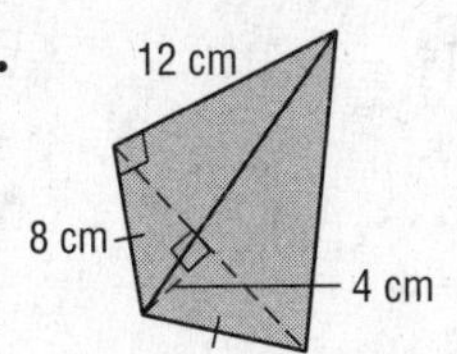

4.

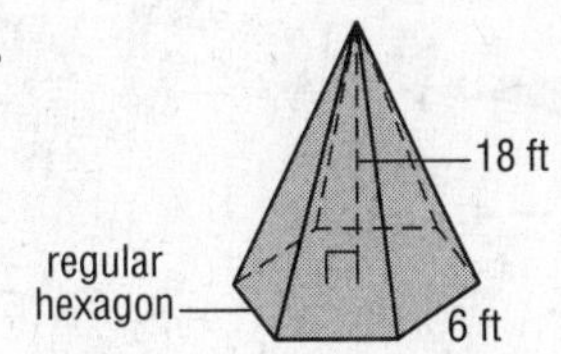

5.

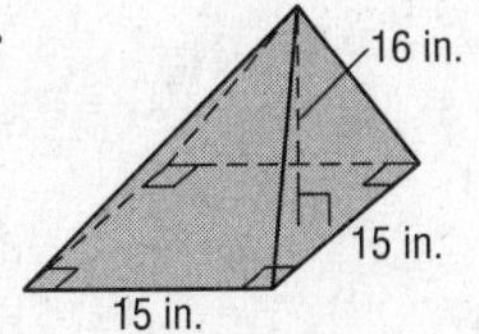

6.

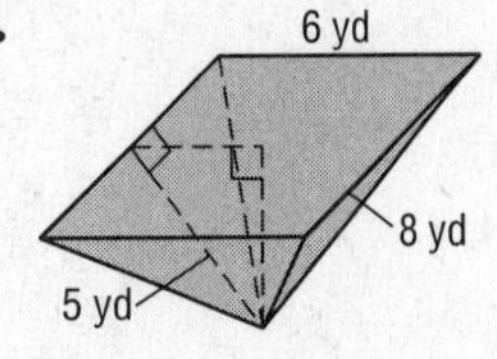

NAME ______________________ DATE ____________ PERIOD ____________

12-5 Study Guide and Intervention *(continued)*

Volumes of Pyramids and Cones

Volumes of Cones For a cone, the volume is one-third the product of the height and the area of the base. The base of a cone is a circle, so the area of the base is πr^2.

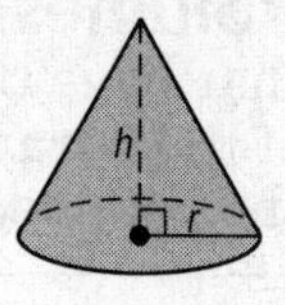

Volume of a Cone	If a cone has a volume of V cubic units, a height of h units, and the bases have a radius of r units, then $V = \frac{1}{3}\pi r^2 h$.

Example **Find the volume of the cone.**

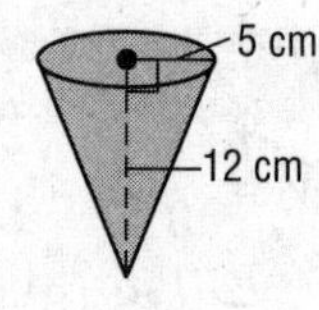

$V = \frac{1}{3}\pi r^2 h$ Volume of a cone

$= \frac{1}{3}\pi(5)^2 12$ $r = 5, h = 12$

≈ 314.2 Simplify.

The volume of the cone is about 314.2 cubic centimeters.

Exercises

Find the volume of each cone. Round to the nearest tenth.

1.

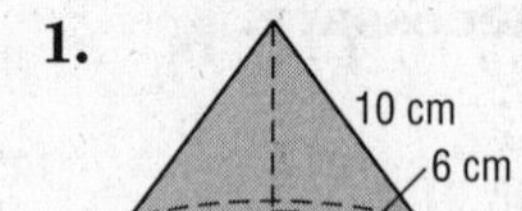

2.

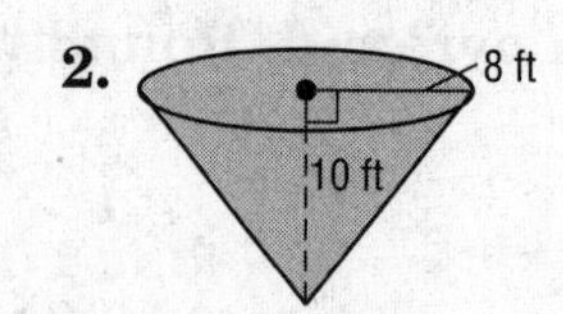

3.

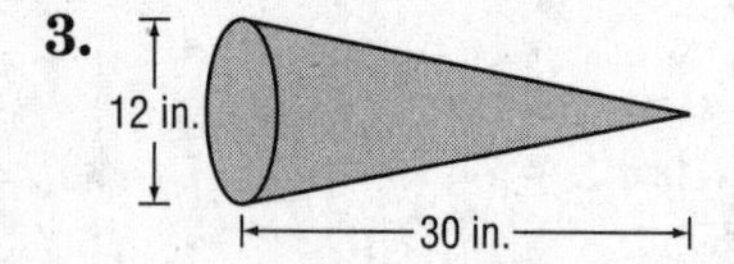

4.

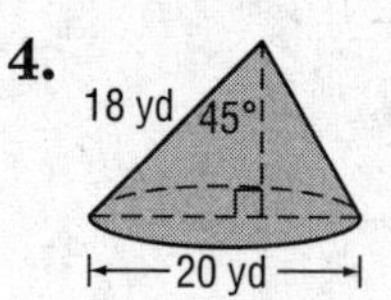

5.

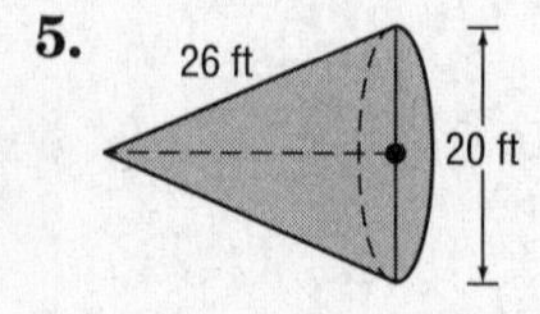

6.

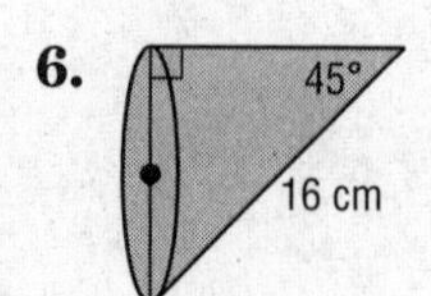

NAME ______________________________ DATE ____________ PERIOD ____________

12-6 Study Guide and Intervention

Surface Areas and Volumes of Spheres

Surface Areas of Spheres You can think of the surface area of a sphere as the total area of all of the nonoverlapping strips it would take to cover the sphere. If r is the radius of the sphere, then the area of a **great circle** of the sphere is πr^2. The total surface area of the sphere is four times the area of a great circle.

Surface Area of a Sphere	If a sphere has a surface area of S square units and a radius of r units, then $S = 4\pi r^2$.

Example **Find the surface area of a sphere to the nearest tenth if the radius of the sphere is 6 centimeters.**

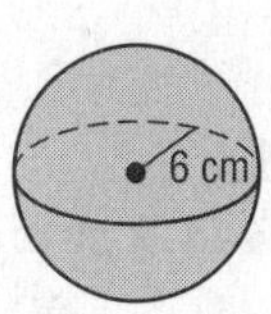

$S = 4\pi r^2$ Surface area of a sphere

$= 4\pi(6)^2$ $r = 6$

≈ 452.4 Simplify.

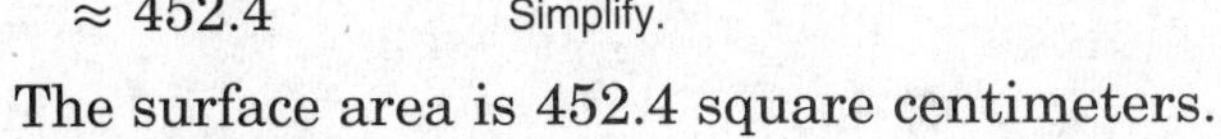

The surface area is 452.4 square centimeters.

Exercises

Find the surface area of each sphere or hemisphere. Round to the nearest tenth.

1.

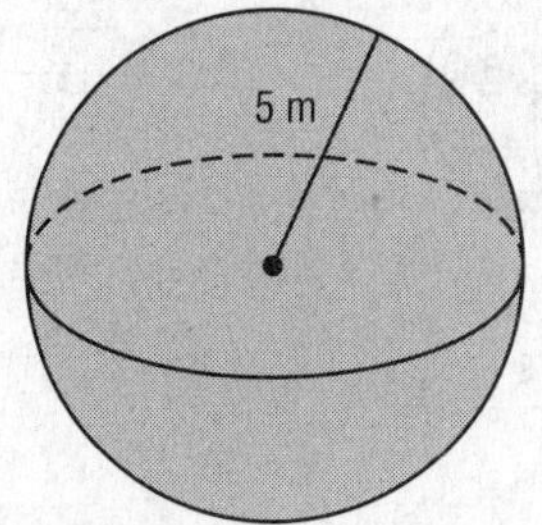

2.

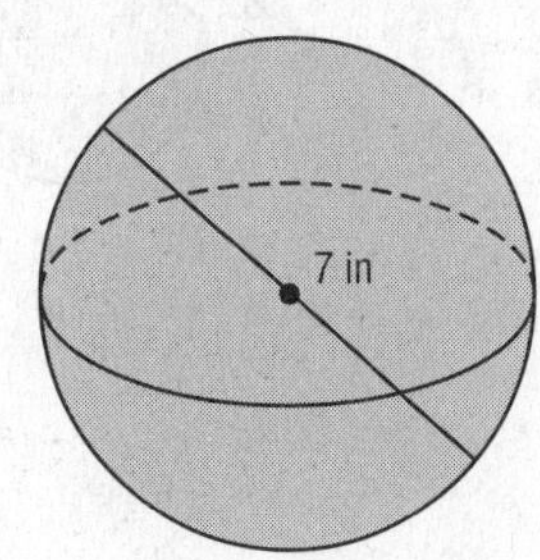

3.

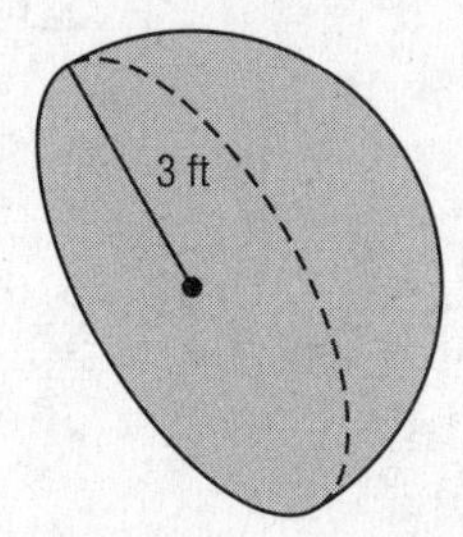

4.

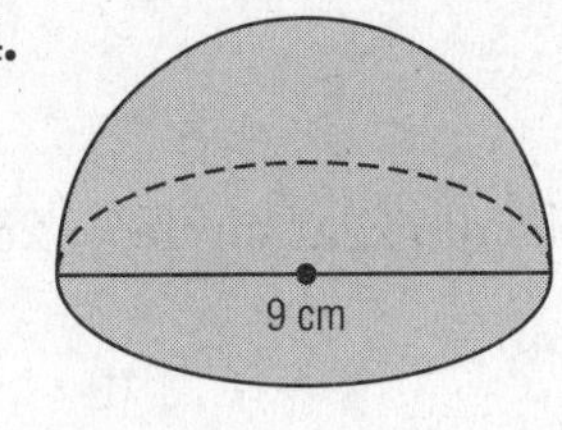

5. sphere: circumference of great circle = π cm

6. hemisphere: area of great circle $\approx 4\pi$ ft^2

NAME ______________________ DATE __________ PERIOD __________

12-6 Study Guide and Intervention *(continued)*

Surface Areas and Volumes of Spheres

Volumes of Spheres A sphere has one basic measurement, the length of its radius. If you know the length of the radius of a sphere, you can calculate its volume.

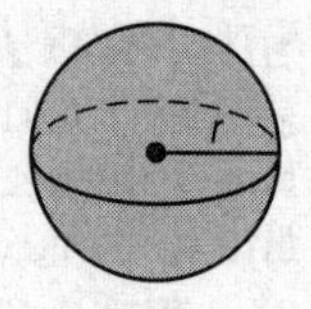

Volume of a Sphere	If a sphere has a volume of V cubic units and a radius of r units, then $V = \frac{4}{3}\pi r^3$.

Example **Find the volume of a sphere with radius 8 centimeters.**

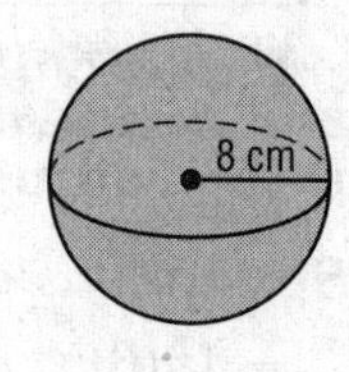

$V = \frac{4}{3}r^3$ Volume of a sphere

$= \frac{4}{3}\pi(8)^3$ $r = 8$

≈ 2144.7 Simplify.

The volume is about 2144.7 cubic centimeters.

Exercises

Find the volume of each sphere or hemisphere. Round to the nearest tenth.

1.

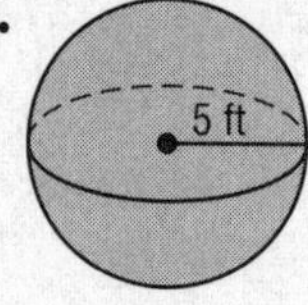

2.

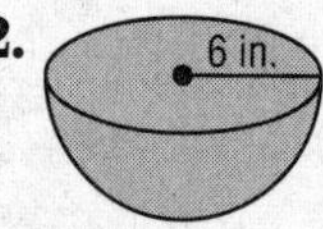

3.

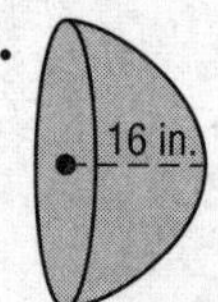

4. hemisphere: radius 5 in.

5. sphere: circumference of great circle ≈ 25 ft

6. hemisphere: area of great circle ≈ 50 m²

NAME ______________________________ DATE ____________ PERIOD ____________

12-7 Study Guide and Intervention

Spherical Geometry

Geometry On A Sphere Up to now, we have been studying **Euclidean geometry,** where a plane is a flat surface made up of points that extends infinitely in all directions. In **spherical geometry,** a plane is the surface of a sphere.

Example **Name each of the following on sphere** $\mathcal{K}$**.**

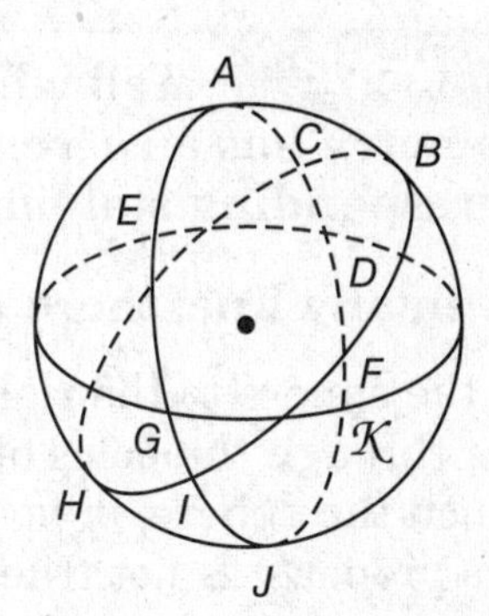

a. two lines containing the point F

$\overleftrightarrow{EG}$ and $\overleftrightarrow{BH}$ are lines on sphere $\mathcal{K}$ that contain the point F

b. a line segment containing the point J

$\overline{ID}$ is a segment on sphere $\mathcal{K}$ that contains the point J

c. a triangle

$\triangle IFG$ is a triangle on sphere $\mathcal{K}$

Exercises

Name two lines containing point Z**, a segment containing point** R**, and a triangle in each of the following spheres.**

1.

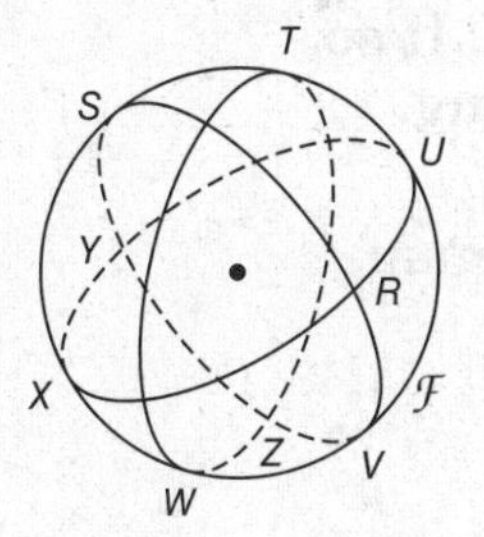

2.

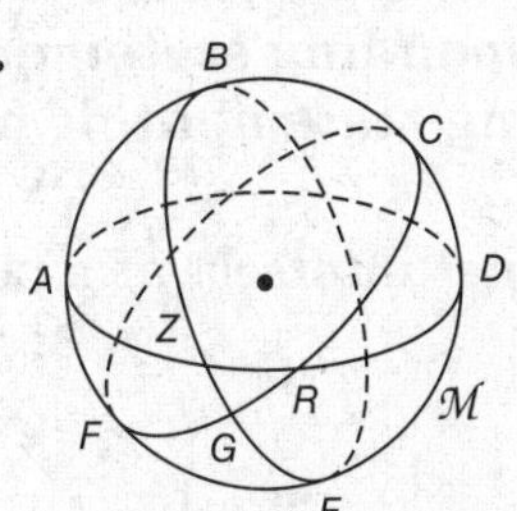

Determine whether figure u **on each of the spheres shown is a line in spherical geometry.**

3.

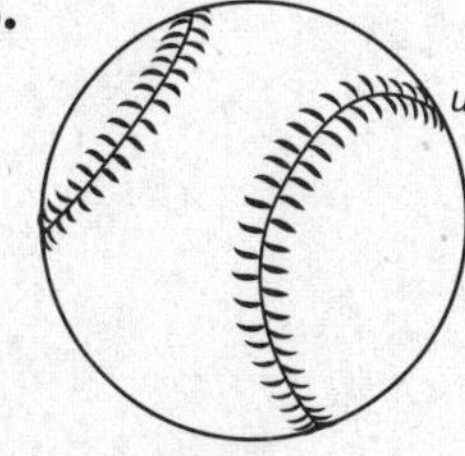

4.

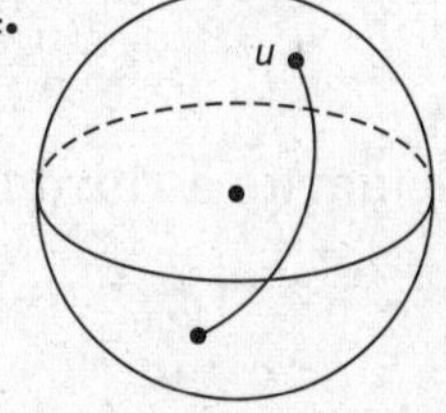

5. GEOGRAPHY Lines of latitude run horizontally across the surface of Earth. Are there any lines of latitude that are great circles? Explain.

12-7 Study Guide and Intervention *(continued)*

Spherical Geometry

Comparing Euclidean and Spherical Geometries Some postulates and properties of Euclidean geometry are true in spherical geometry. Others are not true or are true only under certain circumstances.

Example **Tell whether the following postulate or property of plane Euclidean geometry has a corresponding statement in spherical geometry. If so, write the corresponding statement. If not, explain your reasoning.**

Given any line, there are an infinite number of parallel lines.

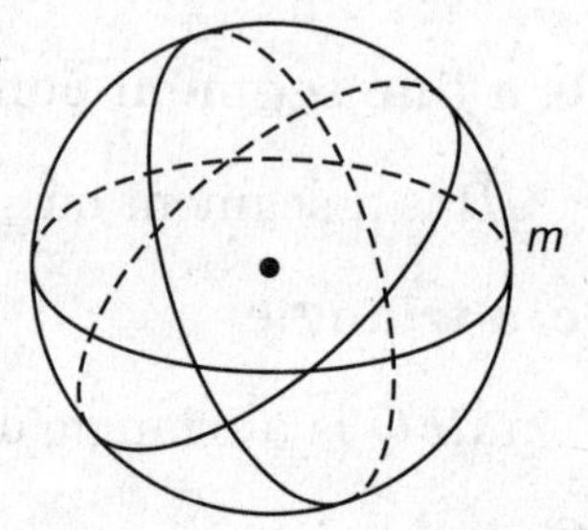

On the sphere to the right, if we are given line m we see that it goes through the poles of the sphere. If we try to make any other line on the sphere, it would intersect line m at exactly 2 points. This property is not true in spherical geometry.

A corresponding statement in spherical geometry would be: "Given any line, there are no parallel lines."

Exercises

Tell whether the following postulate or property of plane Euclidean geometry has a corresponding statement in spherical geometry. If so, write the corresponding statement. If not, explain your reasoning.

1. If two nonidentical lines intersect at a point, they do not intersect again.

2. Given a line and a point on the line, there is only one perpendicular line going through that point.

3. Given two parallel lines and a transversal, alternate interior angles are congruent.

4. If two lines are perpendicular to a third line, they are parallel.

5. Three noncollinear points determine a triangle.

6. A largest angle of a triangle is opposite the largest side.

12-8 Study Guide and Intervention

Congruent and Similar Solids

Identify Congruent or Similar Solids **Similar solids** have exactly the same shape but not necessarily the same size. Two solids are similar if they are the same shape and the ratios of their corresponding linear measures are equal. All spheres are similar and all cubes are similar. **Congruent solids** have exactly the same shape and the same size. Congruent solids are similar solids with a scale factor of 1:1. Congruent solids have the following characteristics:

- Corresponding angles are congruent
- Corresponding edges are congruent
- Corresponding faces are congruent
- Volumes are equal

Example **Determine whether the pair of solids is *similar*, *congruent*, or *neither*. If the solids are similar, state the scale factor.**

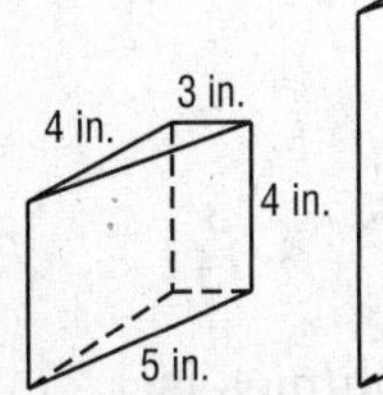

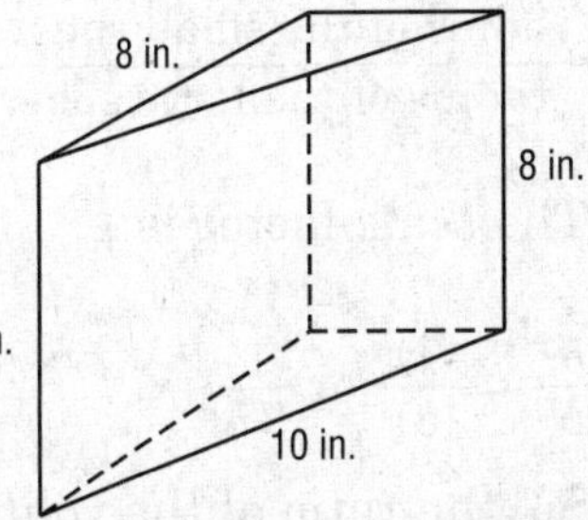

ratio of width: $\frac{3}{6} = \frac{1}{2}$ ratio of length: $\frac{4}{8} = \frac{1}{2}$

ratio of hypotenuse: $\frac{5}{10} = \frac{1}{2}$ ratio of height: $\frac{4}{8} = \frac{1}{2}$

The ratios of the corresponding sides are equal, so the triangular prisms are similar. The scale factor is 1:2. Since the scale factor is not 1:1, the solids are not congruent.

Exercises

Determine whether the pair of solids is *similar*, *congruent*, or *neither*. If the solids are similar, state the scale factor.

1.

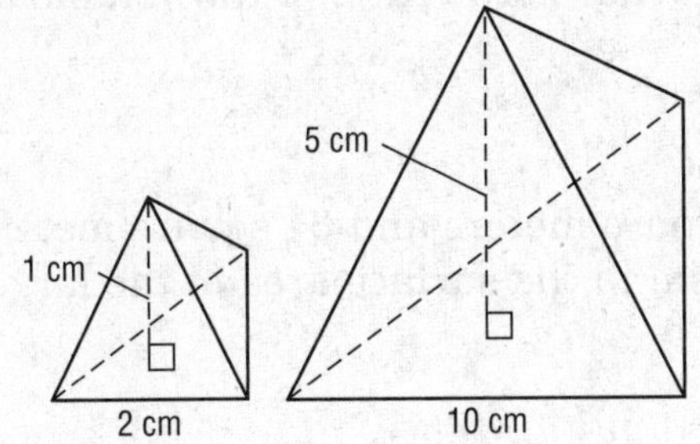

2.

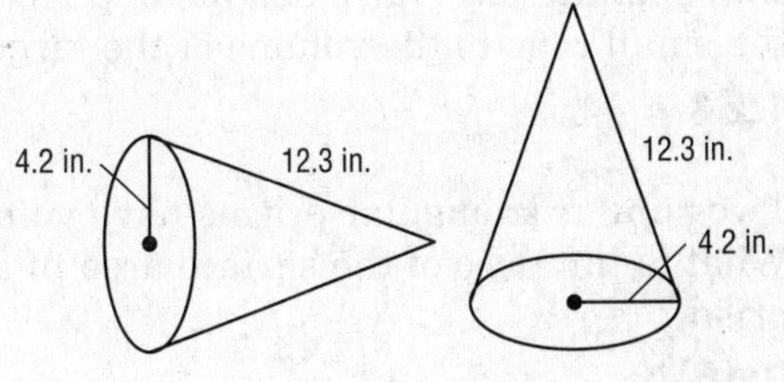

3.

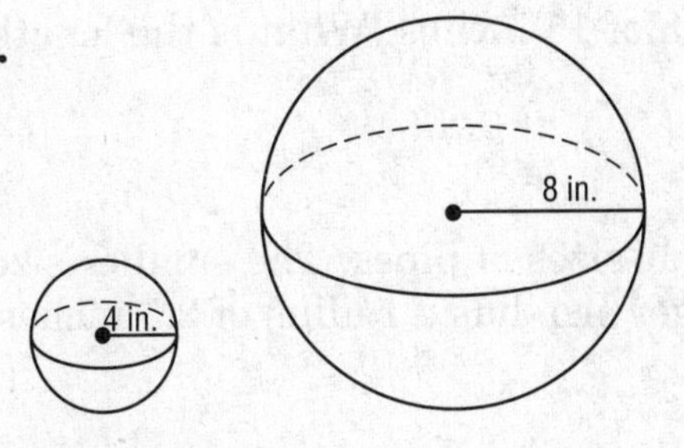

4.

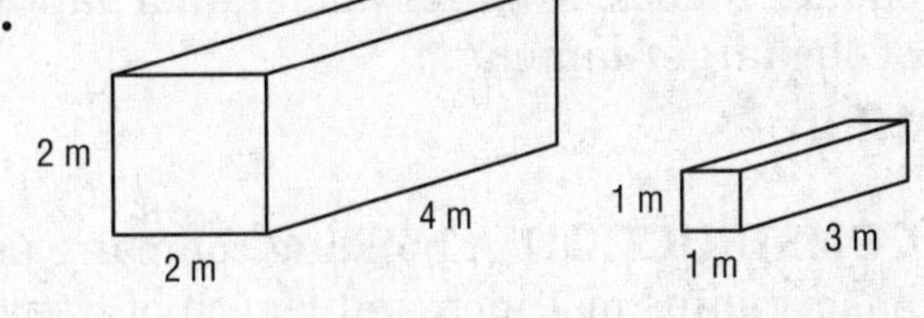

12-8 Study Guide and Intervention *(continued)*

Congruent and Similar Solids

Properties of Congruent or Similar Solids When pairs of solids are congruent or similar, certain properties are known.

If two similar solids have a scale factor of a:b then,

- the ratio of their surface areas is a^2:b^2.
- the ratio of their volumes is a^3:b^3.

Example **Two spheres have radii of 2 feet and 6 feet. What is the ratio of the volume of the small sphere to the volume of the large sphere?**

First, find the scale factor.

$$\frac{\text{radius of the small sphere}}{\text{radius of the large sphere}} = \frac{2}{6} \text{ or } \frac{1}{3}$$

The scale factor is $\frac{1}{3}$.

$$\frac{a^3}{b^3} = \frac{(1)^3}{(3)^3} \text{ or } \frac{1}{27}$$

So, the ratio of the volumes is 1:27.

Exercises

1. Two cubes have side lengths of 3 inches and 8 inches. What is the ratio of the surface area of the small cube to the surface area of the large cube?
   9:64

2. Two similar cones have heights of 3 feet and 12 feet. What is the ratio of the volume of the small cone to the volume of the large cone?
   1:64

3. Two similar triangular prisms have volumes of 27 square meters and 64 square meters. What is the ratio of the surface area of the small prism to the surface area of the large prism?
   9:16

4. **COMPUTERS** A small rectangular laptop has a width of 10 inches. and an area of 80 square inches. A larger and similar laptop has a width of 15 inches. What is the length of the larger laptop?
   12 in.

5. **CONSTRUCTION** A building company uses two similar sizes of pipes. The smaller size has a radius of 1 inch and length of 8 inches. The larger size has a radius of 2.5 inches. What is the volume of the larger pipes?
   125π cubic in.

NAME ______________________________ DATE ____________ PERIOD ____________

13-1 Study Guide and Intervention

Representing Sample Spaces

Represent a Sample Space The **sample space** of an experiment is the set of all possible outcomes. A sample space can be found using an organized list, table, or tree diagram.

Example **Maurice packs suits, shirts, and ties that can be mixed and matched. Using the packing list at the right, draw a tree diagram to represent the sample space for business suit combinations.**

Maurice's Packing List
1. Suits: Gray, black, khaki
2. Shirts: White, light blue
3. Ties: Striped (But optional)

The sample space is the result of three stages:

- Suit color (G, B, or K)
- Shirt color (W or L)
- Tie (T or NT)

Draw a tree diagram with three stages.

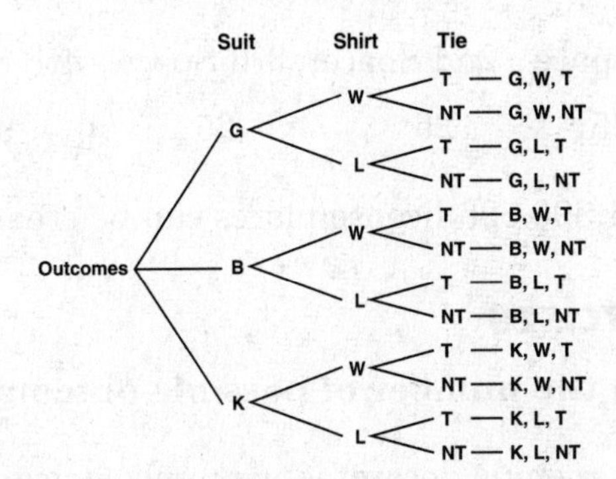

Exercises

Represent the sample space for each experiment by making an organized list, a table, and a tree diagram.

1. The baseball team can wear blue or white shirts with blue or white pants.

2. The dance club is going to see either *Sleeping Beauty* or *The Nutcracker* at either Symphony Hall or The Center for the Arts.

3. Mikey's baby sister can drink either apple juice or milk from a bottle or a toddler cup.

4. The first part of the test consisted of two true-or-false questions.

NAME ______________________ DATE ____________ PERIOD ____________

13-1 Study Guide and Intervention *(continued)*

Representing Sample Spaces

Fundamental Counting Principle The number of all possible outcomes for an experiment can be found by multiplying the number of possible outcomes from each stage or event.

Example **The pattern for a certain license plate is 3 letters followed by 3 numbers. The letter "O" is not used as any of the letters and the number "0" is not used as any of the numbers. Any other letter or number can be used multiple times. How many license plates can be created with this pattern?**

Use the Fundamental Counting Principle.

1st Space		2nd Space		3rd Space		4th Space		5th Space		6th Space		Possible Outcomes
25	×	25	×	25	×	9	×	9	×	9	=	11,390,625

So 11,390,625 license plates can be created with this pattern.

Exercises

Find the number of possible outcomes for each situation.

1. A room is decorated with one choice from each category.

Bedroom Décor	Number of Choices
Paint color	8
Comforter set	6
Sheet set	8
Throw rug	5
Lamp	3
Wall hanging	5

2. A lunch at Lincoln High School contains one choice from each category.

Cafeteria Meal	Number of Choices
Main dish	3
Side dish	4
Vegetable	2
Salad	2
Salad Dressing	3
Dessert	2
Drink	3

3. In a catalog of outdoor patio plans, there are 4 types of stone, 3 types of edgers, 5 dining sets and 6 grills. Carl plans to order one item from each category.

4. The drama club held tryouts for 6 roles in a one-act play. Five people auditioned for lead female, 3 for lead male, 8 for the best friend, 4 for the mom, 2 for the dad, and 3 for the crazy aunt.

NAME ______________________________ DATE ______________ PERIOD ________

13-2 Study Guide and Intervention

Probability with Permutations and Combinations

Probability Using Permutations A **permutation** is an arrangement of objects where order is important. To find the number of permutations of a group of objects, use the **factorial**. A factorial is written using a number and !. The following are permutation formulas:

$n! = n \cdot (n-1) \cdot (n-2) \cdot \ldots \cdot 2 \cdot 1$

$5! = 5 \cdot 4 \cdot 3 \cdot 2 \cdot 1 \cdot = 120$

n distinct objects taken r at a time	${}_nP_r = \frac{n!}{(n-r)!}$
n objects, where one object is repeated r_1 times, another is repeated r_2 times, and so on	$\frac{n!}{r_1! \cdot r_2! \cdot \ldots \cdot r_k!}$
n objects arranged in a circle with no fixed reference point	$\frac{n!}{n}$ or $(n-1)!$

Example **The cheer squad is made up of 12 girls. A captain and a co-captain are selected at random. What is the probability that Chantel and Cadence are chosen as leaders?**

Find the number of possible outcomes.

$${}_{12}P_2 = \frac{12!}{(12-2)!} = \frac{12!}{10!} = 12 \cdot 11 = 132$$

Find the number of favorable outcomes.

$$2! = 2$$

The probability of Chantel and Cadence being chosen is

$$\frac{\text{favorable outcomes}}{\text{total number of outcomes}} = \frac{2}{132} = \frac{1}{66}$$

Exercises

1. **BOOKS** You have a textbook for each of the following subjects: Spanish, English, Chemistry, Geometry, History, and Psychology. If you choose 4 of these at random to arrange on a shelf, what is the probability that the Geometry textbook will be first from the left and the Chemistry textbook will be second from the left?

2. **CLUBS** The Service Club is choosing members at random to attend one of four conferences in LA, Atlanta, Chicago, and New York. There are 20 members in the club. What is the probability that Lana, Sherry, Miguel, and Jerome are chosen for these trips?

3. **TELEPHONE NUMBERS** What is the probability that a 7-digit telephone number generated using the digits 2, 3, 2, 5, 2, 7, and 3 is the number 222-3357?

4. **DINING OUT** A group of 4 girls and 4 boys is randomly seated at a round table. What is the probability that the arrangement is boy-girl-boy-girl?

13-2 Study Guide and Intervention *(continued)*

Probability with Permutations and Combinations

Probability Using Combinations A **combination** is an arrangement of objects where order is NOT important. To find the number of combinations of n distinct objects taken r at a time, denoted by ${}_nC_r$, use the formula:

$${}_nC_r = \frac{n!}{(n-r)!\,r!}$$

Example **Taryn has 15 soccer trophies but she only has room to display 9 of them. If she chooses them at random, what is the probability that each of the trophies from the school invitational from the 1st through 9th grades will be chosen?**

Step 1 Since the order does not matter, the number of possible outcomes is

$${}_{15}C_9 = \frac{15!}{(15-9)!\,(9!)} = 5005$$

Step 2 There is only one favorable outcome—the 9 specific trophies being chosen.

Step 3 The probability that these 9 trophies are chosen is

$$\frac{\text{number of favorable outcomes}}{\text{total number of outcomes}} = \frac{1}{5005}.$$

Exercises

1. **ICE CREAM** Kali has a choice of 20 flavors for her triple scoop cone. If she chooses the flavors at random, what is the probability that the 3 flavors she chooses will be vanilla, chocolate, and strawberry?

2. **PETS** Dani has a dog walking business serving 9 dogs. If she chooses 4 of the dogs at random to take an extra trip to the dog park, what is the probability that Fifi, Gordy, Spike and Fluffy are chosen?

3. **CRITIQUE** A restaurant critic has 10 new restaurants to try. If he tries half of them this week, what is the probability that he will choose The Fish Shack, Carly's Place, Chez Henri, Casa de Jorge, and Grillarious?

4. **CHARITY** Emily is giving away part of her international doll collection to charity. She has 20 dolls, each from a different country. If she selects 10 of them at random, what is the probability she chooses the ones from Ecuador, Paraguay, Chile, France, Spain, Sweden, Switzerland, Germany, Greece, and Italy?

5. **ROLLER COASTERS** An amusement park has 12 roller coasters. Four are on the west side of the park, 4 are on the east side, and 4 are centrally located. The park's Maintenance Department randomly chooses 4 roller coasters for upgrades each month. What is the probability that all 4 roller coasters on the west side are chosen in March?

NAME ______________________ DATE ____________ PERIOD ________

13-3 Study Guide and Intervention

Geometric Probability

Probability with Length Probability that involves a geometric measure is called **geometric probability**. One type of measure is length.

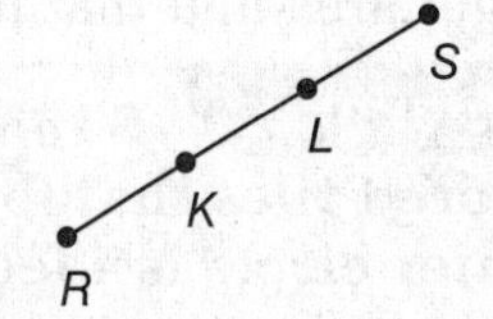

Look at line segment $\overline{KL}$.

If a point, M, is chosen at random on the line segment, then

$P(M \text{ is on } \overline{KL}) = \frac{KL}{RS}$.

Example **Point X is chosen at random on $\overline{AD}$. Find the probability that X is on $\overline{AB}$.**

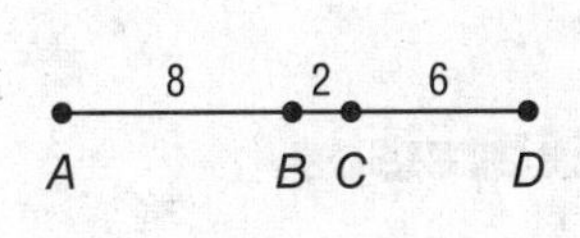

$P(X \text{ is on } \overline{AB}) = \frac{AB}{AD}$ Length probability ratio

$= \frac{8}{16}$ $AB = 8$ and $AD = 8 + 2 + 6 = 16$

$= \frac{1}{2}$, 0.5, or 50% Simplify.

Exercises

Point M is chosen at random on $\overline{ZP}$. Find the probability of each event.

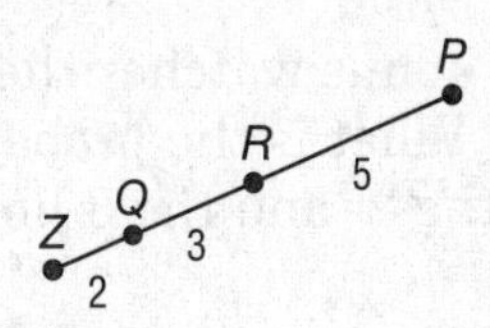

1. $P(M \text{ is on } \overline{ZQ})$

2. $P(M \text{ is on } \overline{QR})$

3. $P(M \text{ is on } \overline{RP})$

4. $P(M \text{ is on } \overline{QP})$

5. **TRAFFIC LIGHT** In a 5-minute traffic cycle, a traffic light is green for 2 minutes 27 seconds, yellow for 6 seconds, and red for 2 minutes 27 seconds. What is the probability that when you get to the light it is green?

6. **GASOLINE** Your mom's mini van has a 24 gallon tank. What is the probability that, when the engine is turned on, the needle on the gas gauge is pointing between $\frac{1}{4}$ and $\frac{1}{2}$ full?

NAME ______________________ DATE ____________ PERIOD ________

13-3 Study Guide and Intervention *(continued)*

Geometric Probability

Probability with Area Geometric probabilities can also involve area. When determining geometric probability with targets, assume that the object lands within the target area and that it is equally likely that the object will land anywhere in the region.

Example **Suppose a coin is flipped into a reflection pond designed with colored tiles that form 3 concentric circles on the bottom. The diameter of the center circle is 4 feet and the circles are spaced 2 feet apart. What is the probability the coin lands in the center?**

$$P(\text{coin lands in center}) = \frac{\text{area of center circle}}{\text{area of base of pond}}$$

$$= \frac{4\pi}{36\pi}$$

$$= \frac{1}{9}, \text{ about } 0.11, \text{ or } 11\%$$

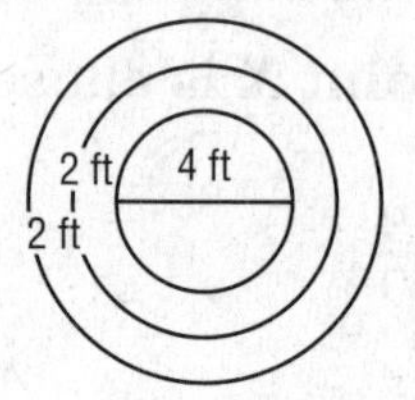

Exercises

1. **LANDING** A parachutist needs to land in the center of a target on a rectangular field that is 120 yards by 30 yards. The target is a circular design with a 10 yard radius. What is the probability the parachutist lands somewhere in the target?

2. **CLOCKS** Jonus watches the second hand on an analog clock as it moves past the numbers. What is the probability that at any given time the second hand on a clock is between the 2- and the 3-hour numbers?

Find the probability that a point chosen at random lies in the shaded region.

3.

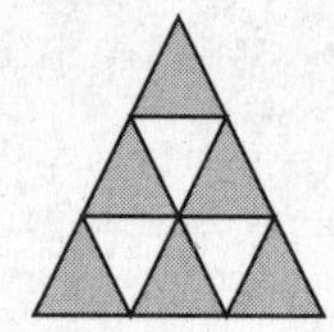

4.

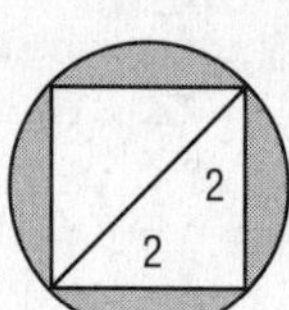

5.

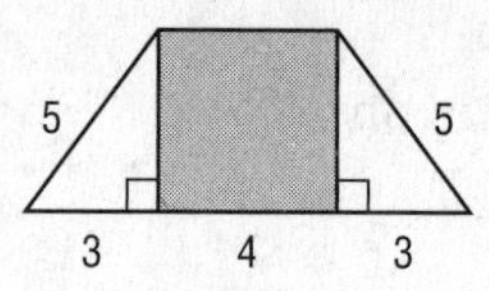

Use the spinner to find each probability. If the spinner lands on a line it is spun again.

6. P(pointer landing on red)

7. P(pointer landing on blue)

8. P(pointer landing on green)

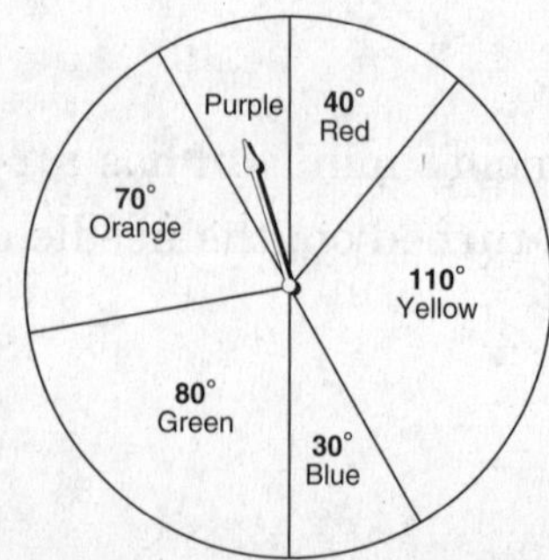

13-4 Study Guide and Intervention

Simulations

Design a Simulation A **probability model** is a mathematical model that matches something that happens randomly. A **simulation** is a way to use the model to recreate a situation to help determine the situation's probability.

To design a simulation:

1. Determine each possible outcome and its theoretical probability.
2. State any assumptions.
3. Describe an appropriate probability model for the situation.
4. Define a trial for the situation and state the number of trials to be conducted.

Example **Joni got on base 40% of her times at bat last season. Design a simulation to determine the probability that she will get on base in her next at bat this season.**

The possible outcomes are Joni gets on base (40%) and Joni doesn't get on base (60%). Assume that Joni will have 90 at bats this season.

Use a spinner divided into two sectors, one containing 40% of the spinner's area, or a central angle of 144°, and the other 60%, or 216°. A trial, one spin of the spinner, will represent one at bat. A successful trial will be getting on base and a failed trial will be not getting on base. The simulation will contain 90 trials.

Exercises

Design a simulation using a geometric probability model.

1. **WRESTLING** Carlos is the star of the wrestling team. Carlos pinned 80% of his opponents in wrestling matches last season.

2. **JEANS** A trendy jeans store sells jeans in 4 different styles. Last year 45% of their sales was straight leg jeans, 30% was boot cut jeans, 15% was low rise jeans, and 10% was easy fit.

3. **MOVIE RENTALS** A local video store has 5 videos in its fairytale section. Last month Cinderella was rented 35%, Snow White was rented 30%, Sleeping Beauty was rented 20%, Rumpelstiltskin 10%, and Rapunzel 5%.

13-4 Study Guide and Intervention *(continued)*

Simulations

Summarize Data from a Simulation After a simulation is created, the results must be reported with numerical and graphical displays of the data. Compare theoretical and experimental probabilities or expected and average values depending on the type of simulation you run.

Example **In a carnival game, a ball is rolled up an incline toward circular regions with different point values. The center circle has a diameter of 6 inches and each successive circle has a radius 4 inches greater than the previous circle.**

Let the random variable X represent the point value assigned to a region on the game. The expected value $E(X)$ is found by adding the products of each region's point value and the geometric probability of landing in that region.

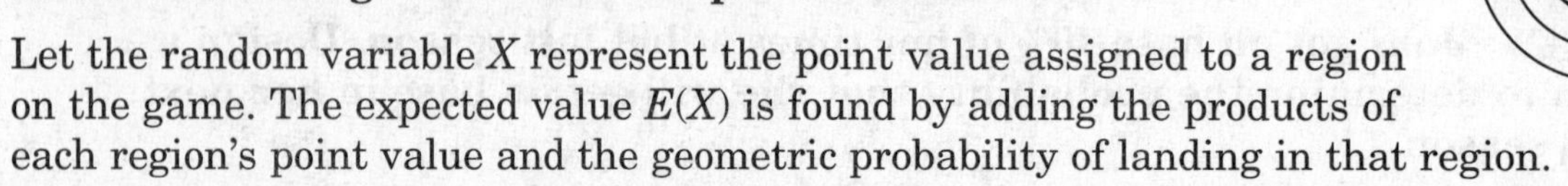

$$E(X) = 100 \cdot \frac{72}{121} + 200 \cdot \frac{40}{121} + 300 \cdot \frac{9}{121} \approx 148$$

The frequency table shows the result of the simulation after using a graphing calculator to generate 50 trials. Use these numbers to construct a bar graph and to calculate average value.

Outcome	Frequency
Region 100	19
Region 200	16
Region 300	15
Total	50

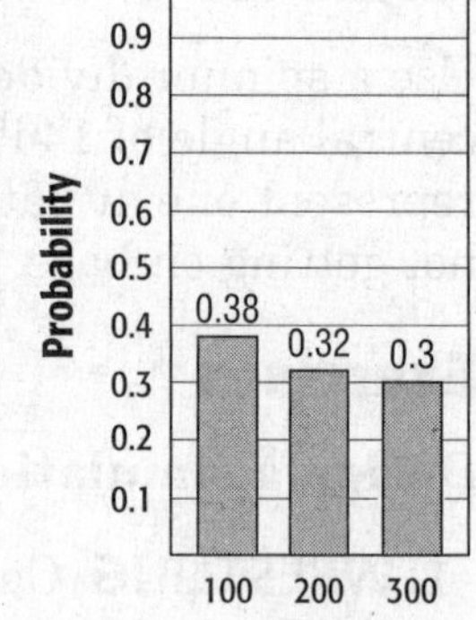

$$\text{Average value} = 100 \cdot \frac{19}{50} + 200 \cdot \frac{16}{50} + 300 \cdot \frac{15}{50} = 192$$

The average value is higher than the expected value.

Exercises

1. BASEBALL For a particular baseball player, out of the total number of times he reaches base he gets a walk 6% of the time, a single 55% of the time, a double 30% of the time, a triple 1% of the time, and a home run 8% of the time. The frequency table shows the results of a simulation. Construct a bar graph and compare the experimental probabilities with the theoretical probabilities.

Outcome	Frequency
Walk	5
Single	60
Double	25
Triple	0
Home run	10
Total	100

2. CARNIVAL In a game similar to the game in the above Example, there are four regions in which the ball can fall. The probability that Jani can get 100 points in a roll is 25%, the probability of 200 points is 50%, of 300 points is 20%, and of 400 points is 5%. Calculate the expected value for each roll.

NAME ______________ DATE ______________ PERIOD ______________

13-5 Study Guide and Intervention

Probabilities of Independent and Dependent Events

Independent and Dependent Events **Compound events,** or two or more simple events happening together, can be independent or dependant. Events are **independent events** if the probability of one event does not affect the probability of the other. Events are **dependent events** if one event in some way changes the probability that the other occurs. The following are the **Multiplication Rules for Probability**.

Probability of Two Independent Events	$P(A \text{ and } B) = P(A) \cdot P(B)$
Probability of Two Dependent Events	$P(A \text{ and } B) = P(A) \cdot P(B\|A)$

$P(B|A)$ is the **conditional probability** and is read *the probability that event B occurs given that event A has already occurred.*

Example **The P.E. teacher puts 10 red and 8 blue marbles in a bag. If a student draws a red marble, the student plays basketball. If a student draws a blue marble, the student practices long jump. Suppose Josh draws a marble, and not liking the outcome, he puts it back and draws a second time. What is the probability that on each draw his marble is blue?**

Let B represent a blue marble.

$P(B \text{ and } B) = P(B) \cdot P(B)$ Probability of independent events

$= \frac{4}{9} \cdot \frac{4}{9}$ or $\frac{16}{81}$ $\quad P(B) = \frac{4}{9}$

So, the probability of Josh drawing two blue marbles is $\frac{16}{81}$ or about 20%.

Exercises

Determine whether the events are *independent* or *dependent*. Then find the probability.

1. A king is drawn from a deck of 52 cards, then a coin is tossed and lands heads up.

2. A spinner with 4 equally spaced sections numbered 1 through 4 is spun and lands on 1, then a die is tossed and rolls a 1.

3. A red marble is drawn from a bag of 2 blue and 5 red marbles and not replaced, then a second red marble is drawn.

4. A red marble is drawn from a bag of 2 blue and 5 red marbles and then replaced, then a red marble is drawn again.

NAME _______________ DATE _______ PERIOD _______

13-5 Study Guide and Intervention *(continued)*

Probabilities of Independent and Dependent Events

Conditional Probabilities Conditional probability is used to find the probability of dependent events. It also can be used when additional information is known about an event.

The conditional probability of B given A is $P(B|A) = \dfrac{P(A \text{ and } B)}{P(A)}$ where $P(A) \neq 0$.

Example **The Spanish Club is having a Cinco de Mayo fiesta. The 10 students randomly draw cards numbered with consecutive integers from 1 to 10. Students who draw odd numbers will bring main dishes. Students who draw even numbers will bring desserts. If Cynthia is bringing a dessert, what is the probability that she drew the number 10?**

Since Cynthia is bringing dessert, she must have drawn an even number.

Let A be the event that an even number is drawn.

Let B be the event that the number 10 is drawn.

$P(B|A) = \dfrac{P(A \text{ and } B)}{P(A)}$ Conditional Probability

$= \dfrac{0.5 \cdot 0.1}{0.5}$ $P(A) = \frac{1}{2} = 0.5$ and $P(B) = \frac{1}{10} = 0.1$

$= 0.1$ Simplify.

The probability Cynthia drew the number 10 is 0.1 or 10%.

Exercises

1. A blue marble is selected at random from a bag of 3 red and 9 blue marbles and not replaced. What is the probability that a second marble selected will be blue?

2. A die is rolled. If the number rolled is less than 5, what is the probability that it is the number 2?

3. A quadrilateral has a perimeter of 16 and all side lengths are even integers. What is the probability that the quadrilateral is a square?

4. A spinner with 8 evenly sized sections and numbered 1 through 8 is spun. Find the probability that the number spun is 6 given that it is an even number.

13-6 Study Guide and Intervention

Probabilities of Mutually Exclusive Events

Mutually Exclusive Events If two events cannot happen at the same time, and therefore have no common outcomes, they are said to be **mutually exclusive**. The following are the **Addition Rules for Probability**:

Probability of Mutually Exclusive Events	$P(A \text{ or } B) = P(A) + P(B)$
Probability of Non-Mutually Exclusive Events	$P(A \text{ or } B) = P(A) + P(B) - P(A \text{ and } B)$

Example **At the ballpark souvenir shop, there are 15 posters of the first baseman, 20 of the pitcher, 14 of the center fielder, and 12 of the shortstop. What is the probability that a fan choosing a poster at random will choose a poster of the center fielder or the shortstop?**

These are mutually exclusive events because the posters are of two different players.

Let C represent selecting a poster of the center fielder.

Let S represent selecting a poster of the shortstop.

$P(C \text{ or } S) = P(C) + P(S)$

$= \frac{14}{61} + \frac{12}{61}$

$= \frac{26}{61}$ or about 43%

Exercises

Determine whether the events are *mutually exclusive* or *not mutually exclusive*. Then find the probability. Round to the nearest hundredth.

1. SHELTER selecting a cat or dog at the animal shelter that has 15 cats, 25 dogs, 9 rabbits and 3 horses

2. GAME rolling a 6 or an even number on a die while playing a game

3. AWARDS The student of the month gets to choose his or her award from 9 gift certificates to area restaurants, 8 CDs, 6 DVDs, or 5 gift cards to the mall. What is the probability that the student of the month chooses a CD or DVD?

4. STUDENT COUNCIL According to the table shown at the right, what is the probability that a person on a student council committee is a junior or on the service committee?

Committee	Soph.	Junior	Senior
Service	4	5	6
Advertising	3	2	2
Dances	4	8	6
Administrative Liaison	1	1	4

13-6 Study Guide and Intervention *(continued)*

Probabilities of Mutually Exclusive Events

Probabilities of Complements The complement of an event A is all of the outcomes in the sample space that are not included as outcomes of event A.

Probability of the Complement of an Event	$P(\text{not } A) = 1 - P(A)$

Example **A school has a photography display of 100 pictures. One of the pictures will be chosen for display at the district office. Lorenzo has 3 pictures on display. What is the probability that one of his photographs is not chosen?**

Let A represent selecting one of Lorenzo's photographs.
Then find the probability of the complement of A.

$P(\text{not } A) = 1 - P(A)$ Probability of a complement

$= 1 - \frac{3}{100}$ Substitution

$= \frac{97}{100}$ or 0.97 Simplify

The probability that one of Lorenzo's photos is not selected is 97%.

Exercises

Determine the probability of each event.

1. If there is a 4 in 5 chance that your mom will tell you to clean your room today after school, what is the probability that she won't?

2. What is the probability of drawing a card from a standard deck and not getting a spade?

3. What is the probability of flipping a coin and not landing on tails?

4. What is the probability of rolling a pair of dice and not rolling a 6?

5. A survey found that about 90% of the junior class is right handed. If 2 juniors are chosen at random out of 100 juniors, what is the probability that at least one of them is not right handed?